The Thirteenth Hussars in the Great War

From a photograph by The Mendoza Galleries.

Lt. Col. J. J. Richardson, D.S.O.
Commanding 13th Hussars from August 1915
to the present time.

The Thirteenth Hussars in the Great War

BY

THE RIGHT HON.

SIR H. MORTIMER DURAND

G.C.M.G., K.C.S.I. K.C.I.E.

WITH MAPS AND ILLUSTRATIONS

DEDICATION

To the Unfading Memory of the

OFFICERS, NON-COMMISSIONED OFFICERS, AND MEN OF THE REGIMENT WHO LAID DOWN THEIR LIVES DURING THE GREAT WAR

1914-1918.

"I vow to thee, my country, all earthly things above,
Entire and whole and perfect, the service of my love;
The love that asks no question, the love that stands the test,
That lays upon the altar the dearest and the best;
The love that never falters, the love that pays the price,
The love that makes undaunted the final sacrifice."

—(*Ode to Duty*, by the late Sir CECIL SPRING RICE, G.C.M.G., G.C.V.O.)

CONTENTS.

LIST OF ILLUSTRATIONS.

PHOTOGRAVURE.

COLOURED PLATES.

HALF-TONE PLATES.

ILLUSTRATIONS IN TEXT.

Thanks are tendered to Messrs Elliott & Fry, to Messrs Gale & Polden, and others, for permission to copy some of the portraits reproduced in this work.

OFFICER OF THE 13TH LIGHT DRAGOONS
1830 - 1836

OFFICER OF THE 13TH LIGHT DRAGOONS
(undress)
1830 - 1836

BEFORE THE CHARGE OF THE LIGHT BRIGADE AT BALACLAVA
25TH OCTOBER 1854

13TH LIGHT DRAGOONS, 1853

FESTUBERT

(From the picture presented to the Regiment by Brig.-Gen. A. Symons, C.M.G.)

MESOPOTAMIA

THE HUSSARS

OF THE

THIRTEENTH NARVA REGIMENT OF RUSSIA

SEND MOST FRATERNAL GREETINGS TO
THEIR VALIANT AND NOBLE COMRADES

OF THE

THIRTEENTH REGIMENT OF HUSSARS

PROUD AND HAPPY IN BEING UNITED
WITH THEM IN CORDIAL COOPERATION
AGAINST A COMMON ENEMY.

THE RUSSIAN FRONT 1916.

XMAS CARD SENT TO THE REGIMENT, 1916

CAPRICE AND FOAL: BORN MAY 1920
(From the picture by GEORGE PAICE, Esq.*)*

TUZ KHARMATLI, 29TH APRIL 1918

THE STORMING OF RICHARDSON'S BLUFF

29TH OCTOBER 1918

The Thirteenth Hussars in the Great War.

CHAPTER I.

INTRODUCTORY.

THE main object of this book is to give an account of the services rendered by the Thirteenth Hussars during the last ten years, especially in the war which has just come to an end.

The earlier history of the Regiment has already been written, and very fully written. On this subject the standard authority must always be Barrett's valuable work, which takes up the story from the beginning and carries it on to 1910, a period of nearly two hundred years. In order that readers of the present narrative may start with a general knowledge of the Regiment and its past, a chapter relating to this period has been introduced. As will be seen, it touches upon most of the wars waged by Great Britain since the days of Marlborough. But it is a mere summary, chiefly drawn from Barrett, and contains little new matter.

In ordinary circumstances this summary would open the book, but any account of the part played by a British Cavalry regiment in the late war must of necessity have some bearing upon the larger question of the part likely to be played by the mounted arm in any wars of the future; and just now this question is of special interest, for it has been freely asserted that recent changes in military conditions, notably the vast increase in the size of armies and the development of the aeroplane, have made Cavalry an obsolete and useless arm; and it is important for us to know whether they have done so, or are likely to do so. Therefore it has been thought desirable to give at the beginning a brief review of the history of Cavalry before this war, and at the close a few remarks upon the lessons of the war with regard to the value of the arm under present conditions.

Perhaps the services of the Thirteenth Hussars will not lose in interest if considered to some extent from this point of view.

CHAPTER II.

CAVALRY BEFORE THE GREAT WAR.

FOR thousands of years the horse has been the companion of man in war.

It is significant that when Job gives us his wonderful description of the strong things of earth and sea and air, he speaks of the horse in this connection, as rejoicing in the sound of the trumpet, and smelling the battle afar off, the thunder of the captains, and the shouting. "He goeth out to meet the armed men. He mocketh at fear, and is not dismayed; neither turneth he back from the sword. The quiver rattleth against him, the flashing spear and the javelin. He swalloweth the ground with fierceness and rage." And in many passages of the Bible, in poetry and in narrative, we have mention of the chariot and the horseman.

Representations of them are to be found in the carvings and tablets of long-vanished dynasties and nations. To take a single instance, they are shown in Assyrian carvings dating nearly a thousand years before Christ, which can be seen now in the British Museum.

Apparently the chariot came into the field earlier than the horseman usually so called, and the first use of the horse in war was to take up to the front in chariots warriors who got down to fight on foot, as the Greek chiefs did in the siege of Troy. But ere long Scythians or other nomads learned to mount the horse himself, and then began that close conjunction and sympathy between man and horse which made the two almost one creature, the Centaur of the fable.

The subject has been touched by many writers. There is perhaps no need to consider here the uses and gradual disappearance of the war-chariot. For present purposes it is sufficient to note that long before the historical age the armed hosts of the great Eastern Empires were composed in part of mounted men, who marched, and often fought, on horseback. The chariots and the people attached to them may have been the first "Cavalry"; but the word as used in this book refers to mounted men only—riders,—and riders who did some part at least of their fighting from the backs of horses.

If the use of mounted men in war began in the East, to which Western nations owe so much, including even their religion, it soon extended to Europe. In the first conflict between East and West on a large scale of which we have any real knowledge, nearly five hundred years before Christ, the Persian invaders of Greece found that the Greeks had little Cavalry to oppose to the thousands of horsemen whom they brought with them. The men of Athens and Sparta fought on foot at Marathon and Thermopylæ. Even at Mount Cithæron, where Masistius in his golden cuirass charged and died, the Greek army was an army of footmen. Nevertheless there were some horsemen in Greece even then, especially on the plains of Thessaly; and the frieze of the Parthenon, of not much later date, shows helmeted Greek soldiers riding spirited horses. The horses are small, apparently not more than thirteen or at most fourteen hands, and are ridden barebacked, but they are evidently war horses. Then we have Xenophon's well-known treatise on Cavalry, a thoroughly practical work, which must have been written in the first half of the next century; and after that the organisation of the Greek Cavalry is fairly well known.

It was Alexander the Great who first showed what horsemen could do in war if properly trained and led. Until his time Cavalry seem to have fought mostly in loose swarms, rather as skirmishers and bowmen than as solid squadrons using the weight of the horse itself to overthrow and destroy bodies of footmen. He saw the value of "shock tactics," and taught his Cavalry to use them, so that when he invaded Persia in 334 B.C. the famous horsemen of Persia went down again and again before his fiery onsets. They had themselves, according to Herodotus, some notion of charging in squadron on the battlefield, but they had never seen Cavalry used in mass, and neither they nor the Persian foot could stand against it.

In the impetuous rapidity of all his movements, especially perhaps in the closeness and vigour of his pursuits, Alexander was in fact a model leader of horse, and his conquests were largely due to his Cavalry, which he not only wielded with dash and power against the Cavalry of the enemy, but kept thoroughly in hand even after a successful charge, and threw into the scale wherever they might be most required to help his foot soldiers.

Ever since those days, for more than twenty centuries, the history of war on land has been the history of a struggle for pre-eminence between horsemen and footmen. The rivalry has been complicated by the invention of Artillery, and of late years by the development of fighting in the air; but it has gone on unceasingly, and can hardly be said to have come to an end even now. In the course of it there has often been a tendency to lose sight of the fact that combined effort for one purpose by all arms, and not rivalry between them, is the secret of success in war. But the long dispute and its vicissitudes form an interesting study.

By the Romans the effective use of Cavalry was for a long time not well understood. Though they had their "Equites" from early days, they got to rely more and more for serious fighting upon their wonderful legions, and it was not until the Punic Wars that they learned their lesson. Hannibal, like Alexander, was a born leader of horse, and when a hundred years after Alexander's death he invaded Italy by way of the Alps, he at once taught Western Europe what Alexander had taught the Greeks and Persians, that in the existing condition of military armament, Cavalry well trained and boldly used in masses could do great things on the battlefield. The successive victories which he gained in Italy, with very inferior numbers, over the proud and confident troops of Rome, were due in large measure to his skilful use of his horsemen. At Cannæ, for example, his wild Numidian light horse, riding without saddle or reins, and his heavier squadrons from Spain and the North, began by driving off the weak Roman Cavalry opposed to them, and then, wheeling inwards upon the rear of the advancing legions, enclosed them in a circle of steel from which there was no escape. Fifty thousand of them are said to have fallen, and for a time Rome seemed to be, perhaps really was, at his mercy. Every one knows the story of his long struggle against hopeless odds, and of his final defeat. When at last he was conquered the superiority in horsemen had passed to the Romans, and he was overwhelmed and crushed by his own methods. He had taught his enemies to fight.[1]

As time went on they forgot in a measure the lesson they had learnt from him, and they suffered some heavy reverses in consequence—for example, in their wars with the Parthians which stopped their expansion eastward; but happily such enemies were rare, and gradually the legions won for Rome the empire of the Western world. It lasted as long as the spirit and discipline of their incomparable Infantry remained unimpaired.

In the closing centuries of Imperial Rome the bulk of her enemies marched against her on horseback, and her own armies came to be composed more and more of Cavalry. Her last great battle was against Attila the Hun, whose people lived on their horses. It was a victory; but it was a Cavalry victory, and won by the help of the Goths. Her Infantry had long since failed her, and the Imperial City had been herself in the hands of the Barbarians. Her fall had been due to the woeful corruption and degeneration of the legions, not to any inherent superiority of the horseman over the footman; but the fact remains that at this time Cavalry was everywhere regarded as the more important arm of the two.

There followed a long period during which the predominance of the horseman grew more and more undisputed. With the collapse of Rome scientific warfare on a large scale became a lost art, and in the disorderly welter of the Dark Ages the fighting power of the footman, which depends so much upon organisation and discipline, sank lower and lower. To deal it a final blow came, a thousand years or so after Christ, the institution of Chivalry, which to a considerable extent undermined national feeling and exalted in its

[1] It may be noted that Hannibal's invasion of Italy led, two thousand years later, to the recognition of the influence of sea power on history. Admiral Mahan himself described to the writer how it one day occurred to him that Hannibal's long march from Spain to Italy had been necessitated by the Roman strength at sea, and that this was the underlying cause of his eventual ruin. Following up the train of thought, Mahan gradually worked out his great conception. Hannibal's march was the germ from which it sprang.

place the individual prowess of the Knight. Having its origin in a praiseworthy attempt to set up a higher standard of right and wrong, to resist cruelty and injustice, to honour woman as she should be honoured, and to make courage and courtesy the aim of men, it did much good, and has left to succeeding ages some noble aspirations and examples. Even now there is surely no better thing one can say of a man than that he is chivalrous—*chevaleresque*—like a knight of old. The horseman had given his name to a new social order and a splendid ideal. In practice Chivalry was not always what it should have been, but the glamour of it lies upon all our poetry and literature. Even the free-lance or the moss-trooper, unprincipled ruffian as he often was, remains to our eyes a picturesque figure. There is still a gleam on his helmet and spear that time cannot take away. The war-horse and his rider had reached in those days the climax of their power and reputation.

Then, very gradually, came a change in the opposite direction. The knights and their retainers had been practically the only fighting men who counted, and were accustomed to ride down with ease and contempt any footmen who ventured to stand against them. Bows and arrows and axes and knives seemed of little avail against the spearman with his almost impenetrable armour and his thundering steed. As Colonel Maude puts it, "the knight in full armour had borne about the same relation to the infantry as an ironclad nowadays bears to a fleet of Chinese junks." But little by little it began to be recognised, first it is said in the Crusades, when the knights had to take or defend fortresses and otherwise fight on foot, that there were operations in war for which the heavily armoured horseman was not well fitted. Bodies of footmen began to be raised again for such purposes, and even to be brought into the open field as archers or cross-bowmen for use in broken ground. They often suffered horribly, but now and then they gained some successes, and as time went on they developed greater skill and confidence. Eventually, at Crécy and Poictiers and Agincourt, the English archers, with their cloth-yard shafts and their bristling defence of pointed stakes, won astonishing victories over the Chivalry of France, and proved to Europe that the horseman was no longer invincible on the battlefield. The lesson had very nearly been taught by the English three hundred years earlier, on the field of Hastings; but the time had not then come. Lured from their stockades, the footmen had been cut to pieces, and the French Cavalry had conquered England. At Crécy the English footmen turned the tables. And elsewhere, about the same period, the Swiss Infantry won almost equal honour.

The Cavalry of Europe nevertheless fought hard for their old pre-eminence, and it was long before they could be brought to see that they would never again be the undisputed masters in battle. But it was a lesson they had to learn. As time went on they found their charges repelled by serried squares of pikemen, from which came showers of arrows and cross-bolts; and later the invention of firearms weighted the scale still further against them. The only offensive weapons of the horsemen were the weight of their horses and the lance or sword; and if the horses failed to break the rows of eighteen-foot pikes, the *arme blanche* could do nothing. At last, after many attempts by the Cavalry to meet these new conditions, by using firearms themselves and other devices, it came to be generally recognised that against confident and steady infantry armed with the pike, deliberate frontal assault by horsemen was practically hopeless, and that for the future Cavalry must depend to some extent upon surprise and stratagem to give them victory. The defence had in some measure triumphed over the attack, and the essentially offensive arm had lost its pride of place.

This is not to say that for the future Cavalry was to be useless on the battlefield—far from it. The range of the unwieldy arquebus, or of the smooth-bore musket which followed it, was not so great as to keep Cavalry out of striking distance; and their speed, if they were led with decision and dash, would yet give them many opportunities of riding down the footmen. They could no longer do so whenever they pleased, but they were still a formidable part of the fighting line.

This was shown very clearly in our own Civil War. The armies of both King and Parliament were largely composed of horsemen, and in fight after fight it was they who were most conspicuous. Finally, the emergence of a great leader of Cavalry turned the scale in favour of the Roundheads. Cromwell's Ironsides, thoroughly trained, and used as in old days the Cavalry of Alexander and

Hannibal had been used, not only with dash but with coolness and self-control, proved too strong for the Royalists, cavaliers though they were. Unlike Prince Rupert, Cromwell kept his horsemen firmly in hand, throwing them into the fight wherever they were most required, and the result was to make him master of England.

On the Continent too Cavalry was still largely used in battle. The Turkish horsemen were numerous and formidable. Before our civil conflicts, in the Thirty Years' War, Gustavus Adolphus had wielded Cavalry with much effect, and while Cromwell was fighting in England the great Condé had sprung into fame by the achievement of his horsemen at Rocroy. Under him and other commanders the French Cavalry gained an enduring reputation, and the same may be said for the Germans under Pappenheim and Montecuculi. The Infantry was now perhaps the leading arm in battle, and it was growing stronger as its firearm improved, while the rise of a more or less effective Artillery was adding to the difficulties of the Cavalry attack; but at the close of the seventeenth century the horseman was still a power in the field.

Throughout the first half of the eighteenth century this state of things continued. In Marlborough's wars Cavalry was used in large numbers, and with great effect. At Blenheim, and other notable fights, his horsemen practically decided the issue between him and the French Marshals. How important the arm was considered may be judged from the fact that at Ramilies the forces on both sides were little stronger in foot than in horse. Between them the opposing armies numbered only 75,000 Infantry to 64,000 Cavalry.

About the same time Charles XII. of Sweden was also using Cavalry in large numbers; and when, under Peter the Great, Russia began to make her mark among the military powers of the world, not the least formidable part of her army was the Cavalry, which, including the afterwards famous Cossacks, amounted at one time to more than 80,000 men.

Then came the crowning period for Cavalry in modern war. In spite of their recognised place on the battlefield, and their many successes, the horsemen of the European armies had not until the middle of the eighteenth century attained to a full comprehension of their possible influence. Awed to some extent by the reputation which the Infantry had gained at their expense in the course of the last three centuries, the Cavalry had become a less swift and dashing arm. They had learnt to rely in large measure upon their fire, and even to fight dismounted as dragoons. "In fact," according to their historian Denison, "the cavalry of all European States had degenerated into unwieldy masses of horsemen, who, unable to move at speed, charged at a slow trot and fought only with pistol and carbine." Even so they were more mobile than Infantry, and had great achievements to their credit; but they had failed to see that a recent change in armaments had thrown the game into their hands. The Infantry, growing over-confident, had discarded the long pike for the bayonet—a very poor substitute—and the Cavalry had once more a chance of riding down their enemy in fair fight by the speed and weight of their horses. Their power was now to be taught them by a keen-sighted soldier, Frederick the Great of Prussia.

When he came to the throne in 1740, and began the career of unscrupulous aggression which was to make Prussia one of the leading nations of Europe, he soon saw that his Cavalry was not all it should have been. "They were," says Denison, "large men mounted upon powerful horses, and carefully trained to fire in line both on foot and on horseback," but they were quite incapable of rapid movement, and never attacked Infantry by the ancient method. "His first change was to prohibit absolutely the use of firearms mounted, and to rely upon the charge at full speed, sword in hand." Marlborough had shown the advantage of using great bodies of Cavalry in mass, and Marshal Saxe had advocated their being taught to move at speed for a mile or more in good order. Frederick now took over both ideas, and by careful and incessant training evolved a Cavalry which was capable of manœuvring in thousands together at full pace, even over rough ground, without disorder or loss of control. Such a force, led by men like Seidlitz and Ziethen, proved to be almost irresistible. Against Austrians and Russians and Frenchmen alike, it had astonishing success. "Out of twenty-two great battles fought by Frederick, his Cavalry won at least fifteen of them. Cavalry at this time reached its zenith."

Frederick's system was copied by all the great military nations of Europe, and at the close of the eighteenth century the influence of horsemen in the field was greater than it had ever been since the battle of Crécy.

Then came Napoleon, and though the Cavalry had not such a pre-eminent place in his armies as in those of Frederick the Great, for it was not as efficient, yet it was used in vast numbers and at times with tremendous effect. Murat was perhaps the most conspicuous figure among all Napoleon's Marshals, and other Cavalry leaders made great names for themselves. At Marengo, at Austerlitz, and in many more of Napoleon's famous battles, the French horsemen won undying renown; and if at last his Cuirassiers had to recoil before the fire of the British squares at Waterloo, every one knows with what magnificent courage and devotion they strove again and again to cut their way to victory.

Among Napoleon's enemies too, Prussian and Austrian, Russian and British, the Cavalry did much fine work throughout; and it is not perhaps too much to say that the Russian horsemen, especially the Cossacks, by destroying his famous squadrons in the great retreat, were among the most notable causes of his downfall. This much is certain, that when he fell the Cavalry of Europe held a high place in the battlefield. Infantry had become the backbone of most armies, and the power of Artillery had vastly increased, but Cavalry was still a powerful and necessary arm.

Then came another marked change in the conditions of war. A generation after the Conqueror's death the rifle took the place of the smooth-bore musket in the hands of the Infantry, and the same principle was applied to cannon. The result was that the power of firearms was greatly increased in range and accuracy, and that the value of Cavalry in battle was proportionately lowered. Soon afterwards the introduction of breech-loading gave the rifled weapons a vastly greater rapidity of fire, which also told heavily against the mounted arm. It was one thing for Cavalry to remain out of range, a few hundred yards away, and then to charge against the slow and inaccurate fire of a smooth-bore musket. It was a very different thing for them to advance from a much greater distance, against a rifle which not only carried three times as far as the musket, but shot straight, and could be loaded in a quarter of the time. From the middle of the nineteenth century it began to be held, at all events in France and England, that the chance of a successful attack by Cavalry armed only with the sword or lance upon Infantry in the battlefield, except under very unusual circumstances, was practically at an end. It seemed a fatal blow to the system of Frederick, and to the hope of the horseman in his long rivalry with the foot soldier.

That conclusion was not shaken by the wars waged by European nations during the remainder of the century. Some successes were gained by Cavalry in various parts of the world outside Europe. For example, the British Cavalry did fine work against the Sikhs in 1846 and 1849; a Persian square was broken and destroyed by a charge of British Indian Cavalry in 1856; and British Cavalry were very useful in the Mutiny soon afterwards, and against the Chinese; but neither in the Crimea, nor in the war between France and Austria in 1859, nor in the war between Prussia and Austria in 1866, nor in the Franco-German War of 1870, nor in the Russian War against Turkey a few years later, could the Cavalry claim to have struck such blows in battle as they had been used to strike in the days of Napoleon. Colonel Henderson in that fascinating book, 'The Science of War,' writing of the "shock tactics" lately prevailing, reviews the achievements of Cavalry under that system. "Such is the record," he says: "one great tactical success gained at Custozza; a retreating army saved from annihilation at Königgratz; and five minor successes, which may or may not have influenced the ultimate issue—not one single instance of an effective and sustained pursuit; not one single instance, except Custozza, and there the Infantry was armed with muzzle-loaders, of a charge decisive of the battle; not one single instance of Infantry being scattered and cut down in panic-flight; not one single instance of a force larger than a brigade intervening at a critical moment. And how many failures! How often were the Cavalry dashed vainly in reckless gallantry against the hail of a thin line of rifles! How often were great masses held back inactive, without drawing a sabre or firing a shot, while the battle was decided by the infantry and the guns!"

Truly, the day of Cavalry seemed to be over, and this was the opinion frequently expressed at the end of the century. Their day was not over.

It will probably have been noticed that so far we have been

dealing only or mainly with the question of Cavalry on the battlefield. But their work lies not only on the battlefield—indeed, it may be doubted whether their work there, however great, has not always been of less value than the services they have been able to render in other ways.

The operations of war are generally treated by military writers as consisting of two distinct branches—those leading up to battle, and those of battle itself. The former are of great variety and scope, involving all the preparations and manœuvres which will result in bringing upon the battlefield an army with "every possible advantage of numbers, ground, supplies, and *moral*" over the army of the enemy. These operations are the province of "strategy." The operations of the battle itself, when the opposed armies have actually come into touch, are the province of "tactics." The latter are the more picturesque, and naturally appeal to the fighting spirit of the soldier; but the former are often, if not usually, of the greater importance to the issue of a war. "Strategy," says Henderson, "is at least one half, and the more important half, of the art of war"; and he says elsewhere: "An army may even be almost uniformly victorious in battle, and yet ultimately be compelled to yield."

Now it may safely be asserted that with regard to strategical operations there has never been any serious question as to the great value of Cavalry in any war confined to the land. To quote Colonel Denison, in "their fitness for scouting, reconnoitring, raiding, &c., Cavalry have always been the foremost arm and without rival. In covering an advance, in pursuing a retreating foe, their capacity has always been unequalled." Henderson, himself an Infantry officer, states that "the Cavalry is *par excellence* the strategical arm," that "it depends on the Cavalry, and on the Cavalry alone, whether the Commander of an army marches blindfold through the 'fog of war,' or whether it is the opposing General who is reduced to that disastrous plight." And Von Bernhardi, discussing the future of Cavalry, says, "It is in the strategical handling of the Cavalry that by far the greatest possibilities lie." He admits that on the battlefield and in retreat their rôle can only be a subordinate one. "But for reconnaissance and screening, for operations against the enemy's communications, for the pursuit of a beaten enemy, and all similar operations of warfare, the Cavalry is, and remains, the principal arm." These passages were written before the aeroplane was used in war, but they show clearly that until then—that is, throughout the nineteenth century—Cavalry was still as necessary as ever for the proper working of a campaign.

And further, it may be pointed out that even with regard to the battlefield, horsemen armed and trained in a different way might conceivably be of greater use than horsemen depending solely or mainly upon shock and the *arme blanche*.

This was proved, though the majority of Continental soldiers would never open their eyes to the fact, by the fighting in the American Civil War. Henderson, with clearer vision, writes of this great conflict: "So brilliant were the achievements of the Cavalry, Federal and Confederate, that in the minds of military students they have tended in a certain measure to obscure the work of the other arms." No doubt many of these achievements were rather of a strategical than a tactical nature, but many were not. The American Cavalry was from first to last constantly used for actual fighting, and in numberless instances its value as a battle arm was amply demonstrated. It would be impossible to enumerate them here, but Henderson expressly declares, for example, that "there is no finer instance . . . of effective intervention (by Cavalry) on the field of battle than Sheridan's handling of his divisions, an incident most unaccountably overlooked by European tacticians, when Early's army was broken into fragments, principally by the vigour of the Cavalry, in the valley of the Shenandoah." The fact was that, adapting themselves to the new conditions brought about by rifled firearms, the Americans had created a mounted service which could fight both on foot and on horseback, with the rifle or the sword or the pistol; "they used fire and *l'arme blanche* in the closest and most effective combination, against both Cavalry and Infantry." Assuredly Cavalry was not yet a negligible arm in battle.

The closing years of the century saw the beginning of another war in which the horse and his rider were again very prominent. The Boers, who made so gallant and protracted a fight against the vast resources of England, were all mounted men, and it was not until the British forces opposed to them also consisted in a large measure of mounted men that their resistance was broken down. They differed in many respects from the American Cavalry. The latter were trained to fight on foot if necessary, but preferred fighting on horseback whenever they could, though they fought with the pistol rather than the sword. The Boers fought mainly, almost entirely, on foot. Their arms and training were inconsistent with fighting from the saddle. They were in fact rather mobile riflemen than anything else. Nevertheless the fact remains that they were mounted men, and that a large part of their value lay in their being so. For many of the essential duties of Cavalry, for scouting and collecting information, for raids on their enemy's communications, for the capture of his trains and guns, for covering a retirement, they were exceptionally well fitted. Henderson, writing of the duties of Cavalry, says: "But most important perhaps of all its functions are the manœuvres which so threaten the enemy's line of retreat that he is compelled to evacuate his position, and those which cut off his last avenue of escape. A Cavalry skilfully handled, as at Appomattox or Paardeberg, may bring about the crowning triumph of grand tactics—viz., the hemming in of a force so closely that it has either to attack at a disadvantage or surrender." The example of Paardeberg is one in which the triumph was due to the British Cavalry, but the Boers had some triumphs of the same kind, for instance at Nicholson's Nek, and they were very near to gaining one which might have shaken the Empire. If Ladysmith had fallen, with its garrison of 12,000 men, as at one time seemed probable, the disaster would undoubtedly have been due in the main to the mobility of the Boers, whose rapid movements on horseback enabled them not only to drive in and besiege White's troops, but afterwards to hold up for months, with inferior numbers, Buller's relieving force, while still maintaining their grip on the starving garrison. In fact it may be said that even on the actual field of battle they fought partly as Cavalry—Von Bernhardi goes so far as to say "exclusively as Cavalry,"—for though they almost invariably dismounted to use their rifles, yet it was by the speed of their horses that they were able to extend their flanks, and, galloping out to any threatened point, form a fresh front against any turning movement. Our slow-moving Infantry had no chance of getting round and enveloping them, but was forced time after time to undertake desperate frontal attacks upon the lines, often more or less entrenched, which their rapidity of manœuvre had made it possible for them to take up. Altogether, the fighting value of the 50,000 Burghers with whom Paul Kruger set out to defy Great Britain, was doubled or trebled by the fact that they were mounted men. It made them in their own country, and perhaps would have made them anywhere, a formidable fighting force.

This was not clearly understood on the Continent of Europe, but it was understood in England. It had a great effect upon the views of our leading soldiers with regard to the future of Cavalry, and the subsequent Russo-Japanese War did not in any way contradict the lessons drawn from the campaigns in America and South Africa.

To sum up this chapter, it may be said with confidence that when the Great War broke out the value of Cavalry, both as a strategical arm and on the field of battle, had been demonstrated by the experience of three thousand years. During that time it had fluctuated, especially with regard to the battlefield, but it had always been great. For some centuries, especially since the development of efficient firearms, the tendency had been for the Infantry to oust the horsemen from their pride of place in the actual shock of armies, and by the end of the nineteenth century the supremacy of the Infantry in this respect had been generally acknowledged. But even so it had not been shown that Cavalry, properly armed and trained, were incapable of joining with effect in the decision of battles, and the American and South African Wars had given reason to believe that it certainly could do so. Its great strategical value was not disputed. Clearly, therefore, Cavalry was still a necessary and important part of any efficient army—one of the most important. Whether for strategical duties or for full victory in battle, the other arms could not do without the horsemen.

No doubt the value of Cavalry might be altered in the future, as it had been in the past, by new developments in the art of war, but such was the position at that time.

We may now turn to the Thirteenth Hussars.

CHAPTER III.

EARLIER HISTORY OF THE REGIMENT.

BEFORE the war of 1914 the Regiment now known as the Thirteenth Hussars had, like most Regiments of the British Army, served in various parts of the world. During the eighteenth and nineteenth centuries it had borne a part in nine wars of one kind or another, and had made acquaintance not only with the Continent of Europe, but with Asia, America, and Africa.

The Regiment was raised in the year 1715. The Duke of Marlborough was then still living, but his long series of victories had been brought to a close by the Treaty of Utrecht two years before, and thirty thousand of the veterans who had won them for him had been ruthlessly disbanded.

After the accession of George I., in 1714, it was seen that this step had been a hasty and dangerous one, for the Jacobite party was strong, and the reduction of the small British Army had given them fresh hopes. It soon became evident that the exiled Stuarts meant to take advantage of their opportunity, and the British Government was obliged to raise fresh troops in place of those so recently thrown away. Among the new Regiments were to be several of Dragoons, and in July 1715 the raising of one of these was entrusted to Brigadier Richard Munden, an officer on half-pay who had served with some distinction under Marlborough.

It appears that Munden had no difficulty in finding recruits, for within three months the Regiment had been raised, and was assembled at Northampton. There it received orders to march to Leeds, and soon afterwards Brigadier Munden was informed that his Regiment, with others, was to be under the orders of Major-

General Wills, whom His Majesty had appointed "to command several of his forces on an expedition."

At this time a Dragoon Regiment in the British Army consisted of 6 troops, and its strength was between 200 and 300, including 19 "Commission" officers. It was not a Regiment of "Horse," though it was mounted, and regarded as Cavalry. The men were armed with the same firearm as the Infantry, or practically the same, and were expected to fight on foot as well as on horseback. This, it will be remembered, was the period when European Cavalry depended largely on their fire, and had not been trained to the system of Frederick the Great, the charge at speed with the *arme blanche*. The officers of Munden's Dragoons, including Munden himself, had almost all served in Regiments of Foot.

The Regiment was "officially declared to be a disciplined force belonging to the regular army on 31st October 1715." It had not to wait long before seeing service, for early in November General Wills learned that the Jacobite "rebels" were over the Scottish border, and marching on Lancaster. He at once drew together his forces at Manchester, and marched thence to Wigan. On the 12th November Munden's Dragoons were in presence of their first enemy, who had advanced as far as Preston, and was in occupation of the town.

It is significant that when General Wills left Wigan with his force to attack the rebels, the order of march was as follows: The advance-guard consisted of fifty musketeers and fifty dismounted dragoons. After the advance-guard came a Regiment of Foot, then three Brigades of Cavalry consisting of one Regiment of "Horse" and five of Dragoons. Evidently Cavalry was not regarded as the eyes of an army.

The action which followed was at first indecisive. The enemy, superior in numbers, and aided by some guns and barricades, repulsed one or two attacks made by Infantry and dismounted Dragoons. But on the following day General Carpenter having come up with three more Regiments of Dragoons, the rebels gave in and surrendered. Their assailants had lost in all one hundred and thirty killed and wounded, so the fighting had not been very severe. Nevertheless Preston was an affair of some importance, for with the indecisive battle of Sheriffmuir, fought the same day by other troops, it sufficed to put an end to the First Jacobite Rebellion and to establish the House of Hanover on the British throne. Munden's Dragoons had only four wounded during the fight, but they seem to have behaved well. Munden himself is said to have led a storming party, and to have been thanked for his gallant conduct. After the fight, the Regiment seems to have been employed in escorting to jail the unfortunate prisoners, whose fate was a sad one.

It may be noted that among the troops who served at Preston was Dormer's Regiment of Dragoons, afterwards the Fourteenth Hussars. Thus began a comradeship between the two Regiments which was afterwards very close.

Then followed for Munden's Dragoons, who about this time became known as the Thirteenth Dragoons, a long period of peace service. In 1718 there was again a reduction of the Army, and some Regiments having been disbanded in Ireland, the Thirteenth were sent over to take the place of one of them. The Irish military establishment was then separate from the British. The pay of the troops was somewhat less, and their circumstances in other respects were very unsatisfactory. It was forbidden to enlist any native of the country, so that men were hard to get, and the barrack accommodation was so scanty that the troops were scattered about in small detachments, to the woeful detriment of their discipline and efficiency. It apparently became the custom for officers to overstay their leave, or absent themselves without leave, and everything got slack in proportion. It was possibly not the fault of the Regiments that their arms were in most cases insufficient and bad; but in every way their condition was deplorable. The Thirteenth Dragoons seem to have suffered like the rest, and probably when their Colonel, Munden, was transferred to another Regiment in 1722, they were not in a very efficient condition.

Munden was one of the officers who followed the body of the great Duke of Marlborough when he was borne to his grave in Westminster Abbey. He died himself, a Major-General, three years later, and Colonel William Stanhope became Colonel of the Thirteenth. This officer, afterwards the Earl of Harrington, was appointed a Secretary of State in 1730.

The stay of the Regiment in Ireland came to an end in 1742, when it was transferred to Great Britain, and in the following year the command of it was bestowed upon Lieut.-Colonel James Gardiner of the Inniskilling Dragoons, then serving in Germany. Thus when the Second Jacobite Rebellion took place, in 1745, the Thirteenth, under this well-known officer, was among the Regiments at the immediate disposal of the Government, and was fortunate, or unfortunate, enough to find itself engaged once more on active service.

When Bonnie Prince Charlie unfurled his standard at Glenfinnan, Sir John Cope, the British General commanding in Scotland, was very weak in the number and quality of his troops. He had no gunners to man his few guns, and the force at his disposal to meet the advancing rebel army, after providing some small garrisons, amounted to about twenty-five companies of foot and two Regiments of Dragoons. One of these two was the Thirteenth. Provisions and transport were very scarce.

It is a curious coincidence that the Regiment came to blows with its second enemy at another Preston, this time in Scotland. Close to it was the house of their Colonel, Gardiner. The Thirteenth had had some trying work during the preceding weeks, when Cope withdrew his small force from Inverness to Dunbar, abandoning Edinburgh to the rebels; and the Regiment was not in good condition, many men and horses being physically unfit for duty.

The result of the battle is well known. The enemy, chiefly Highlanders, attacked on the early morning of 18th September. Cope having no gunners, a Lieut.-Colonel Whiteford and an old Master Gunner of the name of Griffiths fired a few rounds from the guns and cohorns, "none of whose shells would burst," and then the guns were rushed by the Highlanders. It was a fine chance for the Cavalry, as the rebels were in confusion, but the chance was not taken. To tell the simple truth, neither of the two Dragoon Regiments, Hamilton's or Gardiner's, which seem to have numbered six hundred men between them, could be induced to charge, and their only inclination was to gallop off the field. By the exertions of their officers and other gentlemen, about three-quarters of them were stopped, and brought into Berwick next day; but it must be admitted that their behaviour was anything but creditable, and the battle ended in the total defeat of the King's force. This much is to be said in favour of the Regiments, that their officers fought gallantly. The ill-fated Gardiner, who was seriously ill, was wounded at the beginning of the engagement; and later, when his men refused to charge, he received several other wounds, from which he died. His Lieutenant-Colonel, Whitney, was also wounded in trying to rally the men. But the fight of "Prestonpans" was certainly what Brigadier Fowke called it, "an unhappy affair."

After Gardiner's death the command of the Thirteenth was given to Colonel Ligonier, a brave officer who had served under Marlborough, and in the following January it took part in another battle and another defeat at Falkirk Muir. The same two Regiments of Dragoons which had been engaged at Prestonpans, and another, Cobham's, formed at Falkirk a Brigade of Cavalry under Ligonier's orders. This affair was not so discreditable as the former. The Cavalry, very gallantly led by Ligonier, did charge the enemy, and it is said penetrated their first line. But they failed to break the second line, and the charge ended in a confused retreat. Lieut.-Colonel Whitney, wounded at Prestonpans, was killed, and the gallant Ligonier also paid for his courage with his life. Suffering from an attack of pleurisy, he insisted on getting out of bed to command his Brigade in the battle, which was fought in a storm of wind and rain. His exertions in rallying the Dragoons and covering the retreat during the following night were too much for him, and a week later he died.

The Thirteenth saw no further fighting. When the Duke of Cumberland broke the Highland clans at Culloden and put an end to the rebellion, the Regiment was not present. It had been left in Edinburgh to patrol the roads, and intercept any communications between the English and Scottish Jacobites. Its share in the campaign, therefore, had not been a very satisfactory one. Perhaps it was not to be blamed for the second defeat at Falkirk, but certainly it had not won much distinction on the battlefield.

All that can be said is that no troops are likely to do well in the great ordeal of war unless their discipline and general condition have been steadily maintained in peace. History abounds in such lessons. The Regiment was to do great things later under more favourable conditions, and win a fine name for itself as a fighting corps. Its time was not yet come.

In 1748 the Thirteenth was once more transferred to Ireland, and there it remained for a second score of years. A Dragoon Regiment at this time seems to have been very weak in numbers, considerably under two hundred all told, officers and men, with one hundred and fifty horses. The prohibition against Irishmen had apparently been withdrawn, and by 1767 the men were almost all Irish. But none were Roman Catholics, the enlistment of these being still absolutely forbidden. The men were fine, most of them from five foot nine to five foot eleven, and "tollerably well appointed." The officers too were mostly Irish. The barrack accommodation was still very poor, and the Regiment was scattered in detachments as before. The arms were very bad at times.

About 1777 the Thirteenth were converted into Light Dragoons, and much smaller men were enlisted. The example of Frederick the Great was now being followed on the Continent, and Cavalry was being trained for greater speed and hand-to-hand fighting. The Infantry firearm of the Thirteenth gave place to a short carbine, and some changes were made in the uniform, the old three-cornered hat making way for a Cavalry helmet. Bayonets were still carried, but evidently there was some idea of making the Dragoon more of a horseman and less of a foot soldier.

Nevertheless the state of the British Cavalry at that time as to equipment and drill was very antiquated. "The military value of their training," says Barrett, "was practically *nil*." And, to add to their disadvantages, they were now cursed with the system of "proprietary Colonels." How this system came about is not clear, but towards the end of the eighteenth century it was in full force. In Munden's day the Colonel had been "the active officer in command, and always present, unless on leave, whether at home or in the field." Sixty years later, when the old traditions of Marlborough's time had been lost, the Regiment was really commanded by the Lieutenant-Colonel, while the Colonel had become an absentee, seeing the Regiment perhaps once or twice a year.

Yet it was in a sense looked upon as his private property. "The system," says Barrett, "was a bad one. To bad Colonels were due the crying abuses of the pay system as well as those of the clothing system—the systematic robbery of the soldier, the mean frauds by which an income was literally swindled out of Government or sweated off the backs of the men; and the abuse of the power of the lash was owing to the same cause." In 1787 the Colonel of the Thirteenth, a member of Parliament, "lived mainly in London while the Regiment was in Ireland." Arms were bad, desertions frequent, and the duties of the Regiment consisted chiefly of hunting down members of the various lawless societies in Ireland, Whiteboys and Peep-o'-Day Boys, and the like. In spite of all these heartbreaking drawbacks the regimental officers seem to have done something to make the men efficient, for at times the reports of inspecting Generals are good enough, though evidently the standard was not high; and in 1794, no doubt because of the French Revolution and the outbreak of war on the Continent, the strength had been increased to 446 men and 393 horses.

The Thirteenth, however, was not yet to be employed in the Continental war. It was now, after its two campaigns against the Jacobites, followed by fifty years of peace duty, to have its first taste of service abroad, but this was not to be in warfare against a civilised enemy.

In the island of Jamaica the "Maroons," originally runaway negro slaves, had long been giving trouble, and it had now become urgently necessary to suppress them. They held a difficult mountain country, full of densely wooded glens, from which they had been wont for many years to raid the lowlands and plantations, plundering and murdering. After some partial settlements they had again risen, and had openly defied the white men to war. Their numbers were not large, perhaps 1200 all told, but as Great Britain was already fighting the French in the West Indies the complication was serious, and Lord Balcarres, the Governor, was assembling a considerable force to blockade the revolted highlands.

It is remarkable to find, considering the nature of the ground, that in addition to three Regiments of Infantry and some local militia, this force was to consist of five Dragoon Regiments, of which two were the Thirteenth and Fourteenth Light Dragoons.

The Thirteenth was brought over from Ireland to England in 1795, and a couple of troops sailed for Jamaica in advance, the remainder of the Regiment remaining in England until the following February, when, on the 9th of the month, the Headquarters sailed in the *Concord*, which formed part of a fleet numbering more than five hundred sail. In spite of all the circumstances of its peace service, the Regiment seems then to have been in a condition of discipline and efficiency very creditable to officers and men. Fortunate that this was so, for both were soon to be severely tested. A violent storm scattered the fleet three days after sailing, and in the Bay of Biscay the *Concord* took fire, some pitch used for fumigation having been upset by the rolling of the vessel, and blazed up. As the fire was immediately over nineteen casks of powder, the danger was great. It is pleasant to read how the ship's company behaved in this sudden contingency. The Captain, who was writing in his cabin, ran on deck "with his pen across his mouth." An officer was sent down to the hold to cover the powder barrels with wet blankets and mattresses. "Scores of men, with their mattresses held in front of them," threw themselves on the flames and smothered them, while the officer below spread a sailcloth over the barrels and kept it wet under a shower of sparks from the deck above. Eventually, after really heroic exertions, the fire was brought under, and the ship escaped destruction. Soon afterwards she sprang a leak, and had to put back to Cove, but all damage was set to rights in a few days, and on the 26th February the fleet put to sea again. This time all went well, and on the 1st April the fleet was assembled in Barbadoes.

After a short stay there, the Thirteenth was sent on to San Domingo, in which island it remained for some months, helping to put down a rising of brigands. While doing this work the Regiment, which till then had been very healthy, was attacked by the scourge of the West Indies—yellow fever. Much has been written about the awful ravages of the disease in those days. It is only necessary to say here that the Thirteenth suffered as others did. Men died daily, and at last the Regiment was so reduced that it had to apply to the Fifty-sixth Foot for help to bury its dead. How many were left alive does not appear, but by the end of the year the remains of the Regiment had arrived in Jamaica.

It is not easy to follow in detail the course of the campaign against the Maroons; but it seems that though only two troops of the Thirteenth were employed in it, the command of the whole expedition was eventually given to Lieut.-Colonel the Hon. George Walpole of this Regiment, and that after some hard jungle fighting and mutual ambuscades the Maroons surrendered to him, on a promise that they should not be deported. The Jamaica Government broke this engagement, and voted Walpole a sword of honour, which in the circumstances was naturally declined.

The Regiment remained in the West Indies until August 1798, when, after transferring some 95 men to the Jamaica Dragoons, all that were left, 52 in number, chiefly non-commissioned officers, sailed under the command of a Lieutenant for England. Of these 52, many were found on arrival to be totally unfit for service, and were invalided. Most of those not immediately invalided were "completely exhausted and worn out," and were gradually discharged. The Regiment had in fact ceased to exist. During the two years and six months of its absence, though it had lost only one man killed in action, it had left behind it, dead of disease, 19 officers, 7 quartermasters, 2 volunteers, and 287 non-commissioned officers and men. Such were the conditions of service at that time in the West Indies.

But the war with France was now in full course, and Cavalry was necessary, so the Commander-in-Chief gave orders that the Thirteenth be augmented to a strength of 641 men with the same number of horses. As practically nothing remained of the old Regiment but a few officers, this meant raising a new one. Nevertheless, by August 1799, the task had been accomplished, and two years later the strength had reached 902. The short-lived Peace of Amiens in 1802 caused it to be reduced again, after the custom of the times, by about one-half, but the reduction was as short-lived as the peace, and in 1805, when Napoleon had assembled his great army at Boulogne for the invasion of England, the Regiment stood at the highest strength it ever reached, 1064 men, and the same number of horses. From this time on until 1810 the Thirteenth was kept at home. It was then no longer an Irish Regiment, but a trace of its old connection remained in the fact that it now had as one of its squadron commanders Colonel

Patrick Doherty, who had sailed with it for the West Indies in 1796, and that two of his sons were serving in his squadron.

So far the war record of the Thirteenth can hardly be said to have been fortunate. In the ninety-five years of their existence they had served with no special distinction in the two Jacobite rebellions, and in one campaign abroad, where their chief enemies had been climate and disease. But this long period of inglorious and yet trying service was now over. In the next five years, before their first century came to an end, they were to cross swords again and again with the finest soldiers in the world, to learn the lessons of war under the greatest of English commanders, and to win for themselves imperishable renown.

In February 1810 the Regiment was ordered to prepare 8 troops for immediate service abroad, and before the end of the month they were on board ship. They left behind 2 troops in depot at Chichester, and parted with their Commanding Officer, Colonel Bolton, who had done much to raise and shape the new Regiment after the West Indian campaign. He had just been promoted, and was succeeded by Colonel Head from the Twelfth Dragoons. The 8 troops for active service each numbered 85 men and 85 horses, or 680 men with officers. Before the end of March they had disembarked at Lisbon.

The Thirteenth were about to take part in the famous Peninsular War. Wellington had already given the French some rude shocks in this quarter, and was soon to establish his reputation as one of the first soldiers in Europe. He had clearly recognised the power of offence given to Great Britain by her Navy, which was now supreme, and he believed that by clinging on to a foothold in Portugal, he would in time be able to deal a heavy blow to the military strength of Napoleon, which must be strained by a protracted struggle at this distant point of the Empire. It was a fine conception, and the event proved that he had judged correctly. But at the moment his prospects seemed to be very doubtful, if not hopeless. Napoleon had large armies in Spain, fully 300,000 men, commanded by some of his most famous Marshals, while the British force in Portugal was not a tenth of that number, and badly organised. The Spaniards were evidently incapable of defending their country, or of giving any effective help in defending it; and Portugal was not strong enough, or united enough, to do much against such an enemy. Wellington himself was as yet a man of no great weight in Europe, a mere sepoy General, to use Napoleon's words, who was regarded as fit only to fight Asiatics. He was thwarted and decried in England, where such successes as he had gained were minimised by party rancour. Some of his countrymen even wished to omit his name from the vote of thanks accorded to the troops under his command, and the force itself was full of complaints and discontent, chiefly on the part of the officers. It belonged to an Army which had been discredited by almost constant failure since the War of American Independence. Even in its own country it was not highly regarded. And if the British Infantry was now beginning, under Wellington's command, to win some measure of the reputation it was soon to gain as the best in Europe, the British Cavalry was, both in numbers and training, greatly inferior to the magnificent squadrons of France. When the Thirteenth landed in Lisbon there seemed little likelihood of a brilliant future for them. Happily the British soldier is not greatly disturbed by the prestige of his enemies, and individually both men and horses were better than the French. Above all, our troops had now a leader whose indomitable spirit was proof against all discouragements.

The Thirteenth were soon in the thick of the fighting, but at first they seem to have been rather helpless. It is recorded that in July of that year, 1810, the Regiment for the first time found itself in bivouac, "and both the officers and men were perfectly ignorant what to do. . . . Nobody knew what was to be done for food, forage, &c. Provisions were served out to the men by the Commissary, but how to cook them was another matter." They were soon taught how to find shelter and feed themselves, but this was the doubtful beginning of a campaign in which they were to oppose the war-seasoned troops of Napoleon. Nevertheless, within a few weeks of that date some of them had twice successfully encountered the enemy's horsemen, a troop of the Thirteenth on the second occasion charging through and capturing more than fifty French Dragoons.

After this, during the summer, the Regiment suffered severely from sickness, which, however, did not prevent them from being present at the battle of Busaco on the 26th September 1810, when Masséna was met and severely checked in his famous invasion of Portugal. They were not actually engaged, but were observing the plain in the left rear of the force while the battle was fought. As every one knows, Masséna was eventually stopped by the lines of Torres Vedras, and had to retreat. During the autumn and winter the Thirteenth remained in the country not far from Lisbon, watching the French and learning their work in many a rough march.

For some time it is said French and English Dragoons lay on opposite sides of the Tagus, and the retreat being for the time at an end, the Thirteenth used to have frequent field-days on a plain by the river. The vedettes by mutual arrangement refrained from firing on each other, and the French officers used to come and look on, sometimes when the river was low exchanging conversation with their friendly enemies. It was in some ways a chivalrous warfare, in which, however, the unfortunate Portuguese suffered terribly from the wasting of the country and exhaustion of supplies.

Then, in the spring of 1811, the enemy retired to the northward and westward; and a force under Marshal Beresford was sent to intercept communications from the south. The Thirteenth formed part of this force, and while under Beresford's orders it had the luck to be engaged in a brilliant affair which has since formed the subject of much controversy. The town of Campo Mayor had been taken by the French under Latour Maubourg, and was occupied by a force of 1200 Infantry and over 800 Cavalry, with some Horse Artillery and a battery train of sixteen heavy guns. On Beresford's approach this force evacuated Campo Mayor and retreated on Badajos, ten miles away. The British Cavalry was sent in pursuit and overtook the enemy. The action that ensued is not altogether easy to understand; but the Thirteenth charged, and after some very gallant hand-to-hand fighting, broke the opposing French Cavalry, pursuing them up to the gates of Badajos, capturing the whole siege train, with great quantities of waggons and stores, and leaving the rest of the garrison to be followed up and secured by Beresford's heavy Cavalry and guns. The Thirteenth were naturally pleased and proud at their success against a very superior enemy; but, by a mistake which was not fully explained at the time, the advance was stopped, and the Thirteenth given up for lost. They rejoined the force in safety; but Beresford, misled by false information, believed they had shown want of discipline after the charge, and reported in that sense. Wellington, at a distance, and as Fortescue says, "always justly sensitive over the ungovernable ardour of his Cavalry," accepted Beresford's view, and referred to the Thirteenth in stinging words. "Their conduct," he wrote, "was that of a rabble, galloping as fast as their horses could carry them over a plain after an enemy to which they could do no mischief after they were broken. . . . If the Thirteenth Dragoons are again guilty of this conduct, I shall take their horses from them, and send the officers and men to do duty at Lisbon." This threat was not communicated to the Regiment, Beresford having meanwhile learnt something of the truth; but the Thirteenth were nevertheless severely censured for impetuosity and want of discipline. This censure, as may be supposed, they deeply resented. Napier, in his 'History of the Peninsular War,' says that "the unsparing admiration of the whole army consoled them." No doubt to some extent it did, but not entirely.

Fortescue, after a detailed examination of the incident, sums it up as follows: "Of the performance of the Thirteenth, who did not exceed 200 men, in defeating twice or thrice their numbers single-handed, it is difficult to speak too highly. Indeed, I know of nothing finer in the history of the British Cavalry." . . . "But more important than all was the admission of the French that they could not stand before the British Cavalry." Yet, owing to the mistakes of their superiors, the Thirteenth never received for their feat the honour they deserved, or indeed, officially, anything but blame. It was a signal instance of the ill-fortune which sometimes attends upon the noblest conduct.

Whatever may be said of this, the Thirteenth had, at all events, the satisfaction of knowing that they had been thoroughly successful. They were not always to be so, for on the 5th April, less than a fortnight later, a troop of the regiment was surprised by French Cavalry during the night. They were not on outpost duty, having been regularly relieved, and they supposed that their front had been secured by the relieving squadron, a body of Portuguese Cavalry under British officers. The men of the Thirteenth had eaten nothing for two days, and were faint for want of food. After getting a meal, they lay down by their horses, and were sleeping peacefully when the French, who were retiring and came upon them by chance, dashed suddenly among them with the sabre. Two officers and twenty men escaped in the darkness, but the other two officers with practically all the rest of the men were taken prisoners. It is characteristic of warfare in those days that among them was the wife of one of the troopers.

Then there was another turn of the wheel. Ten days after the surprise it was reported that a body of French Cavalry was at Los Santos, levying contributions. The British Cavalry advanced to attack them, and Marshal Beresford himself rode with the Thirteenth, whom he had so severely censured less than a month before. A sharp fight ensued, ending in the rout of the enemy, who were pursued for about nine miles and lost some hundreds of prisoners. The loss of the Thirteenth was very small.

The next month saw the bloody battle of Albuera, which forms the subject of one of Napier's most famous chapters. During the day the Thirteenth were employed in holding off the enemy's Cavalry. They were exposed to severe fire from Infantry and guns, but were successful in carrying out their duty without heavy loss.

There was much hard work for the Thirteenth during the remainder of this year, 1811, and one incident is noteworthy. On the 21st November, Lieutenant King, a fine young officer, was shot by Spanish guerillas when carrying a flag of truce to the fortress of Badajos. His body was recovered by the French and buried with all military honours on the ramparts, General Philippon assembling the whole garrison under arms for the purpose.

During 1812 the Thirteenth again saw some rough service. They shared in the advance to Madrid and Alva de Tormes, and then in the retreat back to Portugal, during which their horses suffered terribly from hardship and starvation.

In April 1813 the British army advanced again, and again reached Alva de Tormes. In June the French took up their position at Vittoria, and the famous battle ensued. The share of the Thirteenth in this combat was interesting. After some sharp fighting they captured King Joseph's carriages and equipment, and then pressed on in pursuit of the beaten enemy, whose losses were great, including over a hundred and fifty guns. Vittoria was in fact the break-up of Napoleon's power in Spain, for many of his commanders and troops had been withdrawn the year before to strengthen his army for the Russian campaign, and he was never able to replace them.

Then followed the march to the French frontier and the battles of the Pyrenees. In November the Thirteenth crossed the border.

The winter was a hard one for the Cavalry. Hilly country intersected by deep ravines, exhausted of supplies, and obstinately defended by Soult and his veterans, was a rough scene for outpost duty. There were many small affairs, especially between foraging parties. The weather was very bad, and the troops had constantly to bivouac in the mud, under torrents of rain, sometimes in snow. There was often no corn or straw for the horses, nothing procurable but gorse, which, pounded and made into a sort of paste, Irish fashion, just kept the poor beasts alive.

One incident which occurred near Orthes, on the 27th February 1814, is striking. The Thirteenth there came in contact with Soult's Cavalry, and charged. At their head rode their Lieutenant-Colonel, Patrick Doherty, with his sons, Captain and Lieutenant Doherty, three abreast. The charge was completely successful, and many prisoners were taken, among them two officers.

Napier has told us how, through the spring of 1814, that fierce fighting went on, in snow and rain and misery—the French, now overmatched, losing battle after battle and many thousands of men, but still, under their indomitable leader Soult, turning to bay again and again. Then at last came the battle of Toulouse, and the white cockade began to show itself, and on the 13th April it was known that peace had been declared. Napoleon had fallen. Soult fought on for five days more, but then it was announced in general orders that hostilities had ceased, and the British Cavalry in pursuit beyond Toulouse desisted from further action.

The Thirteenth had fought almost without interruption for four years, in the long struggle that began at Lisbon and ended at Toulouse. They now had a few weeks' rest, and it was badly needed. Numbers of horses, worn out by want and hard work, had to be destroyed, and the men were in rags. No clothing had been issued during the winter. "Overalls patched with cloth of all sorts of colours, and most frequently of red oilskin—fragments of baggage-wrappers by the way—were universal or almost so." They were indeed "The Ragged Brigade," as they and their old comrades of the Fourteenth had been named. But, starting in May, they marched up through France, and arriving at Boulogne on the 5th July, embarked for England. By the 8th July the Regiment had all been landed in Ramsgate. During an absence of four years and five months the Thirteenth had marched 6000 miles, and had been engaged in twelve battles and thirty-two "affairs," many sharply contested. They had lost by death six officers and 270 men. But the Regiment had now made its mark, and was thenceforward one of the foremost fighting corps of the British Cavalry.

After their return from France the Thirteenth spent some months in England and Ireland, but their enjoyment of peace was short. In February 1815 Napoleon escaped from Elba, and war again broke out. On the 20th April, having meanwhile received royal authority to bear on its guidons and appointments the word "Peninsula," the Regiment was ordered to prepare six troops for immediate service, and soon afterwards the number was increased to ten. In May the Thirteenth were in Ostend (with twenty-eight women and nine children), and by the end of the month they formed part of a force of 6000 Cavalry, under Lord Uxbridge, which was inspected by Wellington and Blücher.

Then followed Quatre Bras and Waterloo. The movements of the Thirteenth up to the morning of the decisive battle are of no special interest, but it seems that having been ordered to join a Brigade consisting of the Seventh and Fifteenth Hussars under Major-General Grant, the Regiment arrived at Quatre Bras on the night of the 16th June, and shared in the retreat of the 17th June to Waterloo. It was a dreary day, for the rain was heavy and they got no food—a bad preparation for the coming battle. Then followed "a dreadful rainy night, every man in the Cavalry wet to the skin," and at four o'clock in the morning of the 18th, the Thirteenth "turned out and formed on the field of battle in wet corn and a cold morning without anything to eat." Their commanding officer, the gallant old veteran Colonel Doherty, had broken down and was lying ill in Brussels, so the Regiment was commanded on the 18th by Lieut.-Colonel Boyse. The Brigade to which it belonged was posted on the right centre of the army, in rear of Byng's Brigade of Guards, who held the house and garden of Hougomont. From this position the Thirteenth witnessed the furious fighting which ensued between the Guards and their French assailants, and they came themselves under heavy Artillery fire, which caused them some loss. Colonel Boyse had his horse killed under him by a cannon-shot, and was severely hurt, the command devolving on Major Lawrence. Two other officers were wounded. There was also severe and repeated Cavalry fighting, in which the Thirteenth did their share, charging more than once the enemy's horsemen, and on one occasion dispersing a square of French Infantry. In this fighting they lost three officers killed or mortally wounded,[1] and two more wounded by sabre cuts. Towards evening the French made another desperate attack with both Cavalry and Infantry, and the Thirteenth charged again, losing three more officers wounded, among whom were both the Doherty brothers. Before the enemy finally gave way almost every officer of the Regiment had lost one horse at least, and Major Lawrence had lost three. When at last the French broke, the Brigade was sent in pursuit, and pressed the routed enemy until nine o'clock. Then it was halted, and the pursuit was handed over to the Prussians. "The last charge," wrote an officer of the Thirteenth, "was literally riding over men and horses, who lay in heaps." And the account goes on to say that "when the Regiment mustered after the action at 10 P.M. that night, we had only 65 men left out of 260 who went into the field in the morning."

Many rejoined later, and these figures do not represent the actual losses as afterwards ascertained, but so far as can be judged the total of killed and wounded was close upon a hundred, of whom eleven were officers.

After Waterloo, the Thirteenth marched to Paris, where they remained some weeks, and then they were sent northward again. At or near Hazebrouck, a name now so familiar, they remained some months. In May 1816 the Regiment returned to England, arriving at Dover on the night of the 13th. During the past year it had lost in killed, died, and discharged, 3 officers and 65 men.

With Waterloo ended the first century of the Regiment's service. If ninety-five years of it had been rather colourless, the

[1] Captain Gubbins, Lieutenants Geale and Pymm.

last five had certainly been as full of fighting as any one could have desired.

For about three years after its return the Thirteenth remained in England. The times which followed the war were bad, and the Regiment was often employed maintaining order among the civil

population, always a detestable duty for soldiers, but nothing of note occurred. On the 9th February 1819 the Regiment sailed for India, and for the next twenty years it rested peacefully in Eastern cantonments.

In India, as well as in Europe, the beginning of the century had been a time of hard fighting in various fields, and when the

Thirteenth went out, the supremacy of the British among the Indian country powers had hardly been established. It was only sixteen years since Sir Arthur Wellesley had routed the Maratha armies at Assaye, and gained his first great victory. After that time other powers had challenged the British, and been with difficulty overthrown. Even in 1819 there remained serious elements of disorder, and it was not until seven years later that a period of complete peace began. Nevertheless, it may be said that the period of general war closed in Asia as in Europe soon after the fall of Napoleon.

The Thirteenth at all events had no fighting to do. They were sent to the extreme south of India, where the name of their old chief was very familiar, and the provinces about Bangalore, where they were quartered, had many fighting traditions; but nothing occurred to test the spirit of the Regiment. In that very pleasant place, and other stations not far distant, the Thirteenth remained year after year, with little to disturb them except inspections and reviews, enjoying plenty of sport, after the manner of British Cavalry Regiments in the East, and maintaining their efficiency in so far as it could be maintained without service in the field. In 1832 a formidable plot was discovered for a native rising in Bangalore. The Thirteenth with a British Infantry Regiment, the Sixty-Second, and a detachment of European Artillery, were to be suddenly attacked at night and massacred, after which the conspirators hoped that a general mutiny of the Native Army would follow. But the plot was revealed by a faithful native officer, and was crushed without any fighting.

Nevertheless it had shown that there was disaffection among the Indian population, and a few years later this came to a head. In 1839 it was found that a certain Mahomedan chief, the Nawab of Karnul, had collected in secret a large quantity of military stores, including some "hundreds" of guns, and that he had in his employ a considerable number of sturdy fighting men, Arabs, Rohillas, and Pathans from the North-West of India—the turbulent mercenaries who had for generations made India a vast battlefield. The matter was considered so serious that a force of 6000 men, of which two squadrons of the Thirteenth formed part, was sent to Karnul. Action had been taken in time, and the fighting on the part of the enemy at Karnul and the neighbouring village of Zorapur, though brave enough, was soon over. A few British officers and men were killed and wounded. The Thirteenth lost more than thirty men, chiefly from cholera, on this expedition, but none by the sword. It was one of the countless forgotten skirmishes upon which the Indian Empire has been built up.[1]

Early in 1840, after twenty-one years spent in the country, the Thirteenth sailed for home. They had seen little fighting, but those were days when India claimed a terrible toll from British troops, and during the short march from Bangalore to the coast at Madras the Regiment lost from cholera forty more men, as well as many women and children. Cholera is no longer the scourge that it was to our countrymen, but the thousands of graves that one finds scattered over the face of the land, often in the loneliest places, are a sad reminder of the price Great Britain has paid for her Eastern dominion.

On return to England the Regiment was very weak, for in addition to its losses from disease, it had left behind many men who had volunteered for other Regiments in India; but it was soon in good order again. It was to be replaced in India by the Fourteenth, and in 1841 the two Regiments, "The Ragged Brigade" of the Peninsular War, met again in Canterbury. There can have been few officers in either who had served together in that war, but the old traditions were still alive, and in remembrance of them the Fourteenth presented to the sister Regiment their mess-table, which had been originally captured by the Thirteenth at Vittoria with King Joseph's household.

During the next ten years and more the Thirteenth served in the United Kingdom, and there is little to record of their doings. In 1852 they formed part of the troops who followed the funeral of their old chief, the Duke of Wellington, and in the next year they attended the first camp of exercise held in England. The Duke had originated the idea. The camp was a success, and proved to be the precursor of many more such gatherings. But something more than camps of exercise was now before the Regiment. In 1854 came war with Russia, and the Thirteenth were warned for service in the field. By the middle of May they had sailed for

[1] Oriental names in this book are not always spelt consistently on one scientific system. For example, it has not been thought proper to alter the original spelling in letters quoted, or to correct well-known names such as Lucknow or Bangalore.

the East. It is memorable that they were now once more commanded by a Lieut.-Colonel Doherty.

The Regiment had some rough work in European Turkey before going on to the Crimea; but in September 1854 it landed at Eupatoria with the Allied forces, and on the 19th of that month it marched towards Sebastopol as part of the Light Brigade under Lord Cardigan. The opening day of the march brought the Thirteenth under fire, and they had a few casualties, which were, it is said, the first in the British army. The enemy withdrew with some loss, and on the following day the whole of the Allied armies, British, French, and Turkish, advanced to the banks of the Alma.

Every one knows the story of the battle which ensued—how the Allies attacked the Russian troops holding the heights across the stream, the British redcoats on the left, the dark masses of the French and Turks on the right; and how, after some stubborn fighting, the Russians were driven off the heights upon the plain beyond. The Light Brigade had little to do during the attack. Their position was on the left front of the British army, where they remained all the afternoon, watching the development of the struggle, and facing large bodies of Russian Cavalry and guns which threatened the flank of our advance. They never did more than threaten, and the Light Brigade had no fighting, though there was some slight loss from Artillery fire. When finally the Russians were driven off the heights, the British Cavalry was sent in pursuit, and crossing the stream at a gallop, pressed up and over the hills, to see below them the beaten enemy in full retreat. Some prisoners were made, and the Russian rearguard was watched as long as daylight lasted, but nothing of much importance occurred. The battle had been won, and the Russians had suffered heavily, but the bulk of their troops maintained some order, and there was no general rout. The Thirteenth spent the night in bivouac, on ground which had to be cleared of many dead.

There followed a month of comparative freedom from fighting, while the Allied armies closed on Sebastopol and took up their position for a siege. The work of the Cavalry was constant, in exploring the country and watching the enemy, but until the 25th October no serious encounter took place. Then occurred the famous battle of Balaclava.

It was a day on which, to an exceptional extent, Cavalry shared in the actual shock of battle, and not only as horsemen against horsemen, but in conflict with other arms. The Russian Cavalry, far more numerous than the British, were the assailants, but it may safely be said that at the end of the day, heavy as our losses had been, and unnecessarily heavy, the British horsemen had established for themselves a personal superiority which was never again challenged.

The action began about daybreak, the Russians advancing in great force to attack certain outlying redoubts held by Turks, which were incapable of much resistance. The Light Brigade and a troop of British Horse Artillery moved out to support the defence, but the redoubts fell quickly and were soon in Russian hands. Then the Russian Cavalry, some three thousand or more in number, with over thirty guns, advanced to complete the Russian success, and to threaten Balaclava itself, which, although of vital importance to our army, was for want of numbers very weakly held. First a body of the enemy's horsemen came on boldly against the 93rd Highlanders, who, with some men of the Guards, received them steadily and drove them off by a couple of volleys at close quarters. Then the main body of the enemy's horse came up unseen by our own Cavalry, and suddenly appeared on the heights within a few hundred yards of Scarlett's Heavy Brigade, which was advancing in imperfect formation through the tents and picket ropes of the camp to support the 93rd. Fortunately the Russians, instead of launching their Cavalry mass upon our greatly outnumbered squadrons, came slowly down the slope, and halted—and then the Heavy Brigade dashed into their ranks. Within ten minutes the great mass, nearly three thousand strong, had been riven by successive charges and scattered over the plain with a loss of 400 men, and Scarlett's troopers, a fifth of their number, rode back in triumph.

This feat was performed under the eyes of the Light Brigade, who sat on their horses, impatiently expecting the order to advance and join in the fight, or at all events in the pursuit of the broken enemy. According to Colonel Tremayne, then a Captain in the Thirteenth, some squadrons instinctively fronted that way; and it seems likely enough that if the Light Brigade had charged the enemy in flank while they were engaged with Scarlett's men, great execution might have been done; but Lord Cardigan considered that his orders forbade him to move, and the opportunity was not taken.

Then followed the separate attack about which so much has been written. It appeared to the British Commander-in-Chief that the Russians were about to carry off the guns captured in the lost redoubts, and he directed the Cavalry to advance rapidly and try to prevent them. What exactly the Light Brigade was ordered or meant to do has been a matter of endless controversy; but this much is certain, that 673 officers and men, of five Regiments, charged a Russian battery in position at the end of a valley a mile or more in length, other Russian batteries and bodies of Infantry being on the sides of the valley, and some Russian Lancers and Hussars in rear of the battery attacked. The Thirteenth, now reduced to a strength of 8 officers and a little more than a hundred men, were with the Seventeenth Lancers in the front line. Behind them came the Eleventh, Fourth, and Eighth Hussars. In front of all, straight into the enemy's guns, rode the Commander of the Brigade, Lord Cardigan. He had been much criticised for habitually sleeping on board his yacht, which lay in the harbour below, and other allegations were made against him, but there was at all events no question as to his courage. The charge was a mad one, due to some misconception. The Brigade reached its objective, but was practically destroyed in the course of the charge and return. When the remnants of the Light Brigade re-formed in rear of the Heavy Brigade, which had not been sent in, it was seen that the losses had been ruinous. The Thirteenth was represented by 1 wounded officer and 14 men. Others rejoined later, but the Regiment lost that day 3 officers and 11 others killed, 12 men taken prisoners, and 30 wounded. The officer in command, Captain Oldham, was among the killed. It was a fatal ride, and the Brigade was sacrificed to little purpose; but officers and men had obeyed their orders with splendid devotion, and it is no wonder that among the Regiments which formed the Brigade the memory of "Balaclava Day" is held in everlasting honour.

The Thirteenth was present a few days later at the bloody battle of Inkerman, where it had no chance of doing anything, and then went through the miseries of the Crimean winter, when men and horses suffered terribly from want of food and clothing. In February the effective strength of the Regiment, exclusive of officers, was 5 mounted men—namely, 1 sergeant, 1 trumpeter, and 3 privates. Lieut.-Colonel Doherty, who had been absent, ill, on the day of the famous charge, was now fit for service again, but that was the strength of his command. During the ensuing spring and summer, successive drafts brought up the number, and when the battle of the Tchernaya was fought in August 1855, the Thirteenth turned out 200 strong. They came, it is said, very near disaster again that day in consequence of an order by the Sardinian General della Marmora, to whom their services had been lent, and were only saved by the interposition of Marshal Pelissier from another hopeless charge at a Russian battery. After the fall of Sebastopol the Regiment had some more hard work in small expeditions, but no more severe fighting. On the 27th May 1856, they were back at Portsmouth.

After their return from the Crimea, the Thirteenth had ten years of peace service in the United Kingdom. There is nothing memorable about this period except that in 1861 or 1862 they became Hussars instead of Light Dragoons.

In 1866 the Regiment was suddenly ordered to Canada, where the Fenian conspiracy had given rise to some excitement; but the invasion proved a fiasco, and the Thirteenth saw no active service. They returned to England in 1869.

In 1870 the Thirteenth were once more ordered out to India, and there they remained for fourteen years. This time they were not sent to the south of the great peninsula, but to the north, to Hindustan proper, where there was more chance of stirring times. India, as one of the best of its Viceroys used to say, is a country where "the bottom is always dropping out of the bucket," and the task of putting matters to rights generally falls to the troops in the north, where the bulk of the British garrison is always kept.

There was in fact some active work for the northern army while the Thirteenth formed a part of it, for in 1878 occurred the Second Afghan War, in the course of which there was much fighting in Kabul and Kandahar. But the Thirteenth had not the good fortune to see it. They were, it is true, sent to Kandahar in 1880, remaining across the border about a year, but in that part of the country the fighting was over, and they returned to India.

In 1884 the Thirteenth left India for home again, but they were sent by way of South Africa, where, in Natal, they spent a year.

From Natal they returned to England. In the United Kingdom they spent the next fourteen years.

Thus it appears that from 1856 to 1899, forty-three years, the Regiment never had the good fortune to see a shot fired in anger. But a considerable part of that time was spent in various parts of the world, in Canada, Asia, and Africa, and the experience gained in such service is not without value. If at times prolonged absence from home may entail a certain loss of smartness, it has its compensations.

In the autumn of 1899 the Boer republics issued their ultimatum to Great Britain, and crossed the border of Natal. The Thirteenth was among the Regiments immediately sent out to strengthen the British forces in South Africa; and as by the time it arrived Sir George White was besieged in Ladysmith, it was ordered to join General Buller's relieving force in Natal. On the 12th December it was included in Lord Dundonald's Cavalry Brigade. Three days later occurred Buller's attempt to force the passage of the Tugela at Colenso, and the Thirteenth was in action throughout the day. The Adjutant, it may be observed, Captain Tremayne, was the son of one of the eight officers who charged at Balaclava. From this time until the 3rd March, the Thirteenth shared all the rough fighting that took place in trying to break through the strong semicircle of hills held by the Boers. Then the relieving force entered Ladysmith in triumph. After that the Thirteenth served throughout the war, and saw much hard work.

It was not until October 1902 that they returned to England, after an absence of nearly three years. In the course of the campaign they had lost eighty men killed in action or by disease, while four officers and forty-six men had been wounded, and a large number invalided home. The Regiment fully maintained its reputation, and received many honours.

Two uneventful years in England followed, and then for the third time the Thirteenth were sent out to India, where they were still serving in 1910. With the beginning of that year Barrett's history of the Regiment ends. It had then been in existence nearly 200 years, and had served in nine wars, among which were the Peninsular War, the Waterloo Campaign, the Crimean War, and the South African War. In all of these it had done well and distinguished itself. Its reputation, whether in war or peace, stood high.

CHAPTER IV.

1910-1914—OUTBREAK OF THE GREAT WAR.

In the beginning of 1910 the Thirteenth Hussars had been more than five years in India, and again in the south, where their first Indian service had passed. The military station of Secunderabad, in the dominions of His Highness the Nizam, the greatest of the Mahomedan Chiefs of India, had long been one of the strategical points at which a considerable force of all arms was kept, and a British Cavalry regiment almost always formed part of the garrison. It is, or was then, as Indian stations go, one of the pleasantest and most sociable, with some sport to be got in the neighbourhood; and, owing to the size of the garrison, there was plenty of amusement, as well as work, in the Cantonment itself. The Nizam and those about him were always friendly and hospitable.

The Thirteenth were not to be in Secunderabad much longer, but in May, while they were still there, occurred the lamented death of King Edward VII., and the accession of King George. On the 9th May the officers of the Regiment, with a party of non-commissioned officers and men, attended at the British Residency at Hyderabad, the capital of the Nizam's dominions, and there heard read the proclamation announcing the beginning of a new reign. It was to prove one of the most memorable in the history of India.

During the remainder of the hot season, which in the East is necessarily the slack season so far as military training is concerned, the regimental records contain notice of little beyond routine occurrences and sport of various kinds, the football and polo and

THE DRUM HORSE—AT THE DURBAR

tent-pegging with which men and officers while away the heat and tedium of an Indian summer. Then, as the heat slackened and another working season began, the Regiment received orders to move from the south of India to the north, to a station nearly a thousand miles away, among a totally different population and surroundings. The Thirteenth left Secunderabad in the middle of October, carrying with them the hearty good wishes of the garrison, and of the General Commanding the Cavalry Brigade, who warmly praised their work and discipline, and expressed his confidence that they would maintain in the north of India the good name they had borne in the south.

Arriving in the northern plains by train, they marched to their new station, meeting on the line of march the Seventeenth Lancers, with whom they had charged at Balaclava more than fifty years earlier. The two Regiments had not met since. The Thirteenth entertained the Lancers to a camp-fire concert, and then they went their ways again.

Meerut, where the Thirteenth were now to be quartered, was a well-known and favourite station. It was memorable as the place at which occurred the first serious outbreak of the Mutiny of 1857, since which time it had, from its central position and nearness to the ancient capital of Delhi, continued to be a large military station. In 1910 the memories of the Mutiny had grown dim, but Meerut was still an important place from a military point of view. It lay in the centre of "Hindustan," the great northern block of territory which has been the seat of countless Empires, Hindu and Mahomedan—the real India upon which the vast Indian Peninsula has in a measure depended for thousands of years. In its broad plains and teeming cities was always concentrated the military power of succeeding conquerors, and the British, when they took the place of the Moghuls, had, like their predecessors, massed their strength on these northern plains.

Meerut, it may be noticed, was also a centre of sport, the site of an annual polo tournament, and within reach of good shooting and "pig-sticking." The Thirteenth arrived just in time to join in the polo tournament, and to be soundly beaten by their Balaclava comrades of the Seventeenth Lancers. They were also beaten soon afterwards at another tournament at Lucknow, this time by the Rifle Brigade; but every one cannot win, and the Thirteenth were at all events to the fore in every kind of sport.

Meanwhile the usual work of military training began again—drill and swimming camps, and marches, and musketry, and inspections, and much more—the steady hard work of which civilians as a rule have no knowledge, but very real and useful work for all that, as the old Army was to show in the dark days which were coming.

Then followed the summer of 1911, and in the autumn the 13th received news of the death of their Colonel-in-Chief, General Sir Baker Russell. He was succeeded by General Sir Robert Baden-Powell.

But this year, 1911, was not to close with another round of customary training. King George had shown from the first, as his father and Queen Victoria had shown before him, a keen interest in his Indian Empire. As Prince of Wales he had visited the country already; now he had decided to visit it again as King-Emperor, and to take his seat in person upon the Imperial throne. It was a momentous decision, and was to have a great effect upon the Chiefs and people of India—how great an effect those only can know who have studied and in some measure understood the traditions and feelings which thousands of years of kingly rule have implanted in the Indian mind. Happily King George understood, and had resolved to take the unprecedented step of leaving England for months to gratify the desire of his Indian subjects. In the whole history of India no such ceremonial had ever been held, for vast as the Empire of the Moghuls had been, it had never embraced the whole of the Eastern dominions now under the British Crown, nor had it formed part of a wider Empire extending to all the continents of the world.

Among the preparations being made to invest the ceremonial with due pomp and splendour, was the assembly at the Imperial Camp of a military force drawn from the Army of India. The occasion was not primarily a military one, and the numbers of the force were limited; but 50,000 troops, British and Indian, were being drawn together to represent the armed might of the greatest power in the East, and to show that if ever he chose, the British Emperor of India would be able to throw into the scale of any

D SQUADRON—AT THE DURBAR

world-conflict an army in which the military efficiency of the West would be blended with the loyal devotion and numbers of the Indian fighting races. Among the Regiments which had the honour of being included in the representative force at Delhi was the Thirteenth Hussars.

The various pageants which took place have been described in detail by Fortescue, the historian of the British Army, who accompanied the King to India. The great Durbar at which the King took his seat upon the throne was a wonderful scene, all classes of the Indian population joining to do him honour, from the humblest to the great feudatory chiefs and their retainers, blazing with jewels and gorgeous clothing and antique armour. The Thirteenth did their part among the soldiers, of whom Fortescue says: "The troops formed the most essential part of the pageant." Besides the Durbar, there were many other interesting ceremonies and amusements—the presentation of colours, receptions, polo and football matches, and so on. But the whole did not last many days. The vast encampment, covering twenty-five square miles, which had risen as if by magic, with its myriads of tents and its luxurious gardens, from the solitude of a barren plain, was gone before the end of the year. The Chiefs of India marched away with their brilliant retinues, the troops and the people were scattered in every direction, and the plains about Delhi relapsed into something like their old lonely peace. But before he went the King had announced with dramatic suddenness, to the astonishment of the great assembly, that Delhi was again to be the capital of India, and that the British Empire, which had risen from the sea, and had hitherto had a seaport for its capital, was for the future to be centred, as former Empires had been, on the plains of Hindustan, surrounded by the territories of the Indian chiefs and the lands of the great Indian fighting races. It was a landmark in the history of India.

To the officers and men of a British Cavalry Regiment the full significance of the ceremonial could hardly perhaps be apparent, and certainly they could not foresee the world-war which was soon to show how fortunate in its consequences had been the King's act in coming to India at the beginning of his reign. Pageants are hardly to the mind of a soldier. Still, the Thirteenth had their part in it, and did well what they had to do. The Regiment was conspicuous among those reviewed by the King, and at the close of the ceremonial it was selected for the honour of furnishing a squadron to escort the Queen during her visit to another ancient capital, Agra. The squadron was under the command of Captain W. H. Eve. Fortescue writes of it: "We had remarked the Regiment at Delhi; but even so we were not quite prepared for what we saw on that Sunday. All the officers of the suite agreed that the escort was the most perfect they had ever seen, so admirably were the distance and the dressing preserved. This may seem to be a small matter, but such details count for much in the discipline of a regiment, for those that are careful in small matters are unlikely to be careless in great. Moreover, it is a real pleasure in this imperfect world to see anything faultlessly done."

Fortescue's words may perhaps seem exaggerated: smartness and discipline are not necessarily the same thing. But they are nearly allied, and there is perhaps no greater mistake made by civilians in judging soldiers than the contempt for drill and "the barrack-yard" which is so readily expressed. Henderson writes in 'The Science of War': "It is unfortunately to be apprehended that few, except professional soldiers, understand the nature or the value of discipline." And he shows very clearly how necessary is the "habit of obedience" for efficient action in war. It was not for nothing that the great American soldier Stonewall Jackson began his career in the Civil War by drilling his undisciplined soldiery until he made himself detested by the officers and men who afterwards learnt to worship him. His brigade stood "like a stone wall" in their first battle when all was melting around them, and earned him the splendid nickname which has become immortal. History teems with instances of the supreme value of the trained soldier in war. Never was it shown more conspicuously than in that wonderful month of the retreat from Mons, when the little army of British regulars went back day after day before the overwhelming numbers of their enemy, only to turn on him at the end and prove to him that in spite of all their losses and sufferings their spirit and efficiency were still unbroken. "It is open to those in whose ears the very name of discipline smacks of slavery, to assert that a powerful instinct of obedience dwarfs the intellect, turns the

THE QUEEN AT AGRA

man into a machine, and rusts his power of reasoning; and in this there is a shadow of truth, but it is only a shadow." It is a question which has been often debated, and in which, *primâ facie*, the contemptuous critic seems to have much right on his side; but to few who have seen war will his view commend itself. The Regiment which shows up well in the manœuvres of the parade-ground will rarely fail to show itself efficient in the field. Like everything else, the principle is capable of abuse, and may be carried too far, but it is a sound principle in the main. Certainly the squadron which won Fortescue's admiration went very straight when it was tried a few years later in something more than escort duty.

The Durbar and its attendant ceremonies at an end, the Thirteenth marched back to Meerut, and the old life of military training and sport began again. There were rifle meetings and inspections, drill and manœuvres, courses in musketry and signalling and machine-guns, polo and races; and then the hot weather of India came once more with its blinding sandstorms and weary nights of heat, when sleep was hard to get and life seemed hardly worth living. There was some sickness too, and the terrible spectre of plague cast its shadow over the Regiment. The men faced the shadow cheerily enough, playing football and hockey and having boxing competitions after the manner of the British soldier; but one or two died, and the Regiment had to be inoculated. The officers kept themselves fit with polo and the swimming-bath. July brought some welcome rain, two or three good showers a week, and the Review report of the General Commanding the Northern Army was received: "A fine regiment, fit for service." But it was a trying time, as an Indian hot weather in the plains always is. India is a picturesque country, full of beauty and romance for those who have eyes to see, but it has its drawbacks. English women face them as well as men. The following extracts are from the letters of a lady who decided to brave the heat with the Regiment.

February 15, 1912.—"The weather has suddenly got very hot. . . . The Inter-Regimental week starts on the 4th of next month, and goes on for about a fortnight. To feel I've got to entertain people for a fortnight is a nightmare!—this place doesn't suit me, and I never feel well. At the last moment —— may be sent up to the hills with the invalid party, but it doesn't look like it, and he's not down for a day's leave of any description."

February 21, 1912.—"We have heard nothing about the Regiment being moved this year, so I suppose we shall stay on here. I have decided to try and stick out the hot weather with ——. I should like to have come home, but if I do —— won't go away at all by himself, and if I have to go away and go somewhere to a hill station he will come too if he can get any leave. Of course every one tells me that no woman can do a hot weather here, but I shall try. . . ."

April 3.—"We have had a nice cool week, for which everybody is very thankful. There was a terrific thunderstorm at the end of last week, and the temperature dropped from 103 to 83, so you can imagine it was a change. We all shivered, but it was lovely. It is warming up again now, and the last two days have been 100 or over in the shade in the middle of the day.

"The early routine has started now and —— has to be up at 4.45, and gets done about 10.30, when he comes in and has breakfast. We generally lie down in the afternoon and try and sleep, getting up about 4 for tea, before going to polo or playing tennis. Nearly every one has gone away on leave, and the place is very empty and desolate."

April 18.—"There is no news to tell you from here—the hot weather is always a dreary time of forced inaction and perpetual discomfort. We are sleeping out of doors every night now with no sheets or blankets to cover us, so you can imagine it is pretty warm. One generally falls into a dead sleep just before the dawn, which is the only cool time during the twenty-four hours. I change my clothes five times during the day—it is one form of exercise. We are both keeping fit, which is the great thing. . . .

"We had a terrific sandstorm here on Tuesday. We could see it coming for miles as the sky was a bright yellow; unfortunately we were caught in it as we were out driving; it was filthy, and we got covered from head to foot with sand. The storm lasted two hours, but we didn't get a drop of rain. If only we had had some rain it would have been cooler for a few days."

So it went on for many months longer, through the blazing

THE ESCORT AT AGRA. 1ST AND 3RD TROOPS OF SQUADRON D

hot weather and the sultry depressing rains. Then began another cold season.

On the 1st November the Thirteenth won the final in the Meerut Polo Tournament, after a desperate struggle with the King's Dragoon Guards. An officer of the Regiment who had been studying the more scientific parts of his profession left for the Staff College at Camberley.[1] There was a Cavalry concentration camp, where a considerable mounted force was assembled for Divisional training, followed by manœuvres of several Divisions together. In the midst of all this soldier work the year was closed by an incident which startled and shocked India. It had been arranged that on the 23rd of December the Viceroy, Lord Hardinge, was to make a State entry into the new capital, and some of the Thirteenth had been sent to join in the ceremonial. The Regimental Diary records very briefly that "a dastardly outrage occurred, a bomb being thrown at the Viceroy, which resulted in his serious injury." It was a painful commentary upon the enthusiastic greeting which had been given to the King-Emperor on the same spot just a year before, and a reminder that in India there exists always a root of sedition and danger which must not be disregarded. Peace in India is armed vigilance. But happily disloyal sentiment is confined to a small minority, and the heart of the great Empire is sound. So Englishmen felt. They showed a fine example of coolness and moderation in face of the treacherous attempt at murder, and all went on again as before. If the traitors had expected to intimidate the white man they were wholly mistaken.

On the 1st of January 1913, ceremonial parades were as usual held throughout the country to celebrate the assumption by Queen Victoria, more than thirty years before, of the title of Empress of India, and over the momentarily troubled waters the ship of Empire sailed forward undisturbed upon its stately way.

During the rest of the year there was from the point of view of the Thirteenth nothing of much importance to record. The Regimental Diary mentions that the English system of messing was introduced for the first time in India; that "C" Squadron won a silver challenge cup for shooting open to all squadrons, batteries, and companies in the Division; and that there was a short spell of "experimental

[1] Captain Norman Neill, afterwards killed on the Western Front.

D

training" in camp, when the Regiment lived entirely on the resources of the neighbouring country. Beyond these incidents, the Diary touches upon little but the doings of the men at cricket and boxing, and "skill-at-arms" competitions, and hockey and football tournaments. Hot work they must have been, for there is this entry referring to the months of July and August: "During these two months the average temperature was about 98. The weather was very trying and injurious to health, mainly due to the rain, followed immediately by sunshine, which caused vapours to rise from the ground." To every one who has served in India this quaintly worded sentence brings back a familiar picture. The British soldier who has "heard the réveillé from Birr to Bareilly" knows only too well the dreariness of the late summer, when the faces of the women and children grow white in the reek from the rain-sodden ground.

On the 25th of October, Balaclava Day, the first "Old Comrades Dinner" was held in London, and the Diary notes that among those present were two Balaclava veterans.

With this month of October 1913, began the last working season of the old order. Everything then seemed peaceful enough, and no one thought that before a year had passed England would be fighting desperately in the greatest war of all time. For the Thirteenth Hussars attention was focussed on the usual incidents of an Indian "cold weather." The Diary records that the regimental machine-gun detachment distinguished itself at the Meerut Rifle Meeting by winning a match open to all India, and that there were some tactical field-days with V Battery of the Horse Artillery. The Regiment was to be associated with V Battery in much hard fighting before they had done with each other. Finally, at the close of the cold season, the Commander-in-Chief in India came down to Meerut, and there was a "Garrison Ceremonial Parade," in which the Thirteenth took part. All went well with them, and the inspection was entirely satisfactory. It was the last they were to undergo before being tested by the ordeal of war.

In the summer of 1914 came the fateful news of the murders at Serajevo, and before long it began to be seen that events were tending towards a great European conflict into which England might possibly be drawn. Every one remembers the excitement of the month that followed. In India, as elsewhere all over the world, it was intense. After so many years of peace, or at all events so many years in which England had looked on at European wars without bearing any part in them, it was difficult for Englishmen to believe that the long-standing German menace had really come to a head, and that "The Day" was upon us. It seemed more probable that England would again stand aside, and that whatever the Continental nations might do, no British Army would be sent to shed its blood on European battlefields. Even when Germany turned upon France, and it became certain that we should see war close to our own shores—war by which our own deepest interests must be endangered—it seemed doubtful whether England would take upon herself the tremendous responsibility of throwing her sword into the scale. Until the 4th of August the issue remained in suspense. Then the doubt came to an end, and on the following day it was known all over the British Empire that the old country had chosen the path of honour.

In no part of the Empire had the suspense been more acute than in India, which was full of martial traditions, and, in spite of local treason here and there, full also of goodwill to the British Crown. The sudden knowledge that Great Britain was at war stilled at once the voice of sedition, and was the signal for an outburst of loyalty on the part of Chiefs and people which astonished our enemies, if not ourselves, though it was no new thing;[1] and it need hardly be said that in the military cantonments scattered over the face of the country, where the soldiers of the King's Army, British and Indian, were gathered in constant readiness for war, the announcement was received with joy and eager hope. They might not be privileged to join in the central conflict on the battlefields of Europe, but surely they would have some share in the fighting, some chance of service and honour.

Meerut was no exception, and among all the King's Regiments there was none which looked forward to the war more eagerly and hopefully than the Thirteenth, with its memories of the Peninsula and Waterloo and Balaclava. Some days before war was declared all officers on leave in the country had been urgently recalled, and

[1] In 1885 England was for a time on the point of war with Russia. The outburst of feeling in India was the same then as in 1914, and should not have been forgotten.

when on the 5th of August the Regiment learnt from a telegram to the Meerut Club that the sword had been drawn, it was ready for immediate service. On the 9th of August the Meerut Division was ordered to mobilise. Then followed some weeks of anxiety, during which the Thirteenth were alternately elated and cast down by contradictory rumours. Early in September they received orders to prepare a large draft of men and horses for the Eighth Hussars, which threw them into the depths of depression; then they got, but could hardly rely upon, private reports that they were not to be left in India. It was a trying time.

Meanwhile it had been raining hard, and this added to the general depression. Polo became impossible, and neither officers nor men had anything to relieve the tedium of waiting. The following extracts from the letters of a junior officer may be worth quoting:—

Lieutenant G. R. Watson Smyth—August 9-12.—"I do not know whether this letter will ever reach you, or where I shall be if it does. At the present moment we are awaiting the order to mobilise: it is sure to arrive at any moment now if the Regiment is to go on service. We don't know if it is decided to take the Meerut Cavalry Brigade, but . . . it is possible that the infantry of the Division may be taken. Whether they will be taken to garrison Egypt or to fight at home is another matter. As I said, though, we are just waiting for the telegraphic order before we start shoeing our horses and sharpening our swords. . . .

"It is now two and a half hours since we should have got our orders, and I am beginning to fear that we shall not get them. . . .

"I have just gone to the Club, and a wire has come in saying that the Brigade is not for it. Rotten luck. . . .

"The Native Regiments here are in a sort of fever of excitement, and are longing to have a go at somebody. . . .

"Skinner's Horse are in Meerut with us now. They are an extraordinarily good and very sporting lot.[1]

"There has only been one day's polo for the last month, as all the grounds are under water, and the rain never stopped long enough to let them dry. . . ."

[1] This Regiment, it may be noted, was named after Colonel Skinner, the founder of the Bengal Cavalry system. He was the son of a Scotch officer and a Rajputni girl, so he had fighting blood on both sides.

DRAFT OF MEN AND HORSES DETAILED FOR THE 8TH HUSSARS
SEPTEMBER 1914

THE BAND AT THE LAST CHURCH PARADE BEFORE LEAVING INDIA
NOVEMBER 1914

August 30.—"We are carrying on in the same way as if there were no war in the world. . . . It really is a bit too thick that here are we, the most efficient Cavalry in the world, stuck in this horrid country. . . . Not a hope of our going to war. We have just heard that they are mobilising three other Brigades, and that the Viceroy is coming with the Court to live at Meerut this cold weather. His escort is one British Cavalry and one British Infantry Regiment with a battery of horse guns. This means that we shall stay here and do escort to him the whole time that the war is on. . . ."

September 17.—"We are becoming deadened to joy or sorrow. It is a perfectly horrible existence, and unfortunately there is no hope of its changing for the better.

"We have had six inches of rain since midnight, and it is still raining—the country will probably be flooded. . . .

"There is a small polo tournament coming off here next week; it ought to give us something to think about, but I am afraid that no one can raise any enthusiasm about anything, as we are all bored stiff."

October 8.—"There is as usual nothing to say this mail except that our chances of getting out look blacker than ever. . . .

"I think I told you that we have been having a little polo tournament on the American system. I am glad to say that we won it. . . .

"We are going into camp with the squadron on Saturday for a fortnight. It will be bad, but a lot better than barracks."

October 12.—"I am writing this in our squadron camp. . . . We have made friends with the local Nabob, and he has lent us an elephant to go out shooting on. It is rather fun shooting off his back, as one never knows what the next shot will be at: it may be a buck or quail or partridge or snipe, or anything. He is a jolly good retriever and will pick up anything that is dead, but he hates to if it is only wounded. . . . The old man who lent us the hathi (elephant), has just come in to complain that two of our men have shot two peacocks, which are sacred birds to Hindus. As there are very strict orders against shooting peacocks . . . I hope that they get it in the neck. They are both in my troop."

That is an old cause of trouble. The British soldier finds it hard to resist at times the temptation to shoot a wild peacock, and add a "turkey" to his rations; but the Government of India is rightly strict on the subject. It is an instance of the care one has to take to avoid hurting Indian feelings.

India, October 25, 1914.—"As perhaps you may guess from the above vague address, we are off to the war. . . . We got the order at 4 A.M. . . . to pack up and come in to barracks at once as the Regiment was mobilising. We had everything packed up by 5 A.M., and the squadron left at 5.30. Considering that this was all done in the dark and that it was raining as well, I think that it is rather a good show. . . . They limit our kit to 35 lbs., which is only two blankets, a change of clothes, an extra pair of boots, and a valise to carry the lot—not very much to sleep in with a temperature of 20 or 30. . . .

"It is rather a coincidence that we got the order to mobilise on Balaclava day, isn't it?"

Balaclava day! Sixty years had passed, and the thought of it was still ready to the minds of those who were now taking the Regiment into another war. That is what a feat of arms in which his Regiment shared means to the soldier—an ever-living memory and example.

The suspense was over. "It is great news," wrote the Captain commanding the squadron, "far better than we dared hope for, and you may imagine how we are all feeling." He was the same officer who had commanded the Queen's escort three years before—the model escort. Now he was going to show whether the men who had won so much admiration in a pageant of peace time would do equally well in the field.

Nothing remained but to complete the number of men and horses, both now below strength in consequence of the draft lately sent to the Eighth Hussars, and to make the final arrangements for a quick departure. Men and horses were found from other regiments, and during the first ten days of November the packing and preparations were completed. Officers disposed of their horses and furniture; many of the polo ponies were taken over by the Remount Department for service as Infantry officers' chargers; the regimental mess was closed; the heavy baggage and valuable books were sent to England; and the Regiment's period of peace service in India was at an end.

CHAPTER V.

THE INDIAN ARMY—BEGINNING OF WAR.

THE Empire of India, with its population of more than three hundred millions, is held by an army which, compared with the hosts of European nations, is a small one. Great Britain has never had in India much more than seventy thousand British troops, not one man to four thousand of the population—a conclusive proof, if any were needed, of the fact that British rule in India is based rather on the goodwill of the Indians than on force. No doubt in the last resort the white soldier is the mainstay of the Government against sedition and revolt; but if sedition and revolt were ever more than partial they would need a much larger garrison to suppress them. Three hundred millions of people would not be indefinitely "kept down" by an army of seventy thousand foreigners, however brave and well disciplined. The truth is that the British supremacy in India, though it has at times involved hard fighting, was founded upon the consent and active co-operation of the Indian races, and is maintained by the same means.

Not only is the number of British troops in India comparatively small, but the British Government has not feared to raise and keep up alongside of them an army of Indian regular troops twice as strong, and to arm and make efficient for war other bodies of men drawn from the population, notably some fine contingents of soldiery in the Feudatory chiefships. Altogether it may perhaps be roughly computed that at the outbreak of the War in 1914 the Crown had at its disposal in India, counting local volunteers, perhaps a hundred thousand armed white men and two hundred thousand Indians. This force had to maintain internal order throughout a country as large as all Europe excluding Russia, and to defend the frontiers against any aggression from without. It was regarded, and organised, not as two armies sundered by the colour-line and mutually suspicious of one another, but as one army in which the white regiments and Indian regiments served side by side, as they had served for many generations in many wars, mutually trusting one another and fighting as comrades against any enemy who might threaten the interests of the Indian Empire.

Some of these enemies had been fought at a great distance from India—in China, in Persia, in Egypt, and in other countries across the sea; but until now Indian troops had not been employed in the battlefields of Europe. More than a hundred years before a great "sepoy General," who had learnt his trade in India, had commanded British armies against the soldiers of Napoleon; and countless other British officers and men had served both in India and Europe. India had, in fact, to quote Henderson's 'Science of War,' been "the great training-ground" of the British Army. And Indian troops had at times, in Asia and Africa, crossed swords with European enemies. Nevertheless, the Indian Army, as such, had not fought in Europe, and the British officers who commanded Indian soldiers had not often served, even individually, in European wars. No Indian soldiery fought in the Peninsular War, or at Waterloo, or in the Crimea, or even in the Boer War, though a contingent of white troops from India did go out to South Africa then, and saved Natal. England, in fact, had hitherto regarded the Indian Army, and the vast reserves of Indian races on which that Army could draw, as a source of strength only for her outlying wars, not as a portion of the Imperial power upon which she could rely if attacked in Europe. That may be said in spite of the fact that on one occasion the far-sighted Beaconsfield had as a demonstration brought a few Indian troops to the Mediterranean.

Unluckily, it may be observed here, this view, and other reasons, prevented the Indian Army in recent times from being brought up to the mark required for scientific warfare in Europe. While the Home Army was being modernised and improved in every way after the Sudan campaigns and the Boer War, the Indian Army

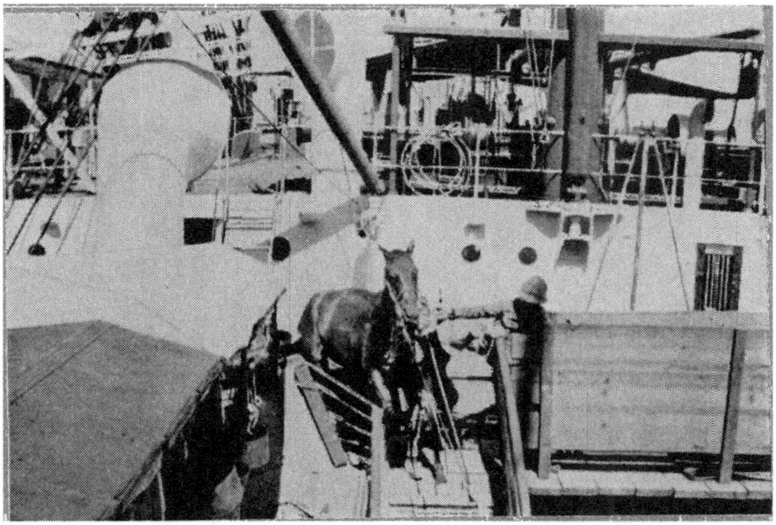

EMBARKING AT BOMBAY. NOVEMBER 1914

was left without similar attention. It was quite fit for Asiatic warfare, but in training, arms, and equipment, its splendid officers and men found themselves at a great disadvantage when employed against European troops of the latest model.

This, however, was not understood by Great Britain.

Now that she found herself involved in a conflict with the greatest military power the world has ever seen, and woefully short of British troops in England to support the comparatively small force she could send to the help of France, her eyes turned to her great dependency; and fully assured of the loyalty of India, in spite of the seditious movements of the past few years, she decided to make use of the reserve of trained strength she had hitherto set aside, and to let the Army in India, British and Indian, have its share in fighting the common enemy on European soil. It was a bold decision, full of important consequences for India and for the Empire; but it was taken, and the call was sent out.

So, when the Thirteenth Hussars received their orders for the front, they were summoned not as an individual Regiment of British Cavalry, but as part of the Meerut Cavalry Brigade, made up of one British and two Indian Regiments, the 3rd and 18th. This Brigade in its turn formed part of an Indian Cavalry Division, the 2nd, and the 2nd Division formed part of an Indian Cavalry Corps.

On the 13th of November the Thirteenth left Meerut by train, in three detachments, and went down to Bombay, where they were to embark. What their destination was they did not know for certain, but it was believed to be somewhere west of Suez. As a fact, their destination was Marseilles, but during the two days they remained in Bombay waiting to embark, they received no definite news of this.

Bombay, the great western port of India, with its magnificent harbour and wooded hills and teeming city, was at this time a very busy scene. It had originally come to Charles II. as a portion of the dowry of Catherine of Braganza, and had been transferred by him to the East India Company for an annual payment of £10, a striking exemplification of the almost magical development of the British Empire in India. Now it was of great value as a commercial port, and as the harbour from which the Indian Government was to carry on the activities entailed by the war. But a Regiment embarking for service had little time for thinking of such matters, for there was much to be done in the two days that elapsed before the troops went on board. On the 17th of November everything was ready, and the embarkation began. Many of the horses were piteously frightened at their novel experience, some of them "screaming like children" as they were slung up into the air and lowered into the hold; but they soon got over their terror, and the men worked splendidly in the Indian heat, the sweat streaming down their faces and through their coats. Before night men and horses were all safely on board, and there had been no mishaps.

The strength of the Regiment when it embarked, under the command of Lieut.-Colonel Symons, was 20 officers,[1] including the Medical Officer, 499 other ranks, including the Assistant-Surgeon, 560 horses, and 1 pony. Several officers were on leave in England, and some of them were expected to join later; but others had already gone to the Front, of whom 2 had been killed and 2 wounded.[2] The Regiment was distributed in two transports—Headquarters and three squadrons, "A," "B," and "D," on board the *Dunluce Castle*, "C" Squadron and the machine-gun detachment on board the *Risaldar*. During the 18th of November the vessels remained at anchor, for they were to form part of a convoy, and some of the other ships were not quite ready to sail; but on the 19th all was in order, and then at 9 o'clock in the morning the whole convoy, to the number of 26, weighed anchor and steamed slowly out over the sunlit waters of the harbour. Outside, the convoy stopped to pick up a few more ships joining from another port, and then the whole formed up, six abreast, and, led by an escorting cruiser, sailed away to the westward. It was a fine sight, though a sad one for the women of the Regiment, who were left behind on shore. Many of them had looked their last upon their men. But that is war.

It was a striking incident that the convoy was escorted from Bombay by the *Dupleix*, a French man-of-war. In the old days, when the French and English were fighting out their long struggle for the mastery of India, the English had no more dangerous enemy than Dupleix, who tried to raise against them a confederacy

[1] See Appendix IV. [2] See Appendix II.

THE DEPARTURE FROM BOMBAY. 19TH NOVEMBER 1914

of Indian powers, and as some believe taught them the use of Indian soldiery trained after the manner of Europe. Sea-power, which he did not understand, baffled all his efforts and decided the struggle in favour of England. Now, if the spirit of the great Frenchman had returned to the shores of India, he would have seen the same sea-power again triumphantly exerted, and would have watched his own countrymen, in a vessel which bore his name, joining with his old enemies to convey to the shores of France, for the help of France, thousands of Indian soldiery drilled and disciplined after his own fashion. If he could have gone with them he might have seen another and even more striking example of the irony of fate. He might have seen on the shores of the Channel the figure of another and greater Frenchman, looking down from his lofty column, not upon the ranks of his veterans gathered together for the invasion of England, but upon the tents of numberless British encampments full of Englishmen assembled on French soil to fight for France. A hundred years before, English sea-power had foiled his vast schemes of conquest. "Those far-distant, storm-beaten ships upon which the Grand Army never looked, stood between it and the dominion of the world." And they had prevailed. Now English sea-power was fighting on the side of the Army of France, and the old enemies combined were to bring down in ruin another scheme of universal empire.

CHAPTER VI.

VOYAGE TO FRANCE.

THE voyage of the Thirteenth across the Indian Ocean was not disturbed by any hostile attack or demonstration, nor by bad weather, and the six-abreast formation was kept until the convoy was near Aden, when a new formation in two lines ahead, or two ships abreast, was taken up. Shortly after passing Aden the *Dupleix* parted company to coal at the French port of Jibuti, and her place was taken by the *Northbrook*, a vessel of the Royal Indian Marine. So far all had gone well.

The wife of an officer of the Thirteenth had left Bombay for England in a passenger steamer a day or two after the Regiment, and she writes in a letter of the time: "On Saturday, 28th of November, we caught sight of our Indian convoy at about nine in the morning. An Admiral asked the Captain to go quite close, as there were so many wives on board whose husbands were in the convoy. So he very kindly altered his course, and we went quite close so that we were able to pick out the different ships, and could actually through glasses see the ships with horses on board." It was a courteous act, and no doubt gave pleasure, if a rather pathetic pleasure, to all concerned.

Though matters had gone well, the voyage had not been altogether without suffering for man and beast. Even at the best season of the year the heat in the Eastern seas can be very trying, and though British troops going on service were no longer exposed to the horrible discomforts of Crimean days, but conveyed in such splendid vessels as those of the Union-Castle Line, the Thirteenth had some unpleasant times. An officer wrote to his wife: "You know

IN THE SUEZ CANAL. DECEMBER 1914

what the weather has been like, but you can have no conception of what it has been below in the horse decks: absolute hell. All across the Arabian Sea it was dead calm and a following wind, and the first day and a half in the Red Sea was even worse. We have worked like slaves with the horses, off and on all day: men stripped, officers in shirt sleeves, and all pouring with sweat; the horses panting for breath, and all we could do by continually moving them, sluicing with vinegar and water, and all sorts of things, to keep them alive. It was heart-breaking. I hope I'll never go through such a time again. . . . All the days in the Arabian Sea seemed to get hotter and hotter, and the horses worse and worse; and the first day in the Red Sea, last Friday the 20th, was worse still, and one of my best horses, No. 133, 4th Troop, a nice bay from Saugar, with pink rings round his eyes, died from heat-stroke. Then one of 'B' Squadron died, and it was desperate. Two or three times they have had the ship round in a circle, to face the wind and try and get some air below for the horses, and it has been a great relief. . . . You wouldn't believe how tame all the horses are now. You can do anything with them. Poor devils, they have had a frightful time. Saturday again was very hot, but the wind gradually came round ahead, and by evening there was a good breeze; and yesterday and to-day has been lovely: a stiff breeze ahead and quite cool. It is like heaven, and the horses are like different creatures and picking up fast. It is sure to last now, I think, right in to Suez, and I hope our troubles are over. . . . The men have worked like slaves, and so have we for that matter."

Another officer, Lieutenant Watson Smyth, writes of the start at Bombay, after five or six hours spent in slinging horses into the hold: "At 8.30 I went down to the horse deck, and never have I met such heat. The horses were packed in pens of five, and were all, all over in a white lather. The temperature was taken by the Vet. and it was 133. This is 6 degrees more than the highest recorded in India, so you can see it was real hot. . . ."

November 29, 1914.—"It has been very hot indeed the last few days, and the horses are feeling it very much. Only two have died so far. . . . I think I said that most of my squadron are in the fore-hold, and the other squadrons are on the decks above it round the hatchway, so that if anything has to be taken out of their decks by a crane it has to be hung over the hold while being hoisted. One of the horses I mentioned died in one of their decks, and when slung up to be dropped overboard, slipped out of the sling and fell forty feet into the hold. Luckily he only grazed one of our horses, another half-inch and it would have been killed. I have decided to take that horse for a charger, as if he can have an escape like that nothing else is likely to hurt him."

It was a rough experience, and not a very good preparation for the cold of a winter in Northern France; but for the moment the discomfort was over, and throughout the voyage not many horses died. The Thirteenth lost four or five in all. Three-quarters of the troop horses were Indian country-breds, and the rest Australians, and therefore also accustomed to some heat. But the country-breds were rather light for British Cavalry, and hardly fitted at best to face snow and wet.

The Thirteenth found the banks of the Suez Canal lined with troops, largely Indian, who were expecting an attack from the Turkish army gathered in the desert to the north, but no attack came while the convoy was in the Canal.

Meanwhile, though still ignorant of their destination and very anxious to know it, they were cheered by a letter from an officer who had seen some fighting on the French Front. "He says the German Cavalry won't face ours at all, and that their Infantry shoot rottenly. He says their Artillery, machine-guns, aeroplanes —anything mechanical, in fact—are perfect—and nearly all the casualties are from gun-fire. He says, man for man they are no match for us, and it is all simply a question of numbers. He says the patrol-work of the German Cavalry is too childish." This confident letter was not altogether wrong in its views, as was afterwards shown by Lord French's despatches and other evidence. Needless to say, the Thirteenth longed to be face to face with the famous Uhlans.[1]

Port Said was full of troops and of French cruisers and destroyers, a very bright and busy scene. There the Thirteenth at last learnt their destination. What they had longed for had come. They were to

[1] The inefficiency of German Cavalry in all but mass manœuvring was only what Henderson's criticisms on their work in 1870 had already pointed out. See the 'Science of War.'

IN THE SUEZ CANAL. DECEMBER 1914

go on to Marseilles, and from there to the Western Front. It was to be real work, against a European enemy.

The passage across the Mediterranean, if rough, was uneventful, and by the middle of December the Regiment was landed on French soil. "We have arrived all fit and well and jolly," Captain Eve wrote, "and have had a very busy day. . . . It is beautifully mild and fine. All the horses are well, and mine flourishing."

The next day the Regiment went on by rail to Orleans. It was an interesting journey, and the French people all along the line gave the Regiment a hearty welcome. "French Red Cross people at all large stations, and lots of soldiers: also lots of enthusiasm, singing, giving the men country wine, and so on. They gave us cigarettes, coffee, tea, flowers, and so on, and were all very nice. Altogether it was very interesting and I enjoyed it. I had to give one badge away to a girl who asked for it, and to kiss another's hand, which I hated. The men made a tremendous noise, but behaved very well indeed, except that two or three of mine got rather drunk on the last night. But it was very difficult for them. I find I can get on a little with my French if I am not hurried. . . ."

That entry was very English, and very English too the thoughts of hunting stirred up by the French *campagne*: "We came a roundabout way, not straight, and at one part came through some awfully nice country just like home, say the Duke's country, enclosed property, and some stone-wall country too, and small coverts, and hilly. I got quite excited looking out at it."

But the journey was soon over. A little after midnight, on the 17th of December, the Regiment arrived at a siding near their camp: "It was bitterly cold, with a white frost and icy wind, and we had to turn out, detrain, and load up all our kit, saddles, and arms on to motor lorries, and then march, leading our horses six miles out to our camp here in pitch darkness. . . . We left the station about 2.45 A.M., and reached camp about 5 A.M., and groped about till we somehow got our lines down." It was not a pleasant beginning to their soldiering in France, a curious contrast to the heat of the Red Sea—"the worst and coldest camp, I think, I have ever seen, about six inches deep in liquid mud, on the top of an exposed hill, with a bitter wind blowing. We are in tents, V.[1] and I sharing an 80-lb. one. We are very warm and comfortable, lots of warm straw on the ground, and our valises on top of it, and the men are in tents too, but the poor unfortunate horses are having a terrible time. . . . They stand always in a bog. The watering-place, about three-quarters of a mile away, is literally up to your knees nearly in liquid mud." Lance-Corporal Bowie's diary says of the arrival at Orleans: "Here we detrained at once in the midst of a terrific hailstorm, afterwards saddling up and leading our horses through the city to the village of La-Source, a distance of nine miles. Our stay at this camp proved to be a very severe test for both men and horses, as we were still clothed in our Indian khaki; at the same time it rained heavily for hours, and was also bitterly cold. The place in which the rough water-troughs had been fitted up, being in a valley, became practically a sea of mud, in places reaching up to our horses' bellies."

At this camp the Regiment found some more of their officers awaiting them, which brought them up to full strength again.

After two or three days they moved to a slightly more sheltered place, and the weather began to change. By Christmas Day it was bitterly cold, but bright and still, with a warm sun, and all was going better. Plenty of warm clothing was being served out to the men, and it was possible to get exercise again; and the food was excellent, good meat and vegetables, and tobacco. The warm clothing indeed was more than the men and horses could carry, and the quantity of blankets and other things had to be reduced to a more reasonable and serviceable scale. To quote Lance-Corporal Bowie again: "On Christmas Day 1914, every one received a post-card photo of the King and Queen, and also a gift from Princess Mary, which consisted of a pipe and an embossed brass box containing tobacco and cigarettes. A majority of us also received a Christmas parcel, which we owed to the generosity of the ladies connected with the Regiment, at the same time being completely overloaded with warm underwear, woollen cardigans, waistcoats, mittens, &c. But the waste of our new kits which we were compelled to obtain before leaving India was disgraceful,

[1] Lieutenant J. V. Dawson.

Major T. H. S. MARCHANT, D.S.O.

Col. A. SYMONS, C.M.G.

Major W. A. KENNARD, D.S.O.
(*Died of pneumonia, December* 1918, *at Etaples*)

Bt. Col. W. PEPYS, D.S.O.

Lieut.-Col. E. F. TWIST
(*Wounded at Lajj, 5th March* 1917)

almost everything being burnt with the exception of some which we had dumped at Marseilles, which, needless to say, we never saw again. On the morning of the 31st of December we were all very glad to march out of this muddy camp, an incident worthy of note being that the men were so overloaded with kit (many of them having on two of almost everything as regards underclothing, having nowhere else to carry it), that they found it an awful struggle to mount, feeling more like a well-dressed Christmas-tree than a cavalryman. However, having all got mounted, we marched direct to Orleans Station, where we at once entrained for Berguette (Pas-de-Calais), where we arrived at 3 A.M. on 1st January 1915. Detraining here, we marched up to a village called Enquin-les-Mines, a distance of some kilometres, where we were allotted billets which consisted of old barns, &c., for the men, whilst we made our horses comfortable under archways, &c."

Certainly the British soldier in this war was equipped and fed as he had never been before, and the Thirteenth ended the year very happily on the whole. It was a contrast to their winter in the Crimea sixty years earlier.

Christmas good wishes and photographs from the King and Queen and Princess Mary came to assure them that they were not forgotten in England. And if the prayer of Their Majesties, "May God protect you and bring you home safe!" was not to be fulfilled for all of them, they faced what was to come with confidence and eagerness, longing only for more stirring work, and a chance of doing their share of honourable service.

It was a pause in the fighting then. The great retreat on Paris and the battle of the Marne were over, and the baffled enemy had made his first attempt to strike out to the westward for the Channel ports. He had been stopped after desperate fighting by the wasted regiments of our little army, and the troops on both sides were settling down into the long trench warfare of the next four years. The British part of the line was woefully short of men, and guns and munitions of all kinds; and to those who knew the real state of affairs the outlook was very dark, for in England there were no trained reserves to send to the Front—plenty of brave men, but no soldiers. Happily the country did not know in what peril its army was, and contingents were coming from India and

Canada and Australia and New Zealand, and the confidence of the men at the Front was unfailing, and all hoped that the worst was over. It seems wonderful now that such confidence should have prevailed at the Front, and so little real anxiety in England; but the fighting men were full of the belief that they were man for man so superior to the enemy that he could never break through. Such gloomy faces as there were could be found only in England, not among the fighting men. In spite of snow and mud and suffering of all kinds, there was no gloom with them.

Bt. Lieut.-Col. E. J. Carter

Major R. F. Cox

Capt. Lord Huntingfield

Capt. Norman Neill
Brig.-Major, 7th British Cavalry Brigade
(Killed at Zwarteleen, 6th November 1914)

Bt. Major R. S. Hamilton-Grace
G.S.O. 2nd Hdqrs. Cav. Corps
(Killed in Motor accident at Burgues, 4th August 1915)

Bt. Major H. Ll. Jones, D.S.O.
(Wounded in France with 4th Dragoon Guards, 28th October 1914)

Capt. F. C. Covell

CHAPTER VII.

1915 IN FRANCE.

THE Regiment was now at full strength, officers and men and horses, and keen for a share in the fighting. The horses had suffered to some extent from the change of climate in the past six weeks, but only required a little rest and feeding-up. The men seemed fit and ready for anything.

But though all hoped for Cavalry work in the near future, and a chance at the Uhlans, this was not to come yet. The enemy's horsemen were no longer to be found in the extreme front, and the fighting was being done by our guns and Infantry, which were deficient in numbers and very hard pressed. The British Cavalry, therefore, though kept as far as possible efficient for their own work in case a chance should occur, had to be utilised to some extent to help the out-numbered foot-soldiers in the trenches; and during the first few days of the new year the officers and men of the Thirteenth, while undergoing Cavalry inspection and training, were hard at work perfecting themselves in their new duties. They had not long to wait.

Before the middle of January they had been taken up to the firing line to be "shot over." "On the 12th," writes Lance-Corporal Bowie, "we were informed that we were to take our places in the trenches as infantry, having been armed with the new H. V. rifle and bayonet, and having had plenty of practice in bayonet-fighting, which was quite a new thing for the Cavalry, we were pretty confident of being able to do anything that was required of us dismounted. So leaving only sufficient men behind to attend to the horses, we started off the next morning in the highest spirits for Béthune, our conveyances being the good old London motor-buses, complete with their own drivers and conductors. Arriving there at 5 P.M., we marched direct to the trenches, just in front of the village of Festubert, a distance of thirteen kilometres, relieving the 6th Inniskilling Dragoons. We remained in these trenches until 6 P.M. the following evening, when we were relieved by the Oxford and Bucks Light Infantry, many of our fellows having to be lifted out of the trenches owing to being cramped with standing in the mud and water for so long. On each man receiving a tot of rum, we at once marched back to Béthune. . . . On arriving at the market square in Béthune, many men fell from sheer exhaustion. Meeting the buses again, we had some hot coffee and returned directly back to our billets, which we were very thankful to reach about 2 A.M. on the 15th of January 1915. One of the most remarkable features of this, our first time in the trenches, was the fact that we did not sustain a single casualty, although we were subjected to a continual bombardment the whole time, the Huns' shooting being fairly good(?), but their shells were very bad, many burying themselves in the mud and failing to explode at all."

Such was the first introduction of the Thirteenth to actual fighting in the Great War. It was very different from what they had hoped—a dreary struggle of endurance against mud and cold, on foot, instead of the stirring hand-to-hand work in the saddle for which a cavalryman naturally longs; but the account shows the cheery spirit in which the men took to their uncongenial duty. Needless to say, the officers set them a good example. One of them, Lieutenant Watson Smyth (14th January 1915), writes: "We got up at Enquingatte, where we were billeted, at 6 A.M. on Wednesday, and at 8 had a three-mile march to another village, Estrée Blanche, where the whole Brigade was concentrated. At about 10.30 along came a fleet of motor omnibuses (London General Company), and halted along the line of troops. We were then told off, and twenty-five men and one officer went in each bus. The buses have the glass out of the windows and the space is boarded up, otherwise they are the same, except that the outside advertisements are painted over, and the whole bus is dark grey (please excuse my writing, but we are under shell-fire—75 mm.

shrapnel—and I am expecting one through the roof any minute). To go on, we left in our buses at 11 A.M. Soon after starting, one skidded into the ditch and had to be jacked and dug out, but this got to be quite a common occurrence later in the trip. At about 1 P.M. we arrived at Béthune, about seven miles from the trenches. We stayed there for an hour, and had our lunch while the men had their dinners. At about 2.30 we got going again, this time on our flat feet, and marched about four and a half miles to a village, Festubert, where we halted. Here we all got a drink of beer, followed by coffee and rum. At 4.45 P.M. we started again, and this time went right on up to just behind the trenches. Here I, with eight men who had volunteered for the job, went on to ——, about 400 yards in front of our machine-guns, which were on the left of our line. . . . When I got up to it we were challenged by the post of the Regiment that we were relieving, and then I went up to them. I asked if they were all right. In a very despondent voice he replied, 'I've two men nearly dead with cold: they are both unconscious, and I don't know how I'll get them back.' Just at that moment one more man went over flop. I thought this was a jolly start, as I was going to be there all night and these fellows had been there in the day. We had great trouble to get them out, as the trench was knee-deep in the most holding mud I had ever met. It beat Wadhurst clay by three stone and a distance. Another difficulty was the fact that the Germans, who were about 600 yards in front, or perhaps a bit more (people are talking all round me, and I keep writing what I hear), kept on sending up 'Very' lights and star-shells, which lit up the whole place far better than it was lit up in the daytime. Owing to the snipers, who were lying up all over the place, we had to drop flat as soon as we saw the light going up, and stay there for about a minute after it had gone. Then I got into the trench, which was bisected by a stream which was just over knee-deep. I put four men one side, and four with myself the near side. I had orders to keep on sniping all night so as to annoy the Germans, so I had one man of each four on sentry for an hour at a time, with orders to shoot about once every five minutes. Of course I could not sleep myself, but I lay down in the wet mud. The trench was over ankle-deep in mud and water, and only just long enough to hold us all. About midnight it got most damnably cold, and I issued the men milk chocolate, and gave them each a tot of rum from a flask I'd got. The snipers kept on shooting at us, but mostly went over, though a few bullets did hit the trench. One horrid fellow, whom we called Bert, was behind us somewhere, and made me very angry. At 3 A.M. we heard the devil of a battle going on a long way off, machine-firing guns going rapid, and a rattle of musketry. This went on for half an hour, and then one or more of our big guns somewhere behind us started firing occasional shots. It made a most colossal row, although it must have been at least half a mile away. At about 5 A.M. we saw the relief coming up, halted it and saw that it was all right, got out of the trench, . . . then we went back to the road behind us and walked along it for about 500 yards till we came to the house that the squadron was billeted in. There we got some tea and more rum, and a bit of bully and biscuit, and the men thawed out. The squadron had been in the trenches all night, and had been relieved, as I was, just before dawn. I do not think I ever appreciated a house and a fire so much before as after that twelve hours of water and mud. . . . The dotted lines show where the snipers were firing. There was one called Fritz who used to fire across the road about every ten minutes. I am sending you one of his bullets. We sat in the house until 10 A.M., when the Germans began to shell the place. The first shell (shrapnel out of captured French guns) burst about 80 feet in front of a group of us, me included, and the bullets went all round us without touching anybody—it was really rather a lucky escape. After that we cleared off to the bomb-proof at the back of the house where I am now. Another shell burst as we were going into the shelter, and scattered all round, but again missed everybody. . . ."

January 15, 1915.—"We are now back in billets, having done only twenty-four hours in the trenches. We stayed in our bomb-proof till about 3 P.M., although they had stopped shelling the village. . . . We found that two shells had gone through the room we had been sitting in and had burst in it. They had only knocked holes in the walls and scattered a lot of plaster and stuff about. We had our transport packed by 4.30 P.M. and fell in at 5 in the dark. . . . I had to wait so as to take the patrol of

Capt. J. N. Lumley, M.C.

Capt. J. I. Chrystall, M.C.

Capt. E. H. Stocker

Lieut. G. R. Watson-Smyth
(*Wounded near Lillers, 14th July 1915*)

Capt. J. H. Hind

Capt. J. A. Jeffrey, M.C.

Capt. J. L. M. Barrett

the relieving regiment down to where I had been. . . . On the way, up went a star-shell, and down I flopped in about six inches of water. As soon as the light had gone—phut!—and a bullet from Fritz hit the ground about 15 yards over. I lay a little flatter, with my back crawling with apprehension—phut!—and another went about 10 yards in front. I lay flatter still—phut!—and another hit the ground about 10 yards behind. I thought this was nice, as he must now be able to see me, and the next shot ought to get me, so I lay very flat and cursed all Germans. But he didn't fire again, so after a bit I got up and splashed (I've never made such a noise before, at least so I thought) forward to the patrol. They also were so cold that they could hardly stand, so I had to stand on the bank and lug them out to the usual accompaniment of star-shell, Very lights, and snipers. . . . We got into our billets at 3 A.M., and I was in bed and asleep at 3.20. We were all in a most filthy mess outside, owing to the mud and water that we had been lying in, and inside our clothes owing to the cod-liver oil that we were anointed with from our feet up to our waists. . . . It is fine stuff to keep the cold out. I was wearing Cording boots with two pairs of socks, the inside pair vaselined, and the outside pair oiled, and puttees over the top of the boots. Although I had been several times in water over my knees, I never got my feet cold or wet. . . . The only casualty in the Brigade was one sowar of the ——, killed. He got scared at a Very light, and stood up in the open staring at it, so of course a sniper shot him and he died. I don't expect we shall do any more trenches for a bit: this effort was only due to the Corps Commander, who wanted to have us shot over. I think it did every one a lot of good: it has certainly taught me that shrapnel is not half so awful as one thinks, and that one can lie out with only a coat on in a puddle all through a winter night, and be none the worse for it, and also that a whack of rum has an entirely beneficial effect."

January 16.—"The patrol of the Regiment that relieved mine saw two dead Germans about 500 feet in front, and so of course all the men who were with me are claiming that they killed them, and the first blood of the Regiment is theirs. . . . The men I had with me were all hard nuts, and when not on sentry lay down in the water and went to sleep. They had their British warms (*i.e.*, coats with a flannel lining that reaches to the knee) and mackintoshes, so that they were fairly warm and dry, except for their legs. Their feet got very cold, though the vaseline helped a lot. . . . It was quite an experience, and although I was most beastly uncomfortable all the time, I wouldn't have missed it for anything. I must say that I never expected that the first time people fired shots in anger at me, and I was retaliating, that my only thought would be how to keep warm, and also not to go to sleep. . . . One rather amusing thing happened while we were in reserve in the village. Our Colonel got an idea that a sniper was concealed in one of the houses (there were no inhabitants left), and so ordered 'X' Squadron to make a house-to-house search. A party consisting of twenty men and one young officer started off with loaded rifles, fixed bayonets, fingers on the trigger—officers waving revolvers. Suddenly they saw a man on a haystack: immediately pandemonium ensued—rifles going rapid, men charging, revolvers going off, wild confusion. Suddenly the fire stopped, and a perfectly *furious* officer leaped off the haystack, rushed at the officer, and started, 'You . . . ' for about five minutes. He then saw the squadron leader, rushed at him, and dragged him off to the Colonel. He then said he'd been on that haystack for five weeks, that all the Germans in Northern France had been shooting at him, as he was in an extraordinarily good place for observing fire, and then these —— did their best—a d—d poor one at only 20 yards—to lay him out. As we had not been warned he was there, I think it was quite natural to plug him. He really was the angriest man I have ever seen."

War has its humours, and it is well to be able to enjoy them.

For a month or so after that first experience there seem to have been no more nights in the front trenches for the Thirteenth, but some parties were told off for trench-digging, and there was much Cavalry-training of one kind or another, with occasional orders to "stand to" and be ready to move at very short notice. These orders of course gave rise at first to much excitement, and eager hopes of some real Cavalry-fighting, but they never came to anything. Perhaps the best way of showing what the Regiment was doing during the remainder of this year, 1915, is to quote some more extracts from letters and diaries.

BILLETS OF CAPTAINS EVE AND JACKSON AT ENQUINGATTE

CAPT. W. H. EVE

CAPT. T. K. JACKSON AND LT. J. V. DAWSON

TRENCHES AT ENQUINGATTE DUG BY D SQUADRON

SCHOOL AT ENQUINGATTE WHERE LT. J. V. DAWSON WAS BILLETED

Lieutenant Watson Smyth — February 6. — "When I got back I found my squadron 'standing to,' and ready to move at fifteen minutes' notice. However, that has now been cancelled, and we are now living in the same old peaceable way. We had a sham fight this morning to practise dismounted action. I and my troop had to run along a dry stream-bed for about three-quarters of a mile. I was nearly dead at the end of it, but my troop were even more done, so on the whole I was rather pleased. . . . I do not think it is likely that we shall move for some time, as it is absolutely impossible for Cavalry to move once they get off the roads. . . . I have just finished my evening task of letter censoring. That is not a nice job as it takes a long time, and I don't much care about reading other people's letters, especially such extraordinarily dull ones as the average soldier writes."

February 17. — "In the afternoon it began to snow, and it snowed as hard as it could all the evening and most of the night. We had been going to have a Divisional route-march the next day (Thursday), but that night the orders were cancelled. On Thursday we found it just possible to ride our horses, but only just as the roads were deep in snow, and it was balling badly. . . . We are rather badly off for water in these billets: I do not mean that there is not enough—the whole place is soaking—but none of it is very good. I rather think that that is one of the causes of our horses not looking as well as they might. Watering is almost as important as feeding, isn't it? . . . Horses are my special care, but it's rather disheartening having these beastly little country-breds to look after."

It may be observed that the Indian country-bred is not accustomed to a Western winter and heavy snow. Nor were the men of the Indian Regiments in the Brigade, to whom such weather was as trying as the extreme heat of India is to English troops.

"I had one horse get his leg broken by a kick from his neighbour two nights ago. It was smashed clean in two about four inches above the knee. Must have been some kick, as the bone is pretty thick at that part. I had him shot where he stood, hitched on one of the draft horses, and pulled him about 200 feet into a field over the way, and the defaulters buried him in the afternoon. A six-foot grave for a horse takes a bit of digging, and fairly made 'em sweat. It nearly killed an old fat reservist, who was doing defaulter for getting drunk on the way up from the Base. However, if he has a few more to bury, he will be an easier man to mount."

February 27.—"To-day we had the coldest day we have had in France. We paraded at 9 A.M. and did a Brigade scheme. I hated every minute of it, and so did our wretched horses. We were out from 9 till 1.45, and most of the time in a snowstorm on the side of a hill. . . .

"My first servant, Farmer, is a tiger for work. I discovered the other day that he had been working at a big butcher's in Jermyn Street before he joined the Army. As I also found some young pigs in one of the farms, I took him down to pick out a nice sucking-pig. He chose one, and I bought it for eight francs, and we are all going to eat it to-night: Farmer was great at cleaning, and scalding, and killing it. It was a most comic affair, as there were about thirteen little pigs, the lady of the farm, Farmer, and self in a covered sty about 12′ × 8′ × 6′ high. We were all talking at once, a child was howling, the pigs were screaming, and we were all trying to catch a different piglet. At length, however, we succeeded in collaring the right one, and I've never heard any animal make such a colossal noise as this little beast did when he was carried off. I nearly died with laughing, as just as we were coming off the road we met the General riding down. He was frightfully tickled. . . ."

It appears from Major Cox's diary that "during the month of February a semi-station routine of Brigade route-marches, Brigade field-days, lectures on various subjects, and squadron schemes, was carried out.

"Quite a lot of snow fell during the month, and cold frosty weather was the rule."

March opened with a very sad accident to the battery of Horse Artillery, V Battery, which formed part of the Brigade.

According to Major Cox's diary, "A trench-mortar bomb exploded during instruction, mortally wounding Major Goldie commanding the battery, two subalterns, and twelve men. Forty-one N.C.O.'s and men were wounded. As bad luck would have it, the whole of the battery was assembled round the trench-mortar

FARRIERS, D SQUADRON

OFFICERS OF D SQUADRON

MAJOR R. F. COX

OFFICERS OF D SQUADRON
TAKEN AT WARNES, MARCH 1915

when the explosion took place." All officers of the Thirteenth who could attend the funeral did so, and it was distressing to think of so many brave men killed and wounded, not by the enemy in fight, but by an accident of the kind.

This happened in Serny, a village adjoining Enquin.

Lieutenant Watson Smyth—March 7.—"To-day we had to find thirteen men a troop to go and dig trenches: they left at 6 A.M., and aren't expected back till 8.30 P.M. This left us, allowing for servants, sick, &c., about six men a troop for duty. We spent our time tidying up and straightening out the billets, and have been at it all day."

Lieutenant Chrystall—March 16.—"We have been on the move and bivouacking every night in a wood, so have had no time to write. We were in the advance to Neuve Chapelle, but were not used. . . . We always travel by night owing to hostile aircraft being about, and the consequence is sleep is impossible."

Captain W. H. Eve—March 16.—"I got your letter in hospital at St Omer. . . . I was in a terrible funk they would send me off home, as I knew what that would mean—two or three weeks perhaps, and then to Aldershot to wait my turn to come out. So I got at the doctors at once, and they said I should be kept there and go straight back to duty as soon as possible. I was very relieved. . . .

"Then rumours began to come through of this forward movement of ours between Armentières and La Bassée, and the hospital had to get ready for one thousand extra cases, though holding five hundred usually. So we knew something was on, and could also hear the big guns at times. At last on Thursday the 11th they told me I could leave hospital next day. Of course this is much too soon really, and would not be done in peace time. But now it is different.

"I went off to get my movement orders and asked 'Any news?' They said, 'Haig has sent for his Cavalry.' We are Haig's Cavalry—1st Army—and you can imagine the state I was in. Next day I left by train—8.24—having slipped out of hospital without even having my things disinfected. . . . All the Indian Cavalry Corps was crowded up there [Berguette?], mostly in billets, but our Brigade in bivouac in a wood—all in reserve. We had done nothing so far, and I was relieved. I was fearfully anxious lest I should be too late. . . . Well, now you will have seen by the papers we have done pretty well, but I fancy somehow we haven't done all we thought we might. I don't understand it, and we don't know the truth; but they said if we had got as far as we hoped, the British Cavalry Corps, which had been brought up too, was to have gone round the north of Lille, and we the Indian Cavalry Corps round the south, and had a cut at the Germans behind. But, anyhow, apparently the thing didn't quite come off, for on Sunday the 14th we got orders to march back here to billets. We were very sick indeed; it looked as though we had missed our chance by so little. But, of course, we really know nothing. We marched back Sunday night and are now about a couple of miles from the station where we detrained when we came back from Orleans, about twenty miles still behind the line. . . . How long we shall be here I haven't the least idea. We have to be ready to move at two hours' notice, but that may not mean anything. It is a dull and trying business this, but we must be patient. We have quite nice billets here."

Another account of the move is given by Lieutenant Watson Smyth: "At 12.30 A.M. on the morning of the 11th we were woke up and told that the squadron was parading at 3 A.M. We were, of course, sleeping in our clothes, as everything was packed, and we had had orders to be ready to move at one hour's notice. On being woke up I went to sleep again till 2.15, when I got up, put my coat and boots on, and went out to hurry up my troop. . . . We started to trot about 4.30 A.M. and trotted steadily until 8.30, except for two very short halts of about three minutes each, when we had just time to look round our horses. On coming to we turned out of the town, and the head of the squadron turned out of the road into a large sand-pit: this was found to be just large enough for a squadron, so the rest of the regiment was bivouacked in the wood. (I forgot to say that the sand-pit was in a wood.) We had easily the best place, as it was quite out of the wind and, better still, entirely free from mud. . . . The horses were perfectly happy, and so were the men. The latter dug holes running into the side of the pit, put a hurdle over the entrance, and were quite warm inside. We had very nice weather, sunny and so warm, and had

IN THE SANDPIT. MARCH 1915

nothing to do except listen to the rumble of the guns at Neuve Chapelle. . . . We stayed in our sand-pit for three days, and then one day got orders to move at 2 P.M.; about 1 P.M., however, these orders were cancelled, so we thought we might get another night in peace. This was rather too much to expect, and we were not very surprised when we were told to parade at 7.45 P.M. We did so, and had a perfectly ghastly march back to where we are now. We walked for hours on our horses, and then dismounted, and led the brutes for three and a half miles. It's no fun walking on one's flat feet when in marching order—*i.e.*, belt, revolver, spare ammunition, compass, haversack, field-glasses, knife, and water-bottle. We then lost ourselves for a bit, and every one lost their tempers, and cursed everybody junior to themselves, and their horses, and the roads, and the staff. Eventually we hit our village about 2 A.M. . . .

"We got orders to-day, and are off into the blue to-morrow.

"Our night march the other day was extraordinarily impressive, as we could see the flashes of the guns, and the searchlights swinging round, and the star-shell, and Very lights lighting up the whole horizon. The noise of the horses' hoofs on the *pavé* was not enough to drown the thunder of the guns, and at one time we distinctly heard the crackle of rifle and machine-gun fire."

March 18.—"We paraded at 8 this morning and started to march to ——, where we are going to be billeted. About 10 we halted and dismounted. . . . I tied up the horses, off-saddled, and let the men fall out to visit the town. At 12.30 I watered and fed the horses, and succeeded in stealing a bale of hay (100 lbs.) off a lorry that foolishly halted about ten yards from the horses. That pleased me and the horses a lot. I am now sitting on a tree-trunk near the horses writing this."

March 27.—"I found a dead motor-cyclist to-day: he had tried to take a corner far too fast in our billets, and had hit a tree and knocked his head in. I am now hoping to be able to 'make' the bike, as except for its front forks and wheel it is in excellent condition and would be very useful."

During this month there was much trench-digging, and Major Cox says, "Brigade field-days and regimental schemes were carried on similar to the routine in an Indian station." It was doubtless necessary, but as instruction in Infantry work was going on at the same time the men were extremely hard worked.

The month closed with a visit from the Honorary Colonel, General Sir R. S. Baden-Powell, who happened to be in France on a short tour. An inspection of the Regiment was held, and a short address was made by Sir Robert, who also presented to the Regiment a large number of cigarette-cases.

Captain W. H. Eve—April 2.—"The Indian Cavalry Corps has been nicknamed 'The Iron Rations,' because they are only to be used in the last extremity. I believe this is all over the place, and am afraid it may be a little true, though let's hope not. Anyhow, it's very funny and very clever of whoever thought of it. You see the iron rations (tinned meat and biscuit) carried by each man is only supposed to be used in the last extremity."

April 19.—"We are very busy all training more or less as in peace, and occasionally digging trenches; but one can find out no news or anything of what's likely to happen, and can only be patient. We are all very fit and flourishing and doing ourselves grand."

Lieutenant Watson Smyth—April 23.—"Wild excitement has possessed us for the last four hours, but it is now dying down, and in fact is nearly dead. It all started by our getting some wild story of Ypres, and asphyxiating gas, and the French, and standing-to. We were just starting out on a Brigade scheme, but this was abandoned, and we came back to billets and commenced furiously to pack. We are now feverishly unpacking. It really is extraordinary the rumours that get about out here; it is only very seldom that one meets any one who really knows anything worth knowing, and will tell what it is.

"It is a very good thing for every one to have these occasional bursts of energy, as one learns a lot about packing, and how things are lost, &c. To-day, of course, I got caught short of forage. Some one had stolen one of my sacks of oats. . . . I had to buy a sack of oats and feed on oat straw instead of hay. That is the advantage of a country-bred, he will eat anything, and his example makes the walers and English join in. I wish we could get a move on: these are excellent billets, but I want to see a German before peace is declared!"

IN THE SANDPIT. MARCH 1915

Captain W. H. Eve—April 30.—"The weather is simply lovely now, has been these last few days, and to-day has been as hot as summer. It is very lucky, for we are (all the men and horses) in the open. We ourselves have got into a barn, where we make ourselves very comfortable with lots of straw. I have celebrated my birthday by having a bath. I always carry the indiarubber one in my saddle-bags, and wouldn't be without it for anything. Well, I can't tell you where we are exactly, but we have moved twice since I last wrote, owing to this scrap which is still going on near Ypres. . . . We are sitting tight again now, listening and waiting. The only thing we see is our own captive balloons, and occasionally a Taube (German aeroplane) coming over and being shelled by our guns. You know, of course, the scrap that is going on now, but the papers make it out a much more important thing than it really is. It is a very weak point round Ypres, because (first) it is a salient, and (secondly) our junction with the French is near here. . . . But all goes well, so much so that I fancy we shall move back in a day or two—where, I don't know in the least. I expect that now we shall be continually up and down the line on this game—mobile reserve until our time really comes.

"In this fight our casualties have been very heavy—ours, I believe, about 18,000, but we have our line all right. . . . It's hard this waiting, but we must be patient. There is nothing in the least to worry about, nor is it the important affair the papers make out.

"We are all tremendously cheery, jolly, and fit. . . . The horses are feeling the benefit of the better weather and are a joy to me now, looking better every day. Of course I seize every opportunity of grazing them in somebody else's fields with the good spring grass coming up. I have never felt fitter in my life. There are crowds all round, but the worst of it is one can't go more than a few hundred yards from one's billet, as we always have to be ready to turn out at once.

"The old lady at our last billet insisted on embracing us when we left with all kinds of good wishes."

The billets, of course, varied greatly, some being very good indeed—one where the officers of the Regiment or squadron were actually provided with beds and "linen sheets,"—others very dirty and bad. Lieutenant Chrystall writes on the 19th of May: "It is very wet and muddy, and we are at present in a coal-mining village, and everything as you may imagine filthy. Last night I slept next my skipper on the road with my head between two spokes of a cart-wheel, and Eve's between the next. Cobbles are not very springy or soft!"

Captain W. H. Eve—May 27.—"You say you are watching for news of the Cavalry. Well, you have it now in the Casualties List. That is the British Cavalry. It is, I think, wicked, for they are men we can't replace, but the fact is they can't trust any but the very best up there at Ypres, and that's why they sent for the Cavalry. They get their chance and are used, but we, poor devils! . . . never get a chance. . . . Our turn will come some day.[1] Up there we are holding on and shall do so, but it costs good men. We sit back here well within sound of the guns, and go on with more or less peace-training and try to be patient. Whenever there is a big show on, off we go up behind the line, and every time we think our time has really come; but every time we come back again in a few days—a sort of mobile reserve, that's all. We are known out here as the 'Iron Ration,' only to be used in the last emergency! We went up like this during Neuve Chapelle into Belgium during the fighting round Ypres, and the other day near La Bassée. We are back here again, now very comfortable in a pretty straggling village. . . . We are all very fit and flourishing, but rather fed up with our own share of the proceedings."

Sometimes the Regiment, or part of it, got a change of work.

Lieutenant Watson Smyth—May 28.—"The day before yesterday the bathing season started and we had (the squadron and officers) gone into the baths at Aire. . . . We were all just nicely in the water, which wasn't very cold, when a heated bicycle orderly dashed up and said, '"B" Squadron to return to billets at once.' We couldn't think what it was for. . . . Yesterday (after a twenty-mile night march) I went for my orders and discovered that I am 'Corps Cavalry,' and that my duties are to send patrols out in the Corps area to look out for spies, collect stragglers, control the traffic at various points, and, in general, be a sort of mounted police. It is, I believe, an excellent job, and quite good fun.

[1] "We" being the Indian Cavalry Corps.

HURDLE SHELTERS, BOIS DU REVEILLON. 15TH MARCH 1915.

BILLETS AT WARNES. APRIL 1915.

Of course the roads are being shelled now and then, but the people I relieved had only one casualty, and he was gas-poisoned by a shell. In case of a push I have to find posts at cross-roads, &c., to direct ammunition and supplies to the various places that they want to go to. It is very nice being on one's own, as I am now. The Squadron Headquarters are about six miles off, so I shan't be worried by any one."

Captain W. H. Eve—May 29.—"The British Cavalry are out [of the trenches] now, came out last night, and I think, though am not quite certain, that the 1st Division of this Corps has gone in. If this is so, we may perhaps get our turn next. No one would choose trench fighting, but there's nothing else just at the present.

"Here we are just peace soldiering. 'A' squadron have their sports on this afternoon and we have ours next week. 'B' squadron has gone off to do Divisional Cavalry—chiefly orderly—duties to various Infantry Divisions. I envy them, as it is a change, and they are nearer the centre of things."

Lieutenant Chrystall—June 15.—"On Sunday last (June 13) I had the honour of acting as Escort to the King of the Belgians when he reviewed the Indian Cavalry here. It was quite a decent show for war-time."

Captain W. H. Eve—June 16.—"We have had rumours of a possible move, but otherwise are carrying on as usual. Lovely weather, and we are playing polo this evening, which is a great thing—having got some sticks and balls out. We have got a Horse-Show (Cavalry Corps) on Thursday 24th, which ought to be very good."

Lieutenant Watson Smyth—July 13.—I may have got the date wrong, but the day is Sunday. As far as I can see, there is no likelihood of our doing anything for a bit. The Regiment is digging trenches about three miles in rear of the line; we go up for four days and then come back for eight. In that eight we do exercise every day except one, when there is a Brigade route-march."

Captain W. H. Eve—July 11.—"There is very little news from here, as you will have gathered from the papers. Kitchener was out here and came and inspected us last Thursday. He told us (what none of us knew before) that his father was in the Regiment. We have been fairly busy, and now have about seventy per cent of the men away trench-digging for a week. Three of my officers and most of my men are gone—went up in motor-buses yesterday, and with the few men left I have my work cut out to exercise and look after the horses."

Lieutenant Watson Smyth—July 13.—We are now up at ——,—at least seventy men and two officers per squadron are—digging the second-line trenches. We are about 2000 yards from the Huns, and they can see us nicely, thank you. . . . They shelled the second party yesterday and killed two men of 'D' Squadron,[1] very bad luck, as they have shelled the trenches lots of times before and never yet hit anybody. I had fever all yesterday. I don't know why I got it, but I'm all right now. . . .

"We go up in two parties; one goes up at 7 A.M. and digs from 8.30 to 12.30; the other goes up at 11 and digs from 12.30 to 4.30. It is not very interesting work, but it is better to be up here than back with the Regiment. There is only one man to about eight horses, and the result is that every one is working all day getting the horses exercised, fed, and rubbed over.

"We are up here for a week, we came up last Saturday and go away next Saturday; as usual, we travel in motor-buses. On a fine day, after a little rain to lay the dust, this is rather a pleasant way of travelling. . . .

"The Colonel (Symons) has just been ordered to the W.O. I suppose he is wanted for the Staff. If he is taken from the Regiment, I don't know who will get command."

The writer was wounded two days later. "It was rather a rotten way of getting hit, standing in a great deep trench, thinking I was quite safe, when suddenly we heard whe—bang, and I found my foot had gone numb. I said, 'Anybody hit?' and all the men said 'No,' so I told them I was." Lieutenant Watson Smyth goes on to describe his various moves until he reached some days later the British Red Cross Hospital at Rouen. "I had quite a good night, and woke up just before arriving here at 6 A.M. . . . I showed the doctor a label tied to my pyjamas, giving details of what was wrong. He told the bearer which ward to take me to, and off I went. . . . At 9 A.M. I was carted off to the X-ray room, and my foot and back were each taken from two positions. . . .

[1] Privates J. Tracey and D. Pitman.

D SQUADRON BILLETS, OCHTEZEELE
MAY 1915

QUARTERS OF D SQUADRON AT WITTERNESSE
6TH TO 19TH MAY 1915

QUARTERS OF B SQUADRON, OCHTEZEELE
MAY 1915

THE CHATEAU—QUARTERS OF A SQUADRON
AT WITTERNESSE. JUNE 1915

"RAGS" AND "STILTS," OCHTEZEELE
3RD MAY 1915

ON LINGHAM RIFLE RANGE
JUNE 1915

"In the operation they got about sixty pieces out of my foot, and a few work themselves out every dressing-day."

Not a word of complaint.

August 2.—"The Regiment's total casualties for the three weeks it was digging were four killed, two died of wounds, eight wounded. No other officers were hit, but one was buried one day, and badly shaken in consequence. . . .

August 3 and 4.—"I had a visitor to-day, a Miss Holt, who is working at Lady Mabelle Egerton's canteen at Rouen. She was very nice and cheery and did me a lot of good. They had one hour's notice a few days ago that 3000 men and 40 officers would want breakfast at 3 A.M. There were only three of them on duty that night, but they managed it all right. That shows that the canteen is useful, and the workers work, I think. . . .

"Dr Augier is exceedingly pleased at the appearance of my foot, . . . but says that I must resign myself to not using it for three months."

August 5.—"Another chap in my Regiment has just come into the hospital. I have never seen him, as he joined the Regiment after I was hit. He tells me that the Regiment have been shifted down south and have taken over, or are going to take over, some French trenches there for a bit. I do not understand it, but he's quite sure of his facts. I should have thought that they had sufficient Infantry out here now to hold the line, but I expect that they are fairly quiet trenches, and they want to make the Indian Cavalry Corps work."

August 5 and 6.—"The Regiment is now either in, or just going into the trenches for a fortnight. . . . It is a pity to have missed that. . . .

"I have discovered that I am one of the show-cases here. They show visitors my X-ray photo, and then bring them along and show them me. I dislike that. All the doctors seem to think it a marvellous case."

Captain W. H. Eve—August 22.—"I understand absolutely what you say about wanting a trophy. I hope too I'll get something some day, but I won't bring anything I haven't got myself, and I have never seen a German yet. Isn't it too dreadful, but it's the truth.

"Richardson has got the Regiment, and I can tell you I am glad."

Lieutenant Chrystall—September 7.—"I have been out on several night patrols, crawling through the grass towards the German trenches; and it is very jumpy work, as you never know when you may bump into one of their patrols."

September 20.—"The powers that be thought that a certain old house in No Man's Land ought to be occupied, and certain snipers caught who used to frequent the place and fire into our lines. Well, we occupied it one day and night and held it all day, when just after dark we were attacked in force. I at the time was holding a conference with the C.O. at his post, and all of a sudden the sentry on the outpost let off his rifle. This was the herald of a dozen bombs being hurled into our place, and the opening of a machine-gun and several rifles. There we were, ten of us, cooped up in an outhouse 12 × 12, firing like billy O through the window and door at the flashes of the Boche guns about fifteen yards away. This was kept up for about ten minutes. When we had just given up all for lost they ceased firing: we did likewise, and as we were in an awkward and tight corner we got out of it and held a bridge on a road. The Germans also retired, and as we afterwards found, left four dead, while we had only two wounded—a really miraculous escape, as bullets were crashing through the windows and splintering the panelling in the door, and bombs hitting the brick-work."

Captain W. H. Eve—October 9.—"We are trying to be patient, knowing very little and hoping for the best. A few days after I got back from leave we were moved off, at the beginning of our offensive, the 22nd [September] to be exact, and were kept in constant readiness for nearly ten days. We were all cramped up in a tiny dirty little village, with all our poor horses in the open, a bad place, with a rotten water-supply, and we had wet cold weather too. Of course we were desperately excited, but we weren't allowed to know much. Then on the 1st we moved on here where we are now, and where we are more comfortable than we have ever been before, in a large village, ourselves, the Third, and Brigade Headquarters, and very nicely situated, and good country. I was very lucky in the area allotted to my squadron, and have got the whole of my men and horses comfortably snug under cover, and it is a real pleasure to see my horses. . . . Of news we know no more

D SQUADRON QUARTERS AND MESS AT BETTENCOURT. AUGUST AND SEPTEMBER 1915

CAPTAIN EVE'S DUG-OUT AT AUTHUILLE
SEPTEMBER 1915

LT. J. V. DAWSON IN HIS DUG-OUT AT
AUTHUILLE. SEPTEMBER 1915

than you do from the papers, and, as I say, have simply to try and be patient, and trust the powers that be, and hope for the very best. . . .

"Meanwhile they seem desperately hard up for officers for some of the Infantry—more senior officers, that is—and have lately been sending all round the Cavalry for fellows to volunteer to take command of battalions and companies. . . . The other day the Colonel asked me if I would care to take promotion to Lieut.-Colonel to command a battalion of infantry, and they have been asking others also. I refused, but for a man who is ambitious irrespective of what brand of the service he gets his promotion in, you can see it is a tremendous chance. . . . But I doubt if they will get many men to volunteer for it. . . . It is extraordinary though, and to me seems such a very short-sighted policy, for when our time comes where will they find *our* officers?"

That was the problem which many Cavalry officers had to face, and it is not surprising that most of them, loving their own branch of the service, and still believing, or at least hoping, that the Cavalry would yet get their chance, should have shrunk from sacrificing their Cavalry training, even for such promotion. It was a clear proof of the great straits to which the losses of the war had brought the Infantry, on whom the brunt had fallen. The letter goes on:—

"News from Russia appears to me all good, and it does look as though the tide were turning at last, and the Germans had shot their bolt, and were up against it. . . . It's such a huge thing this war, and there is such a tremendous lot one wants to know, and so very little one does know. But I feel as confident as ever, as I think we all do out here."

They all did throughout, happily for England.

October 10.—"Nearly every house of a little better class one goes into here has a stuffed fox in it. I can't understand why, and they are such an eyesore to me."

October 12.—"One of the Poona Horse told me he had had a great ride after a really good pig this morning, only couldn't get him as he had only his sword. I had heard there are some in these woods, but hadn't seen any myself so far."

October 13.—"Here we are in our new quarters at Villers. . . .

Coming up here from the river valley we passed the most lovely coverts—all the trees turning fast, and with the sun on them simply looking perfect."

October 14.—"How I long for us to get a real move on. The thought of another winter sitting doing nothing like last is maddening. It can't be. I feel we must push and push and push. And here are we drivelling about doing these silly field-days and 'pretending' always."

October 15.—"A confidential memo. came round yesterday asking for recommendations among N.C.O.'s for the Croix de Guerre and the Médaille Militaire. The first is for gallantry, and the other more like our long-service and good-conduct. It is rather funny for us who have done practically nothing."

October 17.—"Then there was a lecture from the Colonel [Richardson] about drinking in the Regiment, and very well indeed he spoke too."

October 25.—"Exercise. Balaclava sing-song for the men."

November 16.—"This morning there were about four inches of snow on the ground, and it was still falling hard up to about 11 o'clock. During the morning we sent a challenge to 'A' Squadron, saying diplomatic relations were broken off . . . and a state of war would exist from 2 P.M. So at 2 we marched the squadron down, and I divided them into two parties, one doing a frontal attack, which drew 'A' from their village. When they were hard at it with snow-balls I brought the remainder, followed by the M.G., in on their flank, and we had a royal battle, and it was great fun. The men loved it."

But it was far from being all snow-balling and play.

November 25.—"You can see what it is trying to make us do two jobs at the same time, Cavalry and Infantry. The men are simply worked off their legs and haven't a minute all day. . . . We do all our Cavalry parades, all these Infantry ones, route-marches, afternoon parades, fatigues, evening classes, &c., &c., and they complain if the men don't turn out smartly on parade. In spite of all this we are to organise games, and let the men train for cross-country runs and so on. Whenever can they possibly have the time? And . . . I must help the country people in their farming in my spare time."

December 19.—"We marched to our new area on Friday last and got a very bad village to start with, Marival; then yesterday we moved to a neighbouring one, where we now are, and got ourselves very comfortable indeed. . . . We have managed to hire a large room in a farmhouse here, which we are going to fit up as a reading and recreation room for the men. . . . This will make a good deal of difference to them. They have a very dull and wretched time, little else but work and dirt. We shall be able to give them a good spread at Christmas, have ordered puddings, bought a whole pig, and are getting fruit and vegetables, at sales, this week. They will have a good feed anyhow. Your cigarettes will be really welcome and appreciated, there is nothing they want or appreciate so much. . . .

"What the future may hold we don't know; but though there is little news from here, things are very satisfactory all along our front, and we are very much on top of our friends the Germans. I have it first hand from Oakes and Annett of my squadron, who are just back from a month spent in the trenches with different Infantry regiments. . . .

"Haig's appointment out here will, I think, make little or no difference. He is one of French's men, and both are first-class."

December 27. — "It has blown continuously for days with a great deal of rain. . . . I think we made the men pretty cheery and comfortable here. They had a great feed, and very much appreciated all the things that were sent to them. I issued all their presents on Christmas Eve, a parcel for each from the Regimental Comfort's Committee, also from two old comrades, smokes from some newspapers, matches, match-cases and cards from ——, mittens from ——, and cigarettes from ——, so they didn't do so badly altogether. And they have more stuff to be given out later, two more goes of puddings sent by various newspapers, &c., sweets, and so on. I am spreading them out, leaving a week.

"We ran off a Divisional Marathon race, six miles, here on Christmas morning, and the Regiment was second. We went down to eat our dinner with 'A' Squadron, Tom Marchant's crowd, in the evening, and had a cheery time. 'A' Squadron and ourselves share the same village, and they have got the chateau as a mess. It is a large, fine old place, but not properly kept up, like most of them about here. The owner is in the Army, and only the old mother, the Dowager Marquise, is living there now. She came and dined with us. . . .

"We are all very well and flourishing, and are managing to get a lot of football, &c., arranged for the men now—so their life won't be so bad. This is rotten weather for the trenches. Wet is the worst for them, nearly always up to their knees, and often deeper in places. Cold weather they don't feel nearly so much, owing to the depth of the trenches."

So ended for the Thirteenth their first year of war service on the Western Front. It had been in some ways a year of disappointment. Though they had had some turns in the trenches, involving a few casualties, the Regiment had seen no active work as Cavalry. But if hope deferred had brought them some heart-sickness, there was hope still, and they entered upon another year under fairly cheerful conditions.

LT. J. I. CHRYSTALL AT BEMAVILLE. OCTOBER 1915

HOUNDS AT L'ABBAYE. 31ST MARCH 1916

CHAPTER VIII.

1916.

THE year 1916 opened, as the last had closed, without any exciting event for the Indian Cavalry Corps. There was heavy fighting for the Infantry and guns, and the more hopeful spirits in the Thirteenth Hussars still believed, or tried to believe, that their chance would come sooner or later, but there seemed to be no immediate prospect of it.

Meanwhile the old routine went on.

Captain W. H. Eve—February 6, 1916.—"We are still busy, though rather dull, at our ordinary training, . . . but the men get a lot of games, &c., too—football, running, and so on. We have been having pretty good weather, much drier, and nothing to complain of. It has been a wonderfully fine winter so far, but it isn't over yet. Things are very forward—too forward, I am afraid.

"All the land is under crops, which interferes terribly with our field-work.

"Oakes (2nd in command, 'D' Squadron) is at present in hospital with a kick on the shin (football), but otherwise we are all very well and flourishing."

March 19.—"There is little or no news from here except that our proposed move is, I think, postponed for a time. I am sending another digging party away to-day and one comes back. I expect we shall always have a certain number of men away now, digging in second line. We are all very fit and flourishing."

It was a necessary measure of precaution to have lines of trench behind the fighting line. All hoped to go forward, but in case of overwhelming German attacks, strong intrenched positions on which troops could fall back were an obvious need. The letter goes on:—

"I attended a trench-warfare class for a week about a fortnight ago, which was a change, but taught me very little.

"The weather is all in our favour for work, but it means that in this closely cultivated country it is almost impossible to find ground to work over, now that they are so busy with the spring work on the land."

"During this time," writes Lance-Corporal Bowie, "we had many inspections, the chief one being by General Remington, commanding the Indian Cavalry Corps in the new year. . . . The Regiment not being required in the actual firing line at this time, many men were sent on various courses of instruction, such as M.M.P.'s, signalling, pigeon-flying, intelligence, &c. On January 31st a party left by motor-buses for the village of ——, to construct trenches, &c., being relieved on March 14th by another party. On March 26th we proceeded to Feuquières-en-Vimes. From here the Regiment left for St Riquier on April the 4th to commence Divisional training, after which we returned to Feuquières on the 18th, meanwhile continuing to relieve our parties in trench-digging. . . . Later we again proceeded to St Riquier to take part in the Corps training, after which we again returned to Feuquières. The remainder of our stay here was occupied by sports, horse-shows, &c., the chief event being an International Horse Show held at Oisemont in May. This event was marked by the attendance of many noted French Generals and the 3rd French Army Corps from Verdun. Immediately afterwards we again proceeded to St Riquier to take part in the Grand Manœuvres."

It all reads more like peace-time than the middle of the greatest of wars. But such was the condition of things for the Cavalry of the Allied and enemy armies on the Western Front. The extension of the front from the sea to Switzerland, and the development of the great line of trenches which covered it, had put an end to all open warfare. The contending hosts were locked in a desperate struggle, which swayed backwards and forwards over a few miles of devastated and blood-soaked ground, giving no opening for the sweeping movements of horsemen. Month by month it became more and more evident that there was no immediate likelihood of this state of things being changed. Neither side seemed strong enough to tear a gap in the opposing line and let through a flood of horsemen into its rear. Till that could be done there could be no chance for the Cavalry as Cavalry.

"CAPRICE." 1915

AT MARTAINNEVILLE. MARCH 1916

Meanwhile the war was developing all over the world, especially in Africa and Asia, and in some quarters the conditions were very different from those obtaining in the densely-peopled districts of the main front, where the colossal armies stood face to face. On two other fronts especially, where Great Britain was opposed to the Turks, horsemen had room to move and were badly needed. In Egypt and Palestine, and in Mesopotamia and Persia, the armies engaged were not in such numbers as to cover the whole ground and reduce the war to the likeness of a gigantic siege. Even there, on the broad plains of Asia, trenches and aeroplanes were to some extent in use, and did much to limit the operations of Cavalry; but they were not all-important. There was still room for open warfare and turning movements and surprise. And for service in those countries, with their comparatively dry climates and hard soil and great heat, the Indian soldier was specially fitted—much better fitted, to tell the truth, than for the cold and mud of Europe. It was therefore decided in the early part of 1916, when trained white troops were beginning to pour out of England in great numbers, that the work of the Indian Army on the European front was done, and that for the future, with its numerous and excellent Cavalry, it could be more usefully employed in other theatres of war.[1]

It has always been the custom to send on service with Indian troops a certain proportion of white men, and much as the white troops were needed in Europe, it was felt that the custom must be maintained in this case. The Indian Regiments now sent to fight in Africa and Asia were therefore mixed as usual with British Regiments. The officers of the Indian Regiments had every confidence in the troops they led; but it would have been a tremendous test to send Indian soldiers unsupported by British troops to face the renowned fighting men of Turkey, organised and directed by Germans, and equipped with all the latest appliances of modern war. The strain would have been increased by the fact

[1] As to the value of its work in Europe opinions differed. Some seemed to think little of it. Some declared that during the time when English reinforcements were not yet ready to go into line, the Indian contingent had saved the British Army from being overwhelmed. Both these views were perhaps extreme. The Indian soldier fought under serious disadvantage in the climate and surroundings of Europe, but undoubtedly he fought with great devotion and suffered heavy losses, for which England owes him deep gratitude. The matter may well be allowed to rest there.

that the number of British officers serving with an Indian Regiment is always small—probably not half the number serving with a British Regiment.

As to the Thirteenth, they were warned in the middle of June to prepare for service in the East, and there were persistent rumours that the destination of the Brigade would be Mesopotamia, where a serious reverse had lately been inflicted upon British arms by the capture of General Townshend and ten thousand men at Kut.

These rumours were well founded, for the nation rightly felt that such a humiliation must be wiped out, and the Meerut Brigade, including the Thirteenth Hussars, was among the troops selected to retrieve the position.

"It was on the 17th day of June," writes Private C. T. Massey, "that my Regiment, the Thirteenth Hussars, who were at that time digging trenches for cables behind the lines in France, received orders to get ready to return to their billets at Chessy, a little village a few miles distant from the town of Abbeville. We heard that the Regiment was under orders for leaving the country, and rumours were flying all over the place: some said Salonika, others said India, whilst others said Basra, a modern town of 10,000 inhabitants on the right bank of the river Tigris in Mesopotamia, and the last one proved to be correct. . . . On arriving in billets we found great excitement prevailed, and every one said we were leaving for Mesopotamia. Every one was in the best of spirits and glad to be leaving France, where Cavalry were inactive."

That was the right spirit in which to take the move; and though no doubt there had been much disappointment among officers and men with the fact that in France "Cavalry were inactive," it would not be easy to find among all their letters and diaries bearing on this time a word of murmuring or regret at the prospect of turning their backs on England again, and entering upon another term of distant Eastern service. The universal feeling was one of satisfaction at being sent to some other theatre of war where Cavalry would come into its own.

Private Massey's diary goes on to describe the voyage out.

"On the 19th we were officially informed that the Regiment was leaving France for Mesopotamia, and the same day we left for the railhead at Pont Rémy, a fairly large station. . . . Whilst waiting,

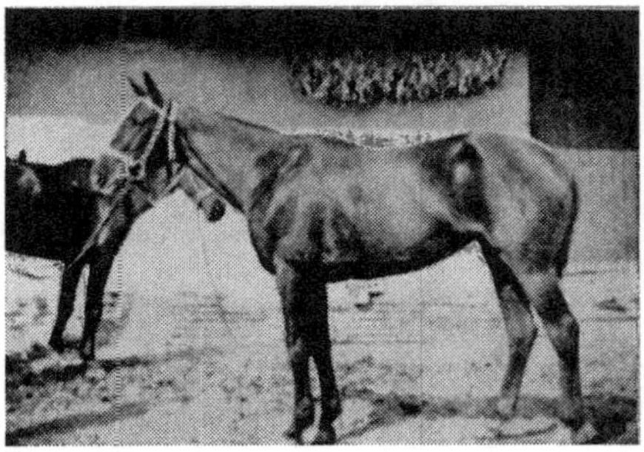

AT FEUQUIÈRES, APRIL 1916

a train-load of Belgian soldiers passed through and we gave them a rousing cheer. . . . All was soon ready, and at 6 the train moved off. We were travelling all day the 20th and the 21st, and on the morning of the 22nd we arrived at Marseilles, and after detraining with our horses, limber, and equipment, proceeded towards La Valentine Camp. We remained in the camp until the 27th, which we left at 7.30 that day, and proceeded to the docks at Marseilles."

It appears from the Regimental Diary that the Regiment embarked 30 officers and 582 other ranks. Private Massey goes on:—

"We handed over our horses to a native regiment. I gave my old horse (No. 62) a few pats on the neck, as he had been a good old horse, and he had carried me hundreds of miles, and I was sorry to lose him. We afterwards went on board the *Kalyan*, and I was put on guard over the steps leading to the bridge; others were put on submarine guard, and every precaution was taken to prevent the ship being torpedoed. . . . On the 31st we were issued out with Indian khaki, as it was getting very hot the farther we went south. The next few days the sea became very rough, and on 2nd July I was sick after tea.

"On the 3rd we arrived at Port Said after a very uneventful voyage. On the voyage across we received the news by wireless every day, so we knew that the great offensive had begun. We watched the coaling of the boat, during the afternoon of the 3rd, and it struck me as a mild form of slavery, to see the natives coaling the ship: they work like devils, and all for a few annas a day. Native boys swim round the ship, diving for pennies which we threw them from the deck.

"When we woke at réveillé on the following morning the boat was already on the move down the Canal. . . . Towards night we sailed past Port Suez, but we could not see much of it; but it looked magnificent with the streets lighted up, and it seemed to stretch a long way along the coast.

"After sailing down the Red Sea for several days we entered the Indian Ocean. We saw large shoals of flying-fish in the Red Sea, also other big fish.

"We had an uneventful voyage across the Indian Ocean, and arrived outside Bombay at about 9 A.M. on the 15th of July. On the 16th we went into dock soon after réveillé, disembarked, and marched to Colaba Camp, about four miles from the docks. I went with several others to the swimming-baths, and for the first time in my life managed to swim a few yards. I was so fond of the salt water that I was in it again that night, aided by the light of the moon."

Lance-Corporal Bowie, already quoted, strikes the same note about the horses. "Here" (at Marseilles), he writes in his diary, "our horses were taken from us, many of us being very sorry to part with some splendid horses which we had brought with us from India, and which had carried us so well all the time in France." In that respect the officers were more fortunate than the men. Though it was not thought desirable to send the troop-horses back again to the East, the officers were allowed to take their chargers—about 100 in all,—a great boon.

June 26, 1916.—"All has gone well," writes the officer commanding "D" Squadron, "and I am taking all my three children with me." And later from Marseilles he mentions "the glad news that we are taking our own children with us. . . . We have handed over a certain number of the best of the walers and country-breds to the 30th, including, I am sorry to say, 'Jean,' and the remainder all go to the depot to-morrow when we depart. . . . All the Regiment will be together, and the 'children' with us too. It could hardly be better, all things considered."

From Port Said again: "All is well. It's the usual dull routine, but better than I remember it before in these parts. The great mercy for me is having the children with us. They give me an interest and are most comfortable—tons of room, two or three stalls for each, and very flourishing so far. 'Caprice' was off her feed and with a temperature yesterday afternoon, but only for a little while, and is as fit as anything to-day."

It is satisfactory to know that for the most part they arrived safely in India, and, though none the better for their voyage, were able to do good service afterwards.

In spite of all it was not an exhilarating voyage. Very little was known to the Regiment about the prospect before them, and not much about what was going on in France.

Captain Eve—Suez.—"As soon as we arrived yesterday we heard the push had started and that we had got through on a front of

Lieut. Basil H. Williams, M.C.
(Wounded in France with 11th Hussars, 31st October 1914)

Lieut. M. H. C. Doll
(Wounded near Authuille, 11th September 1915)

Lieut. T. E. Lawson-Smith
(Killed with the 11th Hussars, near Messines, 1st November, 1914)

Capt. J. O. Oakes

Capt. A. M. Sassoon, O.B.E., M.C.

Lieut. W. P. Crawford-Greene

Capt. S. V. Kennedy. M.C.

twenty-five miles, but we didn't get the telegrams with any details till the afternoon: let us only pray we may do as well as we hope."

Alas! it was a vain prayer. There were to be many "pushes" before the final one two years later.

July 14.—"We get to the end of our march to-morrow. I needn't tell you how glad we shall be. It has seemed desperately long—very different to when going the other way. We are all very fit and flourishing."

A later letter describes the voyage in more detail, and may be worth quoting as an example of war-time voyages. "We went by ourselves, but had various kinds of escorts through the Mediterranean: sometimes a destroyer, sometimes a sham tramp steamer, really of course armed with guns, and sometimes with no visible escort at all; and we went the most peculiar way, all round the west and south sides of Corsica and Sardinia, south of Malta, north of Crete, then south and zigzagged about a lot before making Port Said. I believe a ship was sunk off the north-east corner of Crete not long before us. I think our sham tramp steamers are the most lovely things. Of course we had all lights out, ports closed, no smoking after dark, and great discomfort, and heaps of guards and submarine look-outs, and it was very hot but mostly very calm. However, all this ended at Port Said, where also they took off our gun and gunners. . . . We left at 10 P.M. and went through the Canal very slowly, stopping a long time in the lakes. The camps and trenches had increased enormously, the most perfect trenches now, and huge camps. It is quite a sight, very strong, and there must be a large force in Egypt now. They looked wonderfully fit, and yelled to us to know where we were going, but we would not say."

Arrived at Bombay, the Regiment was informed that it was definitely for Mesopotamia, and the news was received with the greatest pleasure, as doubtful rumours had been in circulation. They had, in fact, been better founded than most rumours, for shortly before the arrival of the Thirteenth, it had been decided that the Brigade was not to leave the country again. The Thirteenth were to return to Meerut, and the Seventh Hussars were to take their place with the Mesopotamia force. Why this decision was altered does not appear, but report said that the Seventh were wanted on the North-West Frontier, where trouble was brewing, and the Thirteenth were taken for Mesopotamia after all. Nevertheless, the old Meerut Brigade was broken up, the 3rd Skinner's Horse and 30th Cavalry being retained in India, to their intense disappointment. The 3rd and the Thirteenth Hussars had lain alongside one another for five years, and it was with special regret that they said good-bye to each other. The place of the two native regiments was taken by the 13th and 14th Indian Lancers. The Meerut Brigade now became the 7th Cavalry Brigade of the Mesopotamia Field Force. It retained with it V Battery of Horse Artillery, which had suffered so severely in France.

Meanwhile the Regiment had lost the services of some officers from illness. Second-Lieutenant Guy Pedder writes on the 4th of July, evidently from the Suez Canal, though after the manner of the time his letter is dated only "on ship going out":—

"Four officers have fallen out now—Covell in France, Kennedy at Marseilles, Munster at Port Said, and Wingfield at the next stop. . . . I wonder where we will get our first letters, &c.; it will be rather interesting. We travel very slowly along here, and stop at each lake as far as I can make out! There are no more submarine guards to be done, thank goodness. I was on eight hours each day, four by day and four by night. Saw a paper yesterday which said we had got the first-line trenches over sixteen mile frontage—wonder if the push is going on as well as was expected."

July 14.—"I expect the first news you will have got of me will be the cable I sent from Port Said; we haven't put into harbour since then, but were able to drop some letters to a small boat just opposite Suez, but I don't know whether you will get that letter. We hope to reach Bombay to-morrow; if so, this letter will just catch the mail, and I will cable to you to-morrow.

"Of course we got out of the danger zone when we got to Port Said. It was very hot, but very interesting in the Canal. Then followed the Gulf, which was calm and dull for four days. Then we got into the Sea, and the so-called terrible monsoon; but it wasn't a bit terrible, although the ship rolled like anything for four days; lots of fellows were sick, and felt very bad, but I was as fit as anything all the time, and smoked cigars and pipes, &c.

"No one knows our final destination up to date, but I fancy

THE "ISLANDA"

we will be in India for a couple of months at least, and then go on. I am looking forward to seeing India awfully, and hope we stay there for a short time at any rate. . . . I have enjoyed the voyage very much, but am just getting bored with it to-day, and want to see a paper and hear the war news: of course we have had none for over a week. This is a topping ship, but we have come along very slowly to save coal. They feed us very well. We had to run for it once in the Mediterranean, but we did not know it at the time, only guessed. It has been very much too rough to play cricket, but to-day it is much calmer. We've had two frightfully heavy rainstorms. Only two horses have died, and both mine are all right. I sang several times after dinner, and on the whole I think the time has flown. I am longing to hear about the Western Front. There is an excellent library on board, and I must have read twenty books, &c."

Bombay, Royal Bombay Yacht Club—July 17.—"This is an extraordinary place, and awfully pretty by night; Daniell (the Staff Captain) gave a large dinner-party here last night, and very nice it was too. I went to the Cathedral yesterday evening after a very busy day, and this morning I have been rushing round shopping. You hardly see two people dressed alike here. This is a ripping Club, and we are all honorary members, Indian Expeditionary Force (D). To-day's paper tells us a little about the part played by what would have been us in the West!"

The hope of staying a short time and seeing India was not to be fulfilled. The state of affairs in Mesopotamia admitted of no delay, and it was soon known that the Regiment was to embark at once for Basra. It was to be carried in two vessels, the *Islanda* and *Chilka*, the former taking "A" and "D" Squadrons, less one troop of "D," under Captain Eve, and the latter the Headquarters and remainder of the Regiment under Colonel Richardson.

By the 19th of July both ships had sailed.

CHAPTER IX.

MESOPOTAMIA.

THE campaign in which the Thirteenth Hussars were now to share was of great importance to England, and not without importance to the decision of the world-war. It may be well to say something here on the subject.

The designs of Germany in bringing about, after many years of elaborate preparation, the conflict of 1914, were not fully understood until they were laid bare in Chéradame's book, 'The German Plot Unmasked,' which was published in 1916, with an Introduction by Lord Cromer. Though England was the greatest of Asiatic powers, and more concerned in the march of events in the East than any other nation of Europe, except possibly Russia, she had to learn from a Frenchman how her Eastern Empire was threatened by Germany —just as she had formerly had to learn from an American sailor, though she was the greatest of maritime nations, the influence of sea-power upon History.

To put it in a few words, Germany aimed, as Napoleon had aimed a century earlier, not only at supreme dominion in Europe, but at supreme dominion throughout the world. To effect this aim she had brought about the alliance with Austria, which placed at her disposal the whole resources of the great central European block, about a hundred and twenty millions of people highly organised for war, and had also succeeded in establishing her paramount influence over Turkey, which meant not only another twenty millions of people, but a vast territory stretching very nearly from the frontiers of Austria across the Bosphorus and far into Asia. The inclusion in the alliance of the comparatively small but powerful state of Bulgaria practically completed Germany's line of communication with Turkey, and made her in fact dominant from the Baltic to Constantinople, and on through Asia Minor to Arabia. A great strategic line of railway had been designed, and in part constructed, which was to

run from Constantinople to Baghdad, and was to be continued to a terminus on the shores of the Persian Gulf.

Italy was a doubtful ally, but was nominally with Germany too, and might at least be regarded as neutral.

The German scheme was to deal first with the two great Continental powers which stood in Germany's way, France and Russia. If they could be attacked and overthrown, as she believed they could be, by a swift onslaught upon each of them in turn, the Continent of Europe would be at her feet. It was calculated that

England would be neither willing nor able to interfere in their behalf before their fate was sealed. Then, with France and Russia powerless, or even possibly enlisted as vassal States, Germany would turn upon the only power which stood between her and the dominion of the world—England. Having a contemptibly small Army, and no allies in Europe, England would either come to heel without fighting, or would be attacked in India and overwhelmed, probably with the help of the Indians themselves. Her fleet would not be able to help her against vast armies, German and other, marching upon the Indian frontier from Asiatic Turkey, and the few hundred thousand trained men she could put into line would be swamped by ten times their numbers. Afterwards, if America or Japan or China gave any trouble, it would be easy enough to deal with them. The only powers that really counted were the three great European powers—France and Russia with their trained armies, and England with her trained fleet.

It was an ambitious scheme, but not one that could be regarded as visionary. It did not take sufficient account of England's sea-power; but undoubtedly if France and Russia had both been struck down, and England had been left standing alone, he would have been a very fearless Englishman who could have faced the future without apprehension. Even supposing that no immediate attack upon England had followed, her prospect of holding her own indefinitely against a Germany rapidly outgrowing her in population and wealth would not have been promising. The silent deposition of the naval power of France by that of Germany in the course of a few years before the War had been a striking lesson. But as a matter of fact a great attack upon England was undoubtedly contemplated. "Der Tag" was to have come, and come soon.

Can any one feel sure that if England had stood by while France and Russia were overwhelmed she could afterwards have successfully resisted that attack? The Boer War had shown that at the beginning of the century a combination of the Continental powers against her was not improbable. Would it have been less probable fifteen or twenty years later, when the sea strength of those powers compared with her own had vastly increased, and when France and Russia had been incensed against her by her failure to help them in their time of need? And if all the navies of Europe had joined against her, could she have drawn for help on India and the Dominions beyond the seas? Would she not have found it hard enough to protect her own coasts? Happily for her she did not stand aside, and that issue was never put to the test. Unready for war as she was, and unwilling, she struck with sure instinct before it was too late.

Even so, though England threw in her lot with Russia and France, the struggle was not an unequal one, and, as every one knows, there were times when it seemed that the Allies might lose the war, or at all events fail to make more than a drawn fight of it. Their latent numbers and resources were greater, but the enemy enjoyed the immense advantage of having chosen his own time, when he was ready and they were not. He had also the advantage of united command and of the central position, whilst the Allies were widely separated. These advantages very nearly outbalanced latent numbers and resources. Eventually they proved insufficient to do so, but they nearly succeeded. Nothing prevented Germany winning but the fact that she had to put out all her armed power at once, and to fight England then, instead of reserving her Turkish strength for a separate duel with England later.

How formidable her Turkish strength was, a glance at the map will show. Not only were the Turks a great military nation, with warlike traditions and a population capable of raising two millions of fighting men, but Turkey stood across the Straits between Europe and Asia, and while guarding them could throw her weight freely upon the East. India was England's most sensitive point, the one where she was exposed to military aggression by land. Strike her there, the Kaiser thought, as Napoleon had thought before him, and the clay feet of the great image would crumble under her.

Between Europe and the Indian frontier lies a stretch of country 2500 miles in breadth, held by three independent powers, Turkey, Persia, and Afghanistan.[1] All these powers are Mahomedan, and of the three Turkey is, or was in 1914, by far the most powerful. Not only was she the strongest from a military point of view, but in the eyes of countless millions of Mahomedans the Sultan of Turkey was the head of the faith, the true successor of the Prophet; and he was entirely in German hands. His power extended over a thousand miles, to the frontier of Persia, which was not only weak, but at the moment unlikely to use such

[1] Beluchistan may for practical purposes be regarded as a part of India.

strength as she had on England's side. Turks and Persians certainly did not belong to the same sect of the Mahomedan faith, and had often been enemies in the past. But the Persians after all were Mussulmans, and their religious sympathies in any quarrel between Mahomedans and Christians were sure to be against the Christians. Persia held a thousand miles more of the space between Europe and India. Beyond her again to the eastward, right up to the Indian border, lay the third of the independent powers—Afghanistan. The external relations of Afghanistan were supposed to be under British control, and her ruler enjoyed a British subsidy. But his people were turbulent and fanatical, and belonged for the most part to the same religious division of Islam as the Turks. They were believed to have little love for the British, who had more than once invaded their country. Finally, along the Indian border itself, and inside India, there were perhaps seventy millions of Mahomedans, some belonging to wild mountain tribes, constantly at war against the British, and most of the rest inclined to acknowledge the religious supremacy of the Sultan. These Mahomedans had, as a rule, served the British Government with fidelity, and formed a considerable part of the Indian Army. But they too were of the faith. Surely the Germans had some ground for hoping that if the Turks made a vigorous push towards India from their own Asiatic territory, their armies, organised and commanded by German officers, and supported by a hot religious propaganda, would succeed in doing much evil to England. They might, perhaps, succeed in sweeping the independent Mahomedan States with them into a great invasion of India. In any case they would seriously disturb the country, and probably stir up a Mahomedan revolt with which England would find it hard to deal. If backed by a great German army they would be irresistible.

The Kaiser was not far wrong. Even though by joining France and Russia in 1914 England disarranged the German calculations, and brought on the Eastern conflict prematurely from a German point of view, it was shown that there had been good reason for the Kaiser's confidence. Turkey under German direction proved strong enough, even without the help of a German army in the East, not only to repulse a great Anglo-French attack upon her in the Dardanelles, but to inflict much loss upon England in

Western Asia, and with the aid of a strong politico-religious propaganda, to cause sensible trouble on the Indian border. In the end she failed, and the blow which was to have brought about the overthrow of England in India resulted in the complete collapse of the Turkish Empire: India, instead of being a source of weakness to England, turned out to be a great addition to her military power. But before this result was reached there were four years of hard fighting, and at times the issue seemed to be very doubtful. Unquestionably, the Anglo-Turkish conflict was a matter of great moment, and the result of it seriously affected the success of the whole German scheme.

It is interesting to consider in some farther detail what was the strategical position of Turkey with regard to war in Asia when she elected to draw the sword. The original home of the Ottoman Turks was on the Asiatic side of the Straits, and it was there that in this century, if not always, the main strength of the Ottoman Empire has lain. Asia Minor was the great recruiting ground for the Turkish armies, and the great central base from which she could strike out eastward. Assuming that her alliances in Europe, and the possession of the immensely strong position on the Bosphorus and the Dardanelles, made her practically secure on the western side, as proved to be the case, the value of the Asia Minor base for action eastward was very great indeed. There she could place the bulk of her large army, and from there she could throw her weight upon the distant possessions of the Allies, where they were incapable of much mutual help,—upon the Russians in the Caucasus on her left—upon Persia, and possibly through Persia upon India in the centre—upon Egypt on her right. The Allies, hard pressed in Europe, and therefore comparatively weak on these extremities of the great semicircle, seemed to be at an almost hopeless disadvantage in meeting the blows she might strike, outwards as it were from the handle of an open fan towards the end of the spokes. Her fronts in Asia were three—Armenia, Persia, Palestine; and it seemed that from her inner position she could act with greater effect upon each of these fronts than the scattered Allies could do, acting from the outside inwards.

On the central of the three fronts the Turks were perhaps in a specially strong position, for they had an established secondary base in Mesopotamia, with its famous capital Baghdad, to which extended, though with one or two gaps, the great strategical railway from Constantinople. Beyond Baghdad they held the lines of the Euphrates and Tigris rivers, and could push eastward into Persia by the highroad which from time immemorial has seen the march of conquering armies eastwards and westwards,—the armies of the Persians for Marathon and Platæa—the armies of Alexander the Great on their way to India—and numberless others before and since.

But what German and Turk alike failed to understand, or at all events to appreciate at its full value, was the sea-power of England. Sea-power had in old days given Rome the mastery over Carthage, and in later times it had enabled England to wear out Napoleon. It was to be the deciding factor now in the overthrow of the Turkish Empire, and with the Turkish Empire, of the great German scheme of world domination.

For recognising at once the great danger to India of letting the Turks push forward into Persia, and possibly into Afghanistan, recognising also the value of the Persian oil-fields and other British interests in that ancient country, Great Britain had determined not to await Turkish and German aggression on its Indian frontier, but to meet the threat with a bold offensive on Turkish soil. Directly it became certain that Turkey had thrown in her lot with the Central Powers, in the autumn of 1914, an expeditionary force sailed from India for the Persian Gulf, and seized the mouth of the Shat-el-Arab, by which the Tigris and Euphrates pour into the sea. The objects of this expedition were at first limited. The protection of the oil-wells, of such importance to our Navy, and the blocking of the German strategical railway through Baghdad, were all that was immediately contemplated. But the comparatively easy success of the Indian force, mainly composed of native Indian soldiery, in defeating the Turkish troops near the coast, encouraged the British commanders to push on up the rivers into Mesopotamia. In 1915, a year after the outbreak of war, a force under General Townshend had taken Kut-el-Amara, three hundred miles from the sea, and the attack on the Dardanelles being on the point of open failure, it was decided that as a counterblast to this failure Great Britain should strike a great blow in the East by marching to Baghdad and conquering all Turkish Arabia.

It was an important decision, and full of interest in many ways. Mesopotamia is the cradle of history, sacred and profane. It is the legendary site of the Garden of Eden, and from its plains, from Ur of the Chaldees, the Patriarch Abraham set out with his flocks and herds for the Holy Land. After his day it was the site of great empires. Babylon lies in the centre of it, Nineveh not far to the north, Shushan a few score miles to the east. It has seen Grecian and Roman armies as well as Asiatic hosts, and the first explosion of the new Mahomedan faith was across its plains to Ctesiphon, and Persia, and Syria. Baghdad soon arose as the Mahomedan capital, and became famous throughout the world. Undoubtedly, to attack Turkey in Mesopotamia was to cover India and Persia from attack on her part; and to beat her out of Baghdad was to strike her a blow which would resound all over the world. She would perhaps exhaust herself in trying to recover her position there, as Napoleon exhausted himself trying to recover from a similar sea-borne blow in Spain. From the time when Townshend was ordered to advance on Baghdad, the Mesopotamian Front became one of the important theatres on which the Great War was being played.

As a fighting ground, Mesopotamia had some advantages for Great Britain, and some great drawbacks. The southern part of it came down to the sea, and communications with India and England were therefore open. Everything required for the conduct of war could be supplied. Moreover, though the climate of Mesopotamia was hot in summer, it was perhaps, as before remarked, better suited for the Indian soldiery, who formed the bulk of the British forces, than what Europeans would consider a more healthy climate, the temperate climate of Northern France. Its plains too were free from the geographical obstacles of mountainous countries. Right up to Baghdad they were flat and bare, very different from the wild fighting grounds of the Indian frontier, with their rocky peaks and forest-clad hillsides and rushing torrents.

On the other hand, the summer heat in Mesopotamia was excessive, even for Indians, and desperately trying to white men, while in winter the wind and cold were at times severe. Moreover, the very flatness of the Mesopotamian plains was a difficulty. The great rivers which wound across them were in the rainy season swollen by the melting snows of their upper courses, until they overflowed their banks, and caused vast inundations and swamps impassable for troops. The march of military forces in the hot season, with the thermometer in tents at 120° or more, was too deadly to face, and movement in the flood season was impossible; so the real fighting season was almost limited to the time from the end of the heat till the filling of the rivers—that is, from about the middle of October to the middle of March.

At all other times, and indeed at all times, the rivers themselves were the chief means of communication for troops and supplies; and boats of any carrying power were few. Even when armed movement on land was possible, any advance against an enemy in position was a formidable task, for the flat ground afforded not a vestige of cover, and troops had often to go forward to the assault of trenches over ground as smooth and bare as a billiard-table, perhaps up to their knees in mud, with deep swamps on each side preventing any attempt at a turning movement. Mud, indeed, proved to be a more formidable obstacle than mountains and ravines. Troops could not advance over it with any freedom or swiftness, and they could not camp in it without misery and loss; nor could they be fed in large numbers, for it made the transport of supplies very difficult. Then the whole country, though not really friendly to the Turks, was inhabited by Arabs who were anything but friendly to an invader. Whether in the marshy lands near the rivers or on the dry plains beyond, they were always hanging on the flanks of an advancing or retreating force, their desert horsemen as elusive as the "web-footed" men of the marshes, swift to gather and as swift to vanish in the mirage of an enchanted land where all seemed fantastic and unreal. With stubborn Turkish soldiery, organised by Germans, intrenched in large numbers along the river lines, and supported by larger numbers of these irregular auxiliaries on every side, the country was no easy field of action for a British army.

Nevertheless, in spite of all difficulties of climate and ground, the British expeditionary force had by the autumn of 1915 established itself in control of the river mouths, with a considerable Turkish province in its hands. Then, in an evil hour, came the decision to advance on Baghdad, and a single British Division was pushed forward. It was a very daring if not an insane project, and it failed. Before the end of the year the unfortunate Division found itself besieged by superior forces at Kut-el-Amara, and in the following April, after a siege of five months, a starving British force of more than 10,000 men, nearly 3000 of them white men, was marched away by the Turks into bitter captivity.

This was the heaviest blow that had ever been dealt to British arms and British prestige in Asia. Not only had 10,000 men been taken prisoners, but the Turks had inflicted upon other British forces trying to relieve them a series of bloody repulses. Struggling forward, time after time, with splendid devotion over the muddy flats, in vain attempts to drive from strong lines of trenches an enemy superior in numbers, our soldiery, white and black, had lost over 20,000 men in killed and wounded, and had been forced to admit that for the time they could do no more. The Turks had won a striking success, the measure of which to Great Britain was the loss of an Army Corps.

But, much to its credit, the British nation refused to accept the defeat in Mesopotamia as a final one. Though staggered by it and the still greater repulse at the Dardanelles, England resolved that the Turks should yet be conquered. Smarting from her defeats, she was not wholly just to the leaders who had done all that men could do to effect impossibilities. Some honourable reputations were sacrificed, and wrong done to brave and capable soldiers. But at least her resolution did not fail. Her legions, rapidly increasing not only on the soil of the British Islands but throughout the Empire, and made available by her sea-power for employment all over the world, were poured upon the Turkish frontiers. The Turks had dealt her two stunning blows; but brave fighters as they had shown themselves to be, they were to learn, as Germany learnt, that it is not prudent for any nation to rouse the English.

In Mesopotamia the military chiefs who had failed in their attempts to reach Kut before its garrison was starved into surrender, were relieved of their commands, and the Mesopotamian force was entrusted to General Maude, who, unlike them, was now given time to collect a large army, properly organised and equipped, and was helped in his task by every possible means both in India and in England. Troops were sent to him in numbers sufficient to let him meet the Turks on at least equal terms, and immense efforts were put forth to give him the necessary equipment for scientific modern warfare, and the transport necessary for effective action. Roads and railways were established, and above all, a great fleet of river steamers was gathered from various parts of the world, in order that he might be able to use to the full the natural highways of the country. During the whole summer of 1916 these preparations were steadily pushed on, with a view to another advance when the hot weather would be over.

It was to this country, and during this pause in the conflict, that the Thirteenth were diverted from their work in France. The diversion was of course a disappointment. The Regiment could no longer hope to join in the coming triumph on the Western Front. Not for them the grand pursuit to the Rhine, and on over German country to the gates of Berlin, and the final march Unter den Linden. It was hard to give up such a prospect. But it has been shown in what spirit the order was received. They were soldiers, and their duty was to do their best wherever they might be most useful to the country. If they were more wanted in the East than on the Western Front, so be it.

And, after all, perhaps it might be as well for themselves. The coming triumph in Europe might be long postponed, might even turn out to be one for the Infantry and guns alone. In the plains of Mesopotamia they might reasonably look for some Cavalry ground—for some chance of striking a blow on horseback and justifying their existence. There, at all events, they would not have the work and the honour altogether taken out of their hands by the airmen, who were to them what the eagle was to the horse, and find themselves chafing in impotence while the enemy defied them from the shelter of his trench lines, against which they were as useless as unarmed men. Mesopotamia held out some hope to the cavalryman who still believed in his arm. He might yet get home with lance and sabre, and take his revenge upon the footmen who had so long held him at a distance with fortifications and "villainous saltpetre." Asia had always been the land of the horseman. Surely it would be so again.

And he was not wrong. Both in Mesopotamia and in Palestine, horsemen were to strike heavy blows before the war ended.

CHAPTER X.

SUMMER IN LOWER MESOPOTAMIA.

THE voyage of the Thirteenth to Mesopotamia was uneventful but not altogether pleasant, as any one can understand who has been on board ship in the Indian Ocean and Persian Gulf in the month of July. A few extracts from diaries and letters will give some idea of it.

Private Massey, "D" Squadron—Diary.—" Réveillé was at 3 A.M. on the morning of the 18th, and I was ordered to stay behind and load the rifles and swords on the motor-lorry, and went down to the docks on it afterwards, so I was saved a march of about four miles. After reaching the docks, the Regiment commenced to embark the horses and mules. All were soon on board, and at 11 o'clock the ship, the *Islanda*, set sail, and it was a voyage I shall never forget. . . ."

July 23, 1916.—" During the voyage we had the horses to water, feed, and clean out their stalls, which involved a great deal of hard work, besides which we had to get their forage out of the hold, and carry it through no end of awkward gangways, and up and down flights of stairs. Added to this was the awful food we had issued: there was only half enough to supply the wants of the men, and what there was was unfit to eat. The mutton we had served up had always to be killed the same morning. The sheep were herded together in the bow of the ship, and though they had shelter from the burning sun the poor animals were nothing but skin and bone: little wonder we could get no meat off the bones at meal-time. Complaints were made, but it was of little use. Often at breakfast the porridge was spoilt in the cooking. I remember we were issued out with presents from the 'Ladies from the Presidency of Bombay,' which included cake, toffee, and games, but there was so few given to each troop that the three sections in the troops decided to toss for it, and our section won the games—deuced little comfort for a man with an empty belly. I myself was so weak that I made sure my knees would give way under the weight of my body, but the worst was yet to come. We continued sailing all day of the 24th, and in the early morning of the 25th July we arrived in close vicinity to the port of embarkation [*sic*], and after waiting for an hour or two moved alongside of the bank, and stopped against a sort of promenade which reached out into deep water. Every one was glad to see land once again, as it meant the chance of a decent meal anyway, for those who had money. The sun was now beginning to make itself felt.[1]

"After we had anchored we commenced to disembark the horses and mules. It was a slow and dangerous job, as horses were led down a gangway from the ship to the land. Finally, the painful job being finished, we next commenced to unload the saddles, after having tied up the horses in a hurdles plot. I myself was stall guard, and I had had to work in the blazing sun all morning. I had to get into an empty stall to attend to the horse next to it, and put my topee down on the deck so that I could get under the entrance, and the horse put his foot on it, smashing the topee to bits.

"After we had finished the saddles it was getting late, but there was a lot of hard work to do yet, and I do not remember stopping for meals; even if we did, it was doubtless the same as the proceeding [*sic*] days.

"Late at night, and lighted by lamps, we commenced to unload limbers and guns, also two aeroplanes, from the hold of the ship. Before commencing, we were issued with a pint of beer each, and what was left over was taken down in a ship's jug: thus we were able to get a refresher now and again, and it did us a great deal of good. It was 4 A.M. of the 26th before we had finished. Réveillé was at 5 o'clock, so we had an hour in which to rest.

[1] In 1854, when the Regiment went to the Russian War, it is recorded that on one of the transports "the beef had made several voyages to and from India, and then been returned into store. Some barrels of peas even bore the date 1828 plainly painted thereon. It was impossible to boil them—also the pork (salt) was as aged as the salt beef, and as bad." The *Islanda* was better than this.

ARAB POLICEMAN

NEGRO WATERMAN

KURDISH WATER CARRIER

PERSIAN POLICEMAN

BASRA

"We lay down on the tables, the benches, and the floor; the hour soon went, when we got up, and proceeded to unload kit and equipment, then had breakfast, and then commenced to water the horses, and immediately prepared to leave for the camp, which was about four miles away."

Such was Private Massey's remembrance of the voyage, and if he grumbled a little, as is the way of the British soldier when he is not fighting, it must be admitted that he did a long day's work for his "shilling and grub." No forty-hours week for him on a six-pound wage.

Lance-Corporal Bowie's account of the voyage is short: "The voyage to Basrah was uneventful, one horse only being lost from the effect of the heat"; and his Colonel sums it up in exactly the same words.

Another officer writes before the start, giving such news as Bombay could supply about Mesopotamia. The port was full of sick and wounded officers and men sent back from there. Their reports were not unfavourable.

Captain Eve.—"I gather it is hot and unpleasant just now, but quite all right. Vegetables and fruits are the great want. . . . There is a lot of shooting, they say, and pig, and there is also excellent fishing, so we ought to have some fun." But evidently the heat was not negligible, for "We were all issued with Cawnpore topees instead of our helmets, as they say they are necessary for Mesopotamia, and I drew one like the men's. We wear the Regimental colours on them. . . .

"To-morrow I start at six, when 236 horses arrive by train from Deolali, where they have been collected from all over the place. I expect most of them will be partly trained anyhow. There will be half for 'D' and half for 'A,' and I shall simply take the first 118 and let 'A' have the other 118. We can pick them out properly if we decide to at the other end. There are also 75 mules arriving—the whole of the Regimental transport—so there will be plenty to do, and I shall be quite content. We shall have to work to-morrow morning. They will be here by six, have breakfast, and then start away, and the ship has to be out of the docks by eleven. . . . There is practically no room for exercise at all, but it is only a short voyage. I do hope we have luck and don't lose many. . . ."

July 23, 1916.—"All is really well, all of us and the horses very fit and flourishing. I am so pleased, and hoping with luck to get all safely ashore. . . . It was a wickedly hot night the one I wrote to you in the docks. I got to bed about 12.45, but could not sleep a wink, with the heat and the noise and thinking about next day. I got up again about 4.45 and was down before 6, and we worked like anything; the men were first-class, and we had no trouble with horses or mules, and were all aboard by 11 A.M. It was extraordinarily lucky I went round myself the evening before, for I found both the ramps leading below for the horses from the upper deck were made too low to let anything but a pony in. Of course I made trouble and had to have both altered. . . . Things worked beautifully. . . . 25 chargers and 235 horses and 84 mules. . . . The mules we put on board first, a very mixed lot, mostly in poor condition, some very big, some small, but I think they will be all right. The horses we simply took straight from the train on board. . . . They seem to be all walers,[1] and a small lot on the whole—some very weedy and light of bone, not many common ones, and a few showing a lot of quality; very few though showing much scope or size, and the majority of them in distinctly poor light condition. They look very healthy, and well in their coats. . . .

"We could hardly be more comfortable than we are on this ship. She is the best for horses I have ever seen—the men are very comfortable, and so are we. . . . There is a lovely head breeze and it is blowing right through the ship, and it isn't at all bad below, and all is as well as possible, and if only it goes on like this we shall, I hope, bring all in safe and well. Every one said it would be terrible. . . .

"To-day we might get wireless news from Aden or from B.[2] How I do hope we may. You can't tell how we want news. . . . This evening we are going to have a men's concert on the boat deck, 8 to 9.30. . . .

"We are as fit as fleas, sweating like anything of course, but I don't seem to feel it and am ever so fit, and never been so comfortable and content on a voyage. . . . Every one wears shorts, and they are a great comfort. . . ."

July 25.—"Everything has worked most awfully well, and we

[1] Australians. [2] Bombay.

BASRA FORT

A CREEK

H. ROBINSON BRIDGE

SINDBAD'S TOWER

BASRA

are now, 7.45 A.M., well on our way up the river. . . . Last night was terribly hot. I went below myself about 1 o'clock, and a lot of horses were blowing badly and we had them out in the hatchways. But they got through the night well. . . . There is a most lovely fresh head breeze, and it is as cool as anything. The river here is wide, and we are able to full-speed ahead, but so far it is the most terrible unhealthy-looking place—palm groves very low lying, mud, and sand. Of course this is only the delta. It ought to be much more interesting farther up. . . .

"I can't write proper letters from here, because I am told all officers' letters are invariably opened and read right through by the Censor. . . .

"We anchored about 1 P.M., and then moved on about 3.15 to our berth, and got tied up about 4.30 to the most ramshackle wooden pier—everything truly Eastern, you know what I mean. Luckily there was a good bridge down the river. I disembarked the horses, all down one gangway, in about 1 hour 20 minutes, and put them in sort of railed paddocks on shore. It was very hot. We then got to work at baggage, all the regimental transport, &c., a terrible long, slow business, and we worked the whole night, and I lay down for three-quarters of an hour. The men and we slept on board, a guard with the horses on shore. They came off all fit and well, only two with any temperature. The rest of the Regiment came in a few hours later and started disembarking too. They lost one horse *en route*. I saw the Colonel, who seemed very content. We were at it from 4 next morning, and finally got all ashore about 7.30."

So the voyage ended, and the Thirteenth were safely landed in Mesopotamia, just twenty-eight days after leaving Marseilles. But the first few months in the country were not agreeable. The heat was great, and there was no news or excitement of any kind, nothing to do but to get the men and the new horses fit for the campaign, if there was to be a campaign, when the weather allowed of movement. For the present the Turks were not giving any trouble. Since the capture of Kut they had seemed content to sit quiet, waiting for the British to try another attack if so disposed. Meanwhile, they strengthened the defensive positions on the Tigris, from which they had inflicted so many bloody repulses upon the invaders of their country, and hoped to inflict more. They did not realise that affairs had changed, that troops and guns and equipment of all kinds were pouring in from the sea, and that the attack when it came would be a very different matter from the hasty frontal assaults in the mud, by small forces, which they had beaten off before. The British War Office had now taken over from the Indian Government the control of the expedition, and the whole resources of England were being set in motion to provide the British force with all the things necessary for modern war, and above all, with ample transport for land and river.

The Thirteenth at first suffered considerably. The letter-book from which I have quoted goes on to speak of the move from shipboard into camp.

Captain Eve.—"It was very hot, and we got in about 11 to find our camp on the edge of the desert, about three miles off and a mile from the river. All of us in the usual E. P. large Indian tents,[1] horses in the open, all pretty uncomfortable at first. The men felt it terribly, and about eight or ten of D went down with heat-stroke on the way up, including Sergeant Hill. Pearson was knocked over the night before on the ship and went to hospital, as did all the men, of course. Next day we spent getting straight more or less, and only led the horses out. They felt the heat terribly, and poor Mam'zelle and one or two horses in the squadron died, and several others in the Regiment too. It was very heart-breaking. Three men in the squadron died, and two or three others in the Regiment. . . . I must go and get inoculated for cholera now. Back again. The horses are a moderate lot, still they were much better than I expected. I had hardly any men the first few days, and we were very hard worked. The men simply went down like flies with the heat. It was partly after the long time without exercise. . . .

"The rest of the day has been wretched—a blinding filthy sand and desert storm, everything smothered in layers of filth. . . . Every one remarks on how well I look. Things were uncomfortable at first, and most people seemed to feel the heat very much; but I never did, and have been ever so fit all along and with a tremendous appetite.

"They have now built long matting-roofed stables for the horses, who are improving visibly. . . . The dust is simply unbelievable. The only thing beyond desert is date-palm groves, but I am certain here at any rate it is healthy. We have a field force

[1] European privates' tents, with double roof and sides—as some protection against heat.

BRIDGE OVER ASHAR CREEK

HUTS UNDER CONSTRUCTION

ASHAR—BULLOCK TRANSPORT

THE SQUARE, BASRA

canteen quite close, so while we are here shan't need any of our weekly supplies, but shall keep everything in reserve. Also the rations are very good, both porridge and bacon and fresh meat quite often, and things like dried apricots, figs, &c., and potatoes. We can buy bread here. The ration is all biscuit. . . ."

The officer commanding the Regiment, Colonel Richardson, dismisses the period in a very few words. "Our arrival," he writes, "coincided with a severe heat-wave, and during the first four days about forty cases of heat-stroke occurred, of which ten proved fatal. For the next three months we were stationary in this camp, training and acclimatising horses and men. The men lived in E. P. tents, and after the first fortnight suffered comparatively little from the heat, the cool nights proving a great boon to every one. The chief maladies with which we had to contend were fever, diarrhœa, and septic sores."

Private Massey's account of the heat-wave is as follows:—

"In the blazing heat of a tropical sun we started, each man leading two horses, besides two haversacks, bandolier containing ninety rounds of ball ammunition, a water-bottle, and a rifle and bayonet.

"On and on we trekked, men falling out by the way with heat-stroke, many stark mad, and men were told off to hold the poor devils down, whilst the motor ambulances raced away with them to hospital at Ashar. . . . On getting into camp we tied up the horses, and after stables the squadron leader, Captain Eve, told us we had had a very hard time getting off the boat, and thanked us for what we had done, telling us that such things were likely to happen on active service. The same night the orderly corporal reported Private Tarr had died in hospital. . . . We were glad to get into the tents, and lie down, and drink lots of iced water.

"Next day, the 27th, Private Killackey was reported dead. Many others went sick the next few days, some of them dying. . . .

"Thus ended the month of July, but from this time onwards things began to get better, and the men were getting better food, but there was a big percentage of sick amongst the Regiment, and men occasionally became delirious."

It certainly was a rough beginning, but the men seem to have acclimatised rapidly. The Regiment had considerable work in training the hastily provided remounts, some of which stampeded and got lost; but by the end of September the training was completed, and the Thirteenth were almost fit for active service again. In October some regimental and Brigade drills and musketry put on the final polish. Doubtless the fine physique of the men had much to do with their quick return to health. Just before they landed to face the Mesopotamian heat, one of their officers had written:—

2nd Lieutenant Guy Pedder—July 24.—"It was 100° in the shade this morning, but they say they often get 118° and 120° where we are going, and of course the trouble is, there isn't much shade, so it is generally 150° to 160°. . . .

"All the men wear next to nothing, probably a thin vest and a thin pair of pants: I never realised before what a magnificent lot of men they are, their physique is marvellous; of course, they ought to be, as they are the original old Army plus K.'s best men. . . .

"There is a decent breeze this morning, and it's just like a heat from an oven striking you in the face! . . ."

Some further letters from officers of the Regiment will show how the summer passed.

Captain Eve—August 3.—"Here we are likely to remain for the next six weeks or so, training our horses and getting them fit. We are under canvas, and the horses in straw-matting stabling. . . . It is of course very hot, but it is very dry heat, and the nights are cool. . . . Dust and sand-storms are the worst thing. . . .

"I have got young Pedder, a very nice fellow, transferred to me, and am now full up again with officers and have a very good squadron. . . ."

2nd Lieutenant Pedder—August 3.—"We are staying here for at least two months in a desert about four miles from where we disembarked. . . . I have got no news: there is none out here, every day is precisely the same as the one before. . . . There's a hot sand-storm raging this morning: we get it pretty regularly every day. I have got hold of one very nice new horse; of course all these horses are untrained, so we have an awful job with them. Yes, Stirling is all right, and Munster, who fell out at Port Said, has rejoined us."

Captain Eve—August 3.—"We all wear neck-shades on our Cawnpore helmets, and all wear spine-pads and short sleeves. Later we shall wear our coats, but now no one does, and the men have khaki shorts too. . . .

"I bought what they call a chágal in Bombay, a canvas bag for water, which you hang up full, and which keeps cool. We have all had them issued to us as well, so we are well off. They are invaluable, and the drinking-water is good, and we mostly live on that, some with lime-juice and tea. . . .

"You have no idea the dust, heat, and discomfort in which I write. If the letter arrives in a mucky state you will know. You don't know what a sand-storm is like, and that with real heat and all the sand turning to dirty mud under one's hands and arms when one was sweating, and one's indelible pencil staining one all over for the same reason."

The heat, apparently, was more than uncomfortable, for in spite of the writer's hearty appetite and contentment, and the sober joys of the "chágal" (which, by the way, rhymes approximately with "gargle," as "jangal" becomes "jungle"), his letters for the next fortnight are written from hospital in Basra. Still they are contented enough.

"I shall be very comfortable indeed here, and shall stop till both my complaints are really right. This is a nice high, big, cool building right on the river, with electric fans, and light, and all that sort of thing—in fact, civilised comforts."

The Thirteenth were certainly fortunate in that way, that they came to Mesopotamia too late to share the horrible discomforts and sufferings endured by the sick and wounded during the campaigns of 1914 and 1915, before the medical arrangements had been fully organised.

"The General came to see me last evening, and sat talking a long time. I thought it so nice of him. But he really is simply charming. . . .

"I feel fairly well this morning—just a bit weak and tired, of course. The great thing is the comfort here. A tent in the desert is not a paradise when you are feeling ill."

Lieutenant Munster—August 4.—"The heat is not as bad as I expected. Dust is the great trouble at the moment. . . .

"I do not think I can compare this place to anything I have ever seen. There are little creeks off the river, and the banks are covered with date-palms, but a little way from the river there is nothing but dust.

"A few Bombay shops have opened branches in the town, and we can get most things that we want—of course, at a high price."

Lieutenant Munster must have had a contented disposition in regard to climate, for another officer writes:—

Lieutenant Chrystall—August 4.—"We are close to the Garden of Eden. We cannot go outside from 9 till 5 owing to the heat, which is now 120° in the shade. The flies and mosquitoes are positively awful, and sand-storms are the order every day; water is at a premium and is rationed out, so you see everything in the Garden is not lovely."

Captain Eve—August 16.—"First all is well—no need to worry. I am convalescent, as I knew I soon should be, and am now (moved yesterday) in the Officers' Convalescent Hospital, about four miles down the river from Basra, and feeling very fit and well. I expect I shall be here a week; it's a really nice place—large two-storied bungalow facing on the river: I don't intend going back till I am really fit.

"There is no news at all here: one is buried in an absolute backwater, and there is nothing at all going on out here, not a shot being fired by us or the Turks, I believe. Occasionally there are Arabs to be dealt with, but that's all. They are always scrapping either among themselves or with some one else. . . . I believe we shall do no more fighting with the Turk—that he'll most probably have chucked it before the time when we could do anything real here comes along. I look upon this just as an exile like India. . . . Sometimes it is hard to be really keen about the training one does, feeling as I do about things here, though there is really tons to be done. . . . I wish one could know more, but I will always be hopeful, and, oh! I pray for the end, though I hardly see it in sight yet. But one never knows."

2nd Lieutenant Guy Pedder—August 17.—"We had a small stampede here the other day, 3 troops clearing off into the desert; however, after many searchings we have retrieved all except 7 horses. The Shemal (strong north wind which covers everything with dust) is still going strong (supposed to last 40 days—it has been on 21 days now). We get a fortnight's intense heat, then it gradually gets cold, and in December there is a freezing wind which goes through you apparently. You would laugh to see us walking about in the daytime (when we have to) in stockings, shorts, shirts, sunshade, spine-protector, goggles, and a large umbrella, khaki!"

Captain Eve—August 21.—"The rest of the Brigade is arriving at once, but I doubt if we shall leave here for a month yet. It is dull, but really a good thing, for there is such a lot to be done, with all new horses and so on, and then one wants to get all one's men well acclimatised. As it is, there are a good many still in hospital, but they'll mostly be coming out soon.

"Lake, who has been in command out here, leaves to-day . . . and I believe we are to get Cobbe, lately on the Staff with us, the Indian Cavalry Corps, in France.[1]

"The worst of the hot weather is about over, and it is slowly getting cooler now. It is anyhow infinitely better than India. Here at the base we are doing ourselves pretty well. It is up at the Front when trouble begins, owing to the great shortcomings in the transport."

August 23.—"First, I am out of hospital, and back at work with the Regiment, and ever so fit and well. Next, I have got Caprice,[2] and was riding her this morning. . . . Caprice is of course looking a bit pulled down and poor, but has still good stuff on her, and is very lively, and bright, and hungry, and searching me for sugar, which I haven't got here."

No, the end was not in sight yet, nor would be for two years longer, and meanwhile the Turk was to do much stiff fighting, and the writer was to ride Caprice yet through some long days of it.

Lieutenant Chrystall—August 24.—"The heat is very bad to-day, and you find me writing this under a mosquito-net at 3 P.M. The flies are awful, and without a net writing would be impossible. I am lying practically stark naked, and am sweating buckets! Bathing is carried on in a very primitive form. I stand on a sack

[1] General Cobbe did in fact command an Army Corps later on, but the command-in-chief went to General Maude.

[2] Caprice, his favourite mare, had been left at Bombay among the sick horses.

(after dark, of course) and simply sponge myself all over from a horse bucket—it is the best one can do, and it really is not half bad."

Captain Eve—August 30.—"Still a very large sick list among the men, but the weather is better and the nights cool. . . . Of course all the middle of the days one can do absolutely nothing. That is one of the great trials of the East to me. . . .

"We have nothing definite about moving yet, nothing but rumours. The great difficulties out here are transport and supply, and at this time of year the river is at its lowest, which, of course, makes great extra trouble."

These Mesopotamian rivers, the only real lines of communication, were in fact very difficult to use. In the hot season they became so shallow that even flat-bottomed steamers of small draught found it hard to avoid sticking on their innumerable loops and sand-banks. At the same time, though there was some dry ground, troops could not march on account of the heat. When, on the contrary, it rained, the dry ground rapidly turned to deep alluvial mud, or was even covered with water, while the rivers became too swift for boats unless very powerful and handy.

Lieutenant Munster—August 31.—"We carry out the same routine —early parade, and then slack about in the tent till about 5 o'clock. Some people shoot in the evening. I believe there are some pigeons about. There is moderate fishing here as well. . . . Perhaps I shall begin to learn to knock a polo-ball about soon. A few people play in the desert in a rough sort of way."

September 7.—"We are still at the base. I have just got a pony. We are each allowed one to carry pack-saddles, and they have been selected with a view to polo: probably we shall play quite a lot later on."

Poor boy. His experience of polo was to be a very short one.

Lieutenant Chrystall—September 7.—"We are getting much cooler weather now—although the desert is not the ideal spot to live in. . . . The nights are generally good."

2nd Lieutenant Guy Pedder—September 14.—"Last Monday it was 112° in the shade in our hut, and yesterday it was never higher than 89°, and dropped to 45° at night; everybody turned into bed early, and most people pulled their beds into their tents and then slept in a blanket. . . . It is much cooler now altogether, and at night one realises how cold it is going to be in six weeks or so. . . ."

Lieutenant Chrystall—September 26.—"We have just got a patent fly-trap in the mess, a Japanese thing which slowly revolves a wooden wheel on which you place sugar and lime-juice: the flies are gradually dropped over into a cage. This is a great invention and catches them by the handful. . . . Our sick list is diminishing now, I am glad to say, as it does make work so hard for the men, having about five horses to look after. This country is remarkably good for horses, and they do very well except for a sort of biliary fever."

2nd Lieutenant J. O. P. Clarkson—Amara, Mesopotamia—October 15.—"I've been sent up here on a course for a few days. I came up by river. We passed Ezra's tomb: there is not very much to see really; we were allowed inside, but had to take our shoes off. The boat was rather wide, rather like a Thames paddle-steamer, except they have a second storey to them, and are open all round. They are boats that used to be on the Irawaddi before the War. We churned up the river, with a barge in tow on either side, and in this manner we went right up-stream. We often bumped into the banks of the river, especially at the turns, and there are some very nasty ones. Some were absolutely hairpin turns, and some were lively S-bends, so between the two we had quite a lovely time bumping into the banks. After we had got a short way up the Tigris we went at a walking pace for a whole day—that was in the narrows. There are plenty of Arab villages, and the whole population would turn out offering us chickens and eggs. We had the band of the 104th going up on the same boat, so we had music every evening. The Arabs used to love the band, and would run alongside the boat and jump and dance and shout and clap their hands. One evening when the band was playing we came across a large band of Arabs building a railway or something; anyway when they heard the music they didn't exactly 'down tools,' but picked them up, stopped work, and came rushing up and down and waved their spades, &c., in the air.

"We are billeted in an Arab house, and on the other side is the Club, which has been well organised. There are several concert troupes here, and they are very good."

Captain Eve—October 18.—"We have not moved yet, but I hope and think we shan't be long. It is still very hot in the afternoons, but the mornings are good now, and so are the nights. We are very busy and all very fit, but very heartily sick of this place, and looking forward to a move, and to the march, which should be great fun—they always are—and to the chance of some good shooting *en route*. We are ready, but have no orders yet. The river is still very low, but should rise a foot next month. This makes all the difference. This whole campaign is a question of transport and supplies—the great difficulty of getting the latter.

"I don't expect we go farther than Amara, a ten days' march, at present, as that is, I fancy, the farthest point which full supplies reach yet. But anyhow that will be part way, and a welcome change. This spot is, of course, the worst out here for climate. It is degrees cooler and healthier farther up. Still we have a very small sick list now, though we have lost a lot of men since we got here.

"Horses do wonderfully well here, and look, and are, as fit as fleas. Of course they are on a full and very good ration here, so they ought to do well. It's for their sakes entirely that I don't want to go farther up than full supplies are getting to. I can imagine no greater misery than seeing one's horses slowly starving on half rations. . . . I go on the river in the evening sometimes, generally in a *bellum* (native boat) . . . something like a gondola, worked by punting or paddles. The river is full of life these days, tremendous activity, and there is always something to see. Also it's a great relief after the desert."

2nd Lieutenant Guy Pedder—October 25.—"Anniversary of Balaclava, 1854. To-day is a whole holiday, as the Regiment took part in the charge of the Light Brigade. This evening there's a hockey match, officers *v.* sergeants. I am playing, but rather funk it, as I'm so stiff after playing 'D *v.* B' Squadron yesterday. It was a draw, 0—0. We went out grouse-shooting this morning."

Lieutenant Chrystall—October 30.—"Our messing arrangements out here are rather funny. I have to beg, borrow, or steal firewood, as there is absolutely not a stick in the country; everything has to come from India, even firewood. Then meat is awfully scarce, and of course tough, like leather. You also see me chasing a poor

MAKINA—SERGT. G. COOK'S GRAVE
(DIED FROM EFFECTS OF HEAT, 27TH JULY 1916)

BREAKFAST ON THE MARCH

unfortunate misshapen chicken, and falling over a tent rope in endeavouring to collar it, in true 'Rugger' style."

2nd Lieutenant Guy Pedder—November 2.—"We really are moving to-morrow and have got a long march in front of us. It is very hot again, but the nights are very cold. Just been given (every officer gets one on going up-country) a sackful of war gifts—thick pyjamas, boots, fly-nets, shaving-soap, books, eatables, &c., &c.—ripping things. . . . Played polo for the Regiment yesterday, and a very good farewell concert last night."

So ended the summer training in camp, and the comparative inaction for the Thirteenth. Much of all this may seem trivial, and no real part of the doings of the Regiment in the Great War. But war, and especially a war of such magnitude, cannot be all fighting. The greater part of any campaign is made up of comparatively peaceful days, during which the soldier sees no shot fired. They are none the less an important time, full of work, and yet not without rest and pleasure, all of which have much to do with his fitness when the fighting days come upon him. The months spent in the desert camp at Basra, far from the front, had not been wasted. The end of them found the Thirteenth in good health and spirits, with men hardened to the climate and horses properly trained. They had two trying years in front of them, years in which they were to see much rough fighting and hardship. It was fortunate for them that they had had this breathing space before being thrown into the actual conflict.

CHAPTER XI.

MARCH TO THE FRONT—MAUDE'S PLAN OF CAMPAIGN.

On the 3rd of November, the 7th Cavalry Brigade marched out of their desert camp and set their faces northward. The Brigade marched in four echelons, of which the Hussars formed the fourth.

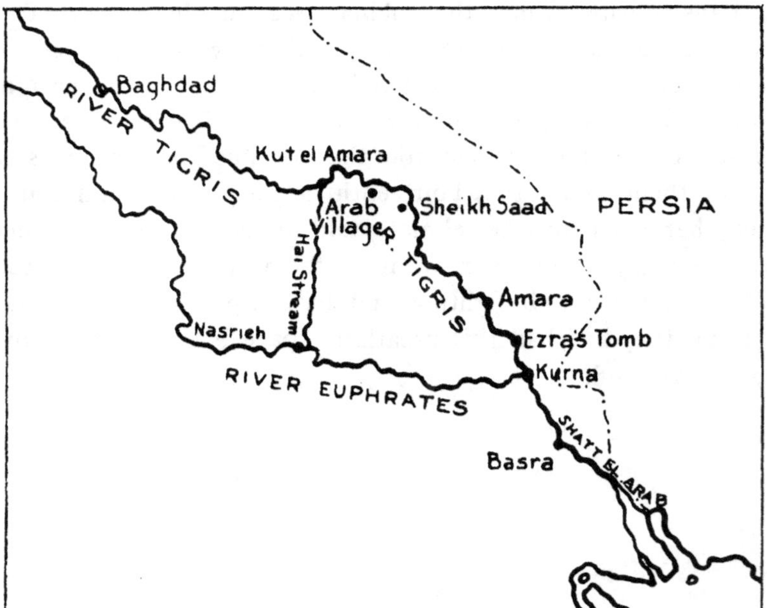

There was nothing particularly exciting about the march. It was a pleasant change from the life of a stationary camp, and the

ON THE TIGRIS

Regiment passed some interesting places on the Mesopotamian rivers, among them the alleged site of the Garden of Eden, near the junction of the Tigris and Euphrates—a dreary spot now. "No wonder the Twelve Apostles deserted," was, according to General MacMunn, a British soldier's comment on seeing it. Farther up on the bank of the Tigris was Ezra's tomb. "The most beautiful of all vignettes is the hedge-sparrow dome . . . set in a small grove of palms." Throughout this country, and Persia, the little blue-tiled domes under a cloudless sky are a familiar feature. And all along the line of march were points which had acquired some fame in the recent campaigns, before the Thirteenth joined the Expeditionary Force. Here and there some enterprising sportsmen found a little game, which went to swell the camp pot. But there was no fighting—the Turks being still to the northward and still inactive, awaiting attack, while the Arab natives of the country gave no trouble beyond occasionally trying to steal rifles at night, which made it necessary to keep a careful watch and form specially arranged night camps.

The following extracts from letters give a more personal touch, and show the daily course of affairs on the way up.

Captain W. H. Eve—November 1, 1916.—"We have finally got our orders. . . . Each day we march in the morning of course, and we camp on arrival in a square—what is called a perimeter camp. The plan for ours is,

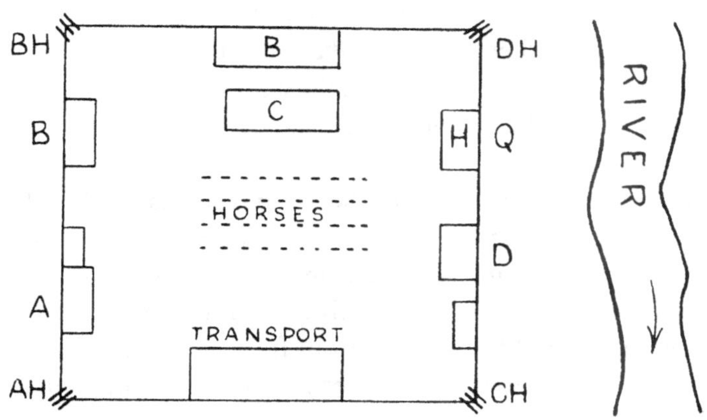

AH, BH, CH, DH show the Hotchkiss guns of each squadron. We take over more than 300 remounts to lead up with us as far as

Amara. They come here immediately before we start. The squadron gets 80 as its share. It will mean pretty hard work for the unfortunate men. We have got any amount of transport—in fact I hardly know how we shall fill it. . . . So we shall travel in tremendous comfort, and cart along all our luxuries, such as tables, chairs, &c. . . . The horses do look well. My own are pictures, and the whole squadron is a pleasure to go round."

November 6.—"I have to write to-day to catch the mail from this place, Kurna, where we are just settling down, 1.15 P.M., with a very nice camp in some palm-trees by the river. I am enjoying myself now, though the first couple of days were uncomfortable. . . . We had taken over a batch of remounts before we started, 80 per squadron—all but half a dozen great cart-horses for gunners. They are very quiet, most of them, but it means nearly double work for the men, and they have all to be led with halters only, and lots of the men are leading two, so you can imagine what it is like. How thankful I shall be when we drop them at Amara. . . . Those that are not heavy draught are ponies for infantry chargers.

"Saturday was a horrid day, cloudy, strong south wind, and trying to rain, and very sticky; . . . yesterday, Sunday, was the same sort of day. . . . We had a long bridge of boats to cross over the old Euphrates into Kurna, and that took time. . . . This is far and away the nicest camp we have had, in fact the only nice one, and we are very comfortable here. . . . We are on the right bank west still, but cross to the other before we get to Amara. The palms end here, and there is corn, &c., on the banks. The new railway is here close beside us, this section apparently about finished, but I don't know how far. We are getting our soda-waters refilled at the hospital here."

The new railway was one of the many works undertaken to strengthen the communications of the army, and make it movable and feedable when the time for the advance should come. With a railway behind him, and a fleet of river steamers, the new General was to be in a very different position from his unfortunate predecessors, pressing on with insufficient numbers and supplies in desperate endeavours to relieve the starving garrison of Kut.

2nd Lieutenant Guy Pedder—Garden of Eden—November 6.—"Just got in and hear there is a mail out at 7 to-morrow, so just

ON THE BANKS OF THE TIGRIS

a short line to tell you I am very fit. This morning we started at 9 and did not get here till 3.30. I had a bathe, and then went for three or four miles with a gun to see if I could find any partridges or snipe, but saw nothing. This is our fourth day's march, and we are at last in Mesopotamia proper (*i.e.*, land in between the two rivers). It's awfully pretty here, and we are right on the Tigris. Yesterday Box (Jeffrey) and I got permission to walk from the second camp to the third in the hopes of getting some shooting; we started off one and a half hours before the Regiment, and got in one hour after it, and walked about fifteen miles. I enjoyed it very much, but we saw only four brace of partridges, and got two and a half brace. At the next camp I believe there is some good duck-shooting.

"The march is rather spoilt as we've got 300 remounts to lead, and so it just doubles the amount of work for the men; however, we drop the remounts in six days' time at Amara.

"Yesterday it actually rained, or tried to, for 2 or 3 minutes. I haven't put up my 80-lb. tent yet this march; it's perfect sleeping out in the open still, as long as one has three or four blankets on one's camp-bed, as I do.

"Dinner!

"Later. Perfect night to-night; have been for a stroll on the river promenade; very tired, so must turn in as réveillé is at 5.30 to-morrow. Boiling hot again to-day, very fit but very tired, so night-night."

Amara, November 12.—"Just a very hurried line, as post goes at 7 to-morrow, to say I am very fit, after 150 miles; we have been just ten days getting to Amara. I have hardly marched with the Regiment at all, as each day I and one or two others have got leave to shoot independently on to the next camp; to-day, for instance, Twist, Jeffrey, and I left the last camp at eight and shot our way here, getting about fifty head. I got 5 brace of partridges, 4 couple of snipe, 5 sand-grouse, and 1 duck—a great day. We arrived here two hours after the Regiment, whom we never saw once on the way. . . . This seems a topping place, but we go on another six miles to-morrow and join the Brigade: how long we shall be there no one knows at present. I have enjoyed the march like anything, bar one or two nights when we struck thousands of mosquitoes. Thank goodness we have handed over all the remounts we had to bring up here and which delayed us so. The last two or three days I have been wading about in shorts after duck and snipe. It is very cold at night now, but still very hot between 12 and 3. Had a tremendous dinner to-night—soup, whole partridge and peas, boiled mutton, onion sauce and beans, tinned peaches and rice, a snipe, followed by a cigar and a bowl of cocoa. . . . The sand-grouse came over to-day in swarms and blackened the whole sky, most of them much too high; must turn in now as I am dead tired."

Captain W. H. Eve—Amara—November 12.— . . . . "To-day we have marched fifteen miles to this place and didn't get in till about 1.30, and then went straight on to the Remount Depot and handed over the remounts—thank goodness! It's been rather a rotten march so far, spoilt by these remounts, which have made a terrible lot of work and caused us to march very slowly, only at a walk, and it has been very hard indeed on the men and very tiring for all of us. . . . The flies and mosquitoes at some of our camps have been wicked. I should think this is quite a nice place [Amara], but have hardly had time to see. Our shooting has been spoilt by our being the last lot of four, and now we can only shoot with an escort, which I shall hate, so I don't suppose we shall do very much. They say there aren't any pig to be found till the rains, when they all get flooded out into the desert. We have been through all sorts of country, a lot very dreary dry marsh, but some very nice, like moorland, short turf and thick scrub. Hardly any just sandy desert since the first few days.

November 14.—. . . "We left the dirty camp at Amara at 8.30 yesterday and marched out here, about 6½ miles up-stream, just on the bend of the river. This is a really nice camp. The country is short heathy turf covered with camel thorn, but all very dry and hard now, and on the opposite bank are gardens and palm groves. . . . The camp is really as in peace-time, and they have trumpet-calls and all that sort of thing. There are no enemy near except Arab rifle thieves. . . . I suppose we shall start regular work here very soon, but we shan't be able to do so much with the horses, as they only get 3 lb. of hay, the rest *bhoosa* (chopped straw), and only 10 or 11 lb. of grain—uncrushed barley and bran. . . . I am so cosy and comfortable in my 80-lb. tent—the same as we had in India.

THE RIVER FRONT, AMARA

PONTOON BRIDGE, AMARA

We have moved the whole of our tents and the mess right up on to the river bank, where all the officers now are, and we have fixed up one mess-tent with the river side of it up horizontally and open to the river, and it is very nice. . . . We are under orders to hold ourselves in readiness to move from to-morrow, but no orders have come, so I'm afraid we are not off yet. But a big native boat has been secured for the Brigade in which some of the heavy kit is being carried."

November 15.—"Away to the east you can plainly see the Persian foothills about forty miles off.

"We are all right so far for rations ourselves, getting fresh meat quite often, and a full allowance; but our unfortunate horses are now on three-quarter rations of grain only, and that uncrushed barley, and hardly any hay, with a little chopped straw in turn. We hope when the railway is finished this may be put right, but it is bad at present, and means we dare do very little with them. The railway is finished in great parts, and they hoped would be through this month. Let's only hope so. . . .

"The nights are cold, but the days still hot, much more so than I expected. That's what makes the climate trying, the tremendous changes during the twenty-four hours. But I think it's very healthy up here, and we are all very fit and flourishing, and hardly any sickness among the men either. My only anxiety is my poor horses.

"They have got canteens going now both here [and] at the Front, so we can replenish always, and are doing ourselves quite well."

Lieutenant Chrystall—November 19.—"We have passed through the Garden of Eden, and a sterile beastly place it looks; and how old Adam existed Heaven only knows, for there is nothing to eat except dates and dust! The next place we passed of interest was the tomb of Ezra, one of the minor prophets? 'Tis a great place of pilgrimage for the Jews. One finds the reading of the Old Testament very interesting, as all the parts round about here are mentioned therein, and also all the customs, &c., and one can see many Abrahams and Ishmaels with the flocks and herds moving over the desert and round the banks of the river."

Captain S. O. Robinson—November 19.—"Since I last wrote we have moved up the river some distance. . . . I believe that we are going up farther in a day or two—*i.e.*, if they can supply us. Our horses are on half-rations at present, but the men are well fed.

"I bought a cheap shot-gun in Bombay before we started, and it has been very useful. There are plenty of sand-grouse and partridge about, which make a very useful addition to the pot. The flies are worse than ever."

2nd Lieutenant Guy Pedder—November 19.—"This letter ought to reach you just before Christmas, so here are the best wishes for a happy Christmas. I have certainly never written those words before on as hot a day as to-day, . . . and though I've written to Bombay for a thousand cartridges I don't know when I shall get them, and I am practically out now, like every one else; it's a rotten state of affairs, as there are now thousands of ducks, geese, grouse, and the farther one gets up-country naturally the shorter the rations get and game is invaluable. . . . From the number of troops coming up-country, I should think there ought to be a fairly good show out here, but it is impossible to say. I am very fit, and am sleeping in my tent on the edge of the Tigris, and have a swim when I get up at 7 every morning. I went into the Bazaar at Amara two days ago and tried to find some curios to send home, and am sending a pair of Arab stirrups and perhaps a bed quilt. . . . Played polo last night and went out shooting this morning, and am going again this evening. . . . I enjoyed the march up here awfully, and am looking forward to going on. Very fit, no news whatever."

November 26.—"A very tiny line to thank you for that ripping waistcoat. I wear it every evening. I am sending home some stirrups, but am keeping the bed quilt as it is so nice and warm. . . . Yesterday I got a beautiful hare, and we are having it to-night . . . in fact, we are pretty well living on game, and have partridges and bacon even for breakfast . . . but it will be more difficult to cater when our cartridges are finished. Am very fit, and have got rid of a filthy cold I had for a week, and have handed it on to Eve. The men are very excited, as they think they are at last going to have a show."

The Regiment marched from Amara on the 28th November and was moving steadily up the river Tigris towards the Front. Captain Eve writes on the 1st December:—

"This march is as nice as the other one was nasty, and I am

THE BAZAAR

VIEW FROM HOUSETOP

MAHEILAS

AMARA

thoroughly enjoying it, and the men are as cheery as crickets. The General and Foster have gone on in front by boat, so the Colonel is commanding the Brigade and Twist the Regiment. Also having no remounts now to lead, we are able to trot along and march a decent pace, and we go largely across country. It has nearly all so far been heathy country with low thorn-scrub and lots of ditches, and it is excellent for the men and horses. . . . We camp in a huge square, always the same way. . . . We usually get in between 11.30 and 12, and then to stables, water, and feed. In the afternoon shooting, &c., but I haven't been out, for I have only one cartridge left. That is my only grouse. . . . Of course we have patrols all round the camp at night and no one is allowed outside. Also at 5 every evening we all parade round the edge of the camp in the places we should occupy in case of attack. The only thing to be carefully watched for is Arab rifle thieves. I sleep with my pistol inside my flea-bag with me. . . . Here we are about as close as we ever get to the Persian hills—about twenty miles—and they are very clear and look so nice when one is in a flat plain, though they look very barren and bare. We aren't on the river bank here, though quite close. There is generally something to be seen on the river, and the monitors look very workmanlike, and I like seeing them. . . ."

Arab Village—December 8.—"Here we are at our destination and all well. . . . Maude, who commands out here, came and saw us march in. . . . General Headquarters is also here and some other Divisions. The trenches are about 8000 yards forward from here. We had an aeroplane over yesterday—a great shooting but no luck, so I expect they know of our arrival by now. We have two pontoon bridges over the river here, and there are other camps on the north side as well. . . . A light railway runs up from Sheikh Saad to the trenches."

At last, therefore, the Thirteenth had reached the real Front. The Regiment was then in excellent health and spirits, and in full numerical strength. The list of officers shows Lieut.-Colonel J. J. Richardson in command, Major E. F. Twist second in command, four Captains, six Lieutenants, and sixteen 2nd Lieutenants—a young lot, but perhaps none the worse for that. At Arab Village the newly-arrived 7th Brigade and the 6th Brigade, which had been in the country over a year, were formed into a Cavalry Division under Brigadier-General Crocker. The 6th Brigade consisted of the Fourteenth Hussars and the 21st and 22nd Indian Cavalry. So, after a lapse of a hundred years, the Thirteenth and Fourteenth, the old Ragged Brigade of the Peninsular War, were again together on service, but some thousands of miles away from Europe.

It has been noted that Lieut.-General Maude, who now commanded the army in Mesopotamia, had met the 7th Brigade as it marched into the camp at Arab Village, and that his own Headquarters were there also.

General Maude had succeeded General Lake some months earlier, and had now made all his preparations for a renewed advance against the Turkish army, which ever since the fall of Kut in April had flaunted its victorious banners in face of the British invaders of Mesopotamia, and not only defied them to retake the place, but threatened to strike out beyond them at Persia and India.

Before giving an account of the memorable campaign that followed, in which the Thirteenth Hussars bore an honourable part, it may be well, at the risk of some repetition, to explain more fully how matters stood when the advance began.

It has been said that the military power of the Turks in Asia was in fact the Eastern wing of the great combination organised by Germany for the conquest of the world. In Europe the Germans, Austrians, and Bulgarians were to overthrow and conquer the main armies of the Allies. In Asia the Turks, aided by German officers and military resources, were to strike out eastward and beat down such forces as the Allies could spare to meet them. Russia was to be attacked in the Caucasus, Russia and England were to be attacked in Persia, which was the highroad to India also, and England was to be attacked in Egypt. The Turkish armies, consisting of several hundred thousand men, securely based upon Asia Minor, were thus to act upon three great Asiatic fronts—the Caucasus on their left, Persia in the centre, Egypt on the right.

Their lines of communication were no doubt long and imperfect, for their railways were not complete; but holding the inner position, the handle of the fan as it were, they were very favourably situated in comparison with the Allies, who had to meet them by acting disconnectedly from outside the semicircle formed by the open fan, while the Turks could strike from inside along the spokes.

In 1914 and 1915 and 1916 there had been fighting on all three fronts of the semicircle — on the Russian frontier towards the Caucasus, the Turkish left; in Persia and Mesopotamia, the Turkish centre; on the frontier of Egypt, the Turkish right. The fighting had fluctuated, but it may be said roughly that on the two wings, towards the Caucasus and towards Egypt, the position was stationary. The Turks had held their own. In the centre they had pushed into Persia and gained some partial success, but as an offset against this, British forces coming from India by sea had landed in the Turkish province contiguous to Persia, and had overrun a considerable part of it. Even here, however, the most recent phase of the war had ended in favour of the Turks. They had repulsed a rash advance on the part of the British, and, shutting up in Kut the force which made it, had beaten off with great slaughter all British attempts at relief, and had finally captured a British Division of 10,000 men. The total loss inflicted upon the British in these operations had been over 30,000. In December 1916, therefore, the prospects of the Turks on their central front were not unpromising. Though they had not conquered Persia, still less succeeded in seriously threatening India, they had made matters very unpleasant for the British in Asia, and inflicted a severe blow upon British prestige. During the hot weather of 1916 both sides had been preparing for a renewal of the conflict upon this front, and the campaign was now about to open.

Judging from a variety of indications, it seems clear that the Turks and their German advisers had decided that the plan of the coming campaign in Asia should be as follows. On their two wings, towards the Caucasus and towards Egypt, the Turks were to content themselves with holding their own, or gaining such success as could be gained without any serious drain on their resources. There was not any vital object to be attained by an advance in force upon these fronts; or at all events a determined advance upon the central front offered a greater chance of decisive results. If Persia could be again invaded, with real success this time, and a Turkish army, or at least a vigorous propaganda, could be pushed on from Persia through Afghanistan to the Indian frontier, the great object of the Asiatic war, which was the overthrow of the British in India, might yet be secured. In comparison with that object nothing else mattered. The Turkish weight, therefore, was to be thrown upon the central front.

But this much being decided, there remained the question how, exactly, the blow was to be struck. Was the British force in Mesopotamia to be destroyed as a preliminary to a further advance into Persia, or were the two operations to be attempted at the same time, or could the British in Mesopotamia be left alone for the moment and an advance into Persia, into their rear, be made without attacking them directly? From the great city of Baghdad, the capital of Turkish Arabia, and the immediate base for operations on the central front, it was possible to avoid the Mesopotamian route, and to strike at Persia by a more northerly line. Which of the three schemes was the best to adopt? The question seems to have been considered in detail.

Eventually it was decided that the third was the most promising. The argument which prevailed with the Turks or their German advisers seems to have been that the British army in Mesopotamia, though lately worsted in its onslaught on prepared positions, was a formidable enemy to attack in the field, and one moreover who was being reinforced from England and India. Such an attack would be a very serious and at best a lengthy operation. It would be better to avoid a direct attack, to make such threats and demonstrations in Mesopotamia as would suffice to keep the British in apprehension of a Turkish offensive, and to leave them facing the positions from which they had suffered so many repulses at the beginning of the year. They would probably be careful about assaulting those positions again, and if in the meantime a Turkish force were to invade Persia, they would probably have to expend their strength in meeting it there. A considerable number of troops was therefore prepared for an advance on the Persian frontier by northerly routes, while the Turks in Mesopotamia were reinforced to such an extent only as seemed sufficient for the maintenance of their main positions on the Tigris, and for threatening demonstrations on the Euphrates.

It must be admitted that this reasoning was strategically not unsound, and that against a timid or over-cautious commander it might well have succeeded. Happily for Great Britain, the new British leader, General Maude, was a man who combined reasonable caution with the knowledge that war cannot be successfully waged without incurring some risks; and happily also, the summer months when active warfare was impossible had been utilised by the British War Office to reinforce and equip his army with such vigour and thoroughness that it had become a much more formidable weapon than the Turks imagined. Not only had additional troops been poured into Mesopotamia from France and elsewhere, until the numerical superiority had passed to the British, but in other respects the force had been completely reorganised. By the end of the summer light railways had been pushed forward, river steamers in great numbers had been collected from various parts of the world, stores of food and supplies of all kinds had been sent up the Tigris and Euphrates, the ports and the rivers themselves had been vastly developed for traffic. By the end of October General Maude had been able to move his headquarters from the base at Basra to the neighbourhood of the Turkish positions, in the knowledge that the difficulties of transport had been overcome, and that he had now under his hand a force of troops superior in numbers to his enemy, and sure for the future of food and all necessary supplies. It had been a great effort, and his own exertions had been incessant, but the worst was over. In a few weeks more, when the weather became fit for campaigning, he would be able to go forward with every hope of success. Early in December, when he brought together his Cavalry Division on the Tigris, the time had almost come.

What General Maude had then to consider, and had doubtless considered very carefully during the three months which had elapsed since he took over command in Mesopotamia, was his own plan of campaign. He knew that the country looked to him to retake Kut and re-establish the reputation of British arms in Asia, which the surrender of a British Division, and the bloody repulses we had suffered in trying to relieve it, had undoubtedly tarnished. That meant a renewed attack upon the Turks in their strong positions on the Tigris, which the army under his command was eager to undertake. And he now knew, or believed he knew, that the enemy intended to advance into Persia in his rear, where the British forces were small and the Russians not much stronger, while the Persians themselves were in very doubtful mood. He could hope for little co-operation on the part of the Russians, either there or on the side of the Caucasus, for Russia was in serious difficulties; nor could he hope for any help from the British forces in Egypt. They apparently had enough on their hands, and in any case they were separated from him by the Arabian desert. On both flanks of their great Asiatic battle-front the Turks were practically safe. For success against them he must depend solely upon the forces under his own command in the centre of Asia. And since the Allies in Europe were barely holding their own, he must have felt as Jervis felt when he sighted the Spanish fleet off St Vincent, that England had great need of a victory at that moment.

It is evident from what General Maude has left on record that he had from the first contemplated the action he eventually took. On this point it is well to let him speak for himself.

Despatch of April 10, 1917.—"Briefly put," he says, "the enemy's plan appeared to be to contain our main forces on the Tigris, whilst a vigorous campaign, which would directly threaten India, was being developed in Persia. There were indications, too, of an impending move down the Euphrates towards Nasariyeh. To disseminate our troops in order to safeguard the various conflicting interests involved would have relegated us to a passive defensive everywhere, and it seemed clear from the outset that the true solution of the problem was a resolute offensive, with concentrated forces, on the Tigris, thus effectively threatening Baghdad, the centre from which the enemy's columns were operating. Such a stroke pursued with energy and success would, it was felt, automatically relieve the pressure in Persia and on the Euphrates, and preserve quiet in all districts with the security of which we were charged.

"This, then, was the principle which guided the subsequent operations, which may be conveniently grouped into phases as follows:—

"*First.*—Preliminary preparations from 28th August to 12th December."

ON THE MARCH

It would be confusing to follow up at this point General Maude's summary of his operations, but enough of it has been given to show that from the time he took command he contemplated a "resolute offensive" on the Tigris, threatening Baghdad, and that his preliminary preparations for that movement were steadily pushed on until the 12th December, when all was ready.

The immediate field of conflict on the Tigris, and the positions occupied by the conflicting armies, are shown in the accompanying sketch-map.

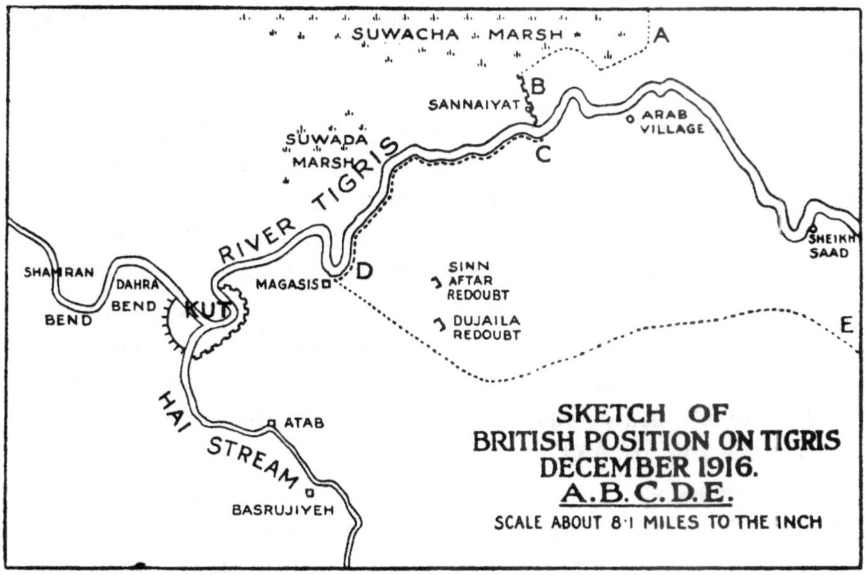

The Turks were astride the river. On the north or left bank they held the same positions as they had occupied since the fall of Kut. At Sannaiyat the enemy awaited attack in the same formidable labyrinth of trenches, flanked by marsh and river, from which in the early part of the year he had three times repelled the desperate onslaught of our troops. "Since then he had strengthened and elaborated this trench system, and a series of successive positions extended back as far as Kut, fifteen miles in the rear. The river bank from Sannaiyat to Kut was also intrenched." On the south or right bank of the river the enemy was not so far forward. He had, on this side, withdrawn to a line of intrenched defences which curved from a point on the Tigris, only about three miles east of Kut, to a point on the Hai stream, and thence round again to the Tigris west of Kut. The Hai stream was also held for some miles southward with posts and mounted Arab auxiliaries.

The British troops held the north bank of the Tigris up to the Turkish trenches at Sannaiyat, and the south bank for about eleven miles farther up-stream. Thus, as General Maude points out, the British were strategically better situated than the enemy, for while their flanks were secure the withdrawal of the enemy's troops on the south bank seemed to offer a chance for a blow sooner or later at his communications on the north bank, which would mean the retreat or capture of the force at Sannaiyat.

In these circumstances it was decided that the proper course was, first, to secure possession of the Hai stream, then to clear the enemy's trench systems on the right bank of the Tigris, and finally to cross the Tigris as far west as possible. This idea of a turning movement by the south was not a new one, for during the British operations for the relief of Kut more than one attempt had been made to seize the Hai; but the British forces had been too weak, and the attempts had failed. Now, as shown above, General Maude's army was better fitted in numbers and equipment to make the attempt with success.

It may be as well to note here what was the composition of the army. Speaking generally, it may be said that about two-thirds of it consisted of Indians, drawn from various races, the remaining third being British. The Indians were not regarded in Europe, or by the Turks, as equal to the British, nor were they—for European warfare at all events. Nevertheless, they had faced the Turks well in previous fighting, and as shown in an earlier chapter, they had won some credit even on the European Front, under great disadvantages. They were, in fact, excellent soldiers, and the Cavalry had a special reputation. The British troops were as good as possible, largely drawn from the old Regular Army, with additions from "Kitchener's men." They were all in high spirits, and eager to get at the enemy. This was certainly the case in the Thirteenth Hussars, where the men were very keen to go forward.

Lieut B. E. H. JUDKINS

Lieut. R. GORE

Lieut. R. C. HILL, M.B.E.

Bt.-Major CHARLES STEELE

Lieut. M. C. KENNEDY

Lieut. J. W. BLYTH, D.C.M.

Capt. C. H. GOWAN, M.C.

On the 12th December the final orders were given. Lieut.-General Cobbe, with a strong force of Infantry and Artillery, was to hold the enemy to his positions on the north bank of the Tigris, and picket the south bank nearly up to the Turkish positions on that side, while the Cavalry and a force under Lieut.-General Marshall were by a surprise march to secure and intrench a position on the Hai. Everything was to be got ready that night for the opening of the campaign on the morrow.

It is curious to note, by the way, that on the 12th December, the day before the advance, a detail which seems to have given special satisfaction to the British troops was the permission to shave the moustache. Private Massey's diary has the following entry: "On the 12th it came in the orders from the Regimental Office that we could shave the hair off our top lip if we wished. Many a time have I heard the men grousing and grumbling because it was against orders to shave the top lip. A great many took advantage of this order, and Captain Eve appeared on parade with his moustache shaved off. I fetched a pair of scissors, and after cutting the hair on my top lip quite short, I shaved it off, and I felt much healthier and cleaner." So the Thirteenth, or many of them, went into the Mesopotamian fighting after the manner of their forefathers in the days of Napoleon—"bien rasés."

CHAPTER XII.

DECEMBER 12, 1916–FEBRUARY 24, 1917—FIGHTING ON THE TIGRIS.

DURING the night of the 12th December the last preparations were completed. General Marshall concentrated his troops in the forward area from which he was to march on the Hai stream, and General Cobbe got ready to bombard the Turkish trenches on the north of the Tigris, so as to give the impression that the British intended once more to attack Sannaiyat.

On the 13th the bombardment opened, and the new campaign had begun. General Maude himself moved his Headquarters some miles forward, and after dark General Marshall's force, with the Cavalry on their left, struck out across country for the Hai.

The night march was carried out without mishap. At 6 A.M. on the morning of the 14th the column had reached its objective unobserved; and the enemy, taken by surprise, made no stand in defence of the stream. It was crossed by General Marshall at Atab, and by the Cavalry a little farther south at Basrujiyeh. The first move of the campaign had been a complete success.

The Thirteenth had borne their part in the night march, and owing to the completeness of the surprise had sustained no casualties. They got over the stream unopposed, and without any difficulty, for in their Colonel's words it was "almost a dry bed," and after an hour's rest for watering and feeding the horses they marched northwards up the right bank of the stream with the Cavalry Division, while General Marshall's Infantry marched up the left bank, the Turks retiring upon their intrenched position covering Kut. The force was, according to Private Massey, shelled by the

ON THE TIGRIS—JANUARY AND FEBRUARY 1917

gunboat *Firefly*, which the Turks had taken during Townshend's retreat on Kut, but without result.

But if the enemy had permitted thus easily the seizure by the British of a point on the Hai, they were none the less determined to cling tenaciously to the several strong positions which they still held on the south bank of the Tigris, and two months of severe fighting were yet to take place before they could be dislodged from the last of these. During that time the main fighting had, of course, to be done by the Infantry and guns, but the Cavalry was constantly engaged in covering their flanks, "in reconnaissances, in harassing the enemy's communications west of the Hai, and in raids, capturing stock and grain." The letters and diaries of the Thirteenth will be quoted as before to show the work of the mounted arm, and their share in it.

2nd Lieutenant Guy Pedder—December 19, 1916.—"We are standing-to at the moment, all packed up ready to move, securing the first opportunity I have had of writing since we started scrapping. We left that camp from where I last wrote[1] on the evening of the 13th, and marched all night and came under Arab fire about 8.30 next morning. We marched on and on, but my squadron did not get any show, as we were on flank guard out of it all. We got to a very famous spot in this part of the world and we had got the Turks pushed back a few miles. We got to bed very cold and very empty about 10 o'clock, the wretched horses having had their saddles on for 27 hours. Next morning I woke up at 5 and found our transport had got up and a ripping hot breakfast was ready for us. We had had practically nothing since midnight the night before. We marched off again about 8, but did not do much all day (a very long and tiring one). The Regiment only had a couple of casualties, and we were shelled a bit. Next day was much the same, getting back after dark. Then Sunday we were told we were going to have a day's rest; however, about 9 o'clock, Pearson and I were sent out with two troops to report on some mounted troops right away in the distance. We galloped out and got as close as we dare and reported about 800 Arabs, and galloped back, the battery having got our range firing over our heads, to find the whole Regiment were out, and as D had turned [out] quickest we formed the advance-

[1] Arab Village.

guard. We went about five miles at a tremendous pace but could not catch them, and then were told to act as rearguard to the Regiment, which was moving back to camp, and then the fun began. Directly we turned homewards on came any number of what we thought were Arabs shooting at us from long range, and making it very unpleasant; however, we did our job and got back all right with only three men hit, two slightly; five horses, three killed. We found out afterwards that we were up against a mixed force of Turks and Arabs. I thought the firing was too unpleasantly good for Arabs. Yesterday we went out in force and waited about all day, but never got a show. Directly we turned homewards we were shelled. The Regiment was extremely lucky, no one hit. Two shells burst only a few yards from me and my troop, but no one was hit and the horses only made a slight fuss for a second or two. My men were excellent under fire. . . . This sort of fighting is better than in France, as though there is not so much cover there is more room and space, and of course there is not nearly so much shelling. We could not possibly camp at night so close to the line in France as we do here. I hear our English mail is on its way from its last standing camp from where I last wrote, and is being brought up by camels. I am more than grateful for that waistcoat, especially in the evening and first thing in the morning, when we get an icy cold wind. During the day the weather is perfect, just like spring at home."

Captain W. H. Eve—December 19.—"These Arabs and Turkish Cavalry are wretched brutes to fight. They won't let you get near them. Then directly you start home away from them they attack and worry. . . . It always seems to be the way. Luckily as a rule they are apparently pretty bad shots. I am rather hopeless of ever being able to get at them mounted, but I shall try, and we may get the Infantry out of their trenches one day. But there's nothing to worry about, we are right on top of them. The job is to get near them."

Lieutenant Munster—December 19.—"We left our camp a week ago and have been skirmishing about ever since. So far we have had only very slight casualties, and it does not appear to me as though we Cavalry should become heavily engaged. Up to now we have been occupied in keeping the Arabs from interfering with operations against the Turks. We are quite comfortable, and on full rations. There seems to be no difficulty about supplies. The weather has not broken yet. Our horses are having a hard time just now."

Private Massey writes in his diary regarding these casualties: "As the bullets which the Arabs use are as big as a man's little finger, with flattened nose just rounded off, they make a big noise going through the air, as well as making terrible wounds." The horses suffered most.

Lieutenant Chrystall.—"We have been doing, as you will no doubt have seen by the papers, a little fighting, and have been doing a lot of night marching on 'operation scale,' which means waterproof sheet, one blanket, and biscuits and bully, and out all day. Our horses at times have been forty-four hours without water, which of course is very hard. Well, we were bemoaning our fate when we were ordered out on Christmas Eve morning for an all-night show, and thinking we will never have any Christmas dinner at all. We returned to camp about 3 P.M. on Christmas Day and found a much-belated Christmas parcel awaiting us, and all your splendid parcels. We had a meal fit for a king, and we had to thank all you good people for it."

Captain W. H. Eve—December 27.—"Personally I see no show ever for us, and am sorry I ever joined the Cavalry. . . . I have no delusions about it now. . . . I shall never get a gallop with my squadron. . . ."

2nd Lieutenant Guy Pedder—December 30.—"We are back again in the same camp as we started out from for this show,[1] and got back just in time, as it has rained almost ever since. We are pretty comfortable in our tents, after much digging. I mean, we dig out the inside of our tents, and make a trench all round to keep the water out. You achieve three things by doing this: (1) your tent is much warmer; (2) you have much more room and do not bump your head every time you turn round; (3) you keep dry. . . . We can't possibly move up again for another show until the sun comes out and dries the mud, which is ankle-deep now. (Thank goodness! I was able to buy a pair of gum-boots from our Ordnance here.) Ormrod, Hill, and Lord arrived with a draft. They marched up-country and brought us some horses which we wanted badly.

[1] Arab Village.

One night the Arabs scored over them well. Ormrod had his gun stolen out of his tent, Hill and Lord had all their kit taken, including bedding, valises, &c., and the sentries were on duty all round and within a few yards of the spot! The Arab is a marvellous thief. . . . The actual Christmas parcels have not arrived yet; all the same, we had a wonderful Christmas dinner only about three miles from Kut—soup, fish, mutton, and vegetables (we got the mutton after raiding an Arab Fort on Christmas Eve), champagne, two bottles for six of us (gift from Lord Curzon, I believe), an enormous plum-pudding which I bought at the E.F.C.[1] before we left here, also there was a Christmas cake and pudding given by telegraph. There is an E.F.C. up here now, but owing to transport difficulties it runs out of everything you want very quickly; but the best thing of all is our regimental coffee-shop (the only one in Mesopotamia, I believe). Even right up here it is open again, and you can buy anything from getting soda-water bottles refilled to buying assorted chocolates. . . . A tremendous thunderstorm is raging as I write, and it is coming down in buckets; to-morrow the mud will be knee-deep."

2nd Lieutenant J. O. P. Clarkson—December 30.—"Here's a good story and true. One of our monitors had been up to an advanced position to shell the Turks, but had got heavily shelled itself. After a few days of this they tried to pull the Turks' leg a bit. They rigged a *mahailla* (Arab boat) with funnels and mast to represent a monitor, towed it up during the night, and the next day put it into position, burning oily rags and brown paper to represent its being under steam. It was not shelled. They towed it up higher still. It was not shelled. The next day a notice appeared from the Turkish trenches, and it ran, "Your real monitors amuse us, but your dummy one is superb."

Lieutenant Munster—December 30.—"Still here and still raining. There can hardly be any doubt that we shall not be moved at present, the mud is so awful. I used to think the mud at Aldershot could not be surpassed, but now I am inclined to think Mesopotamia beats it. We have to build little mud walls round our tents to keep the water out. I did not build mine deep enough, and as a result I think I had a foot and a half of water in my tent. It came

[1] Expeditionary Force Canteen.

just two or three inches short of the level of my bed. I woke up and saw my boots and clothes floating about."

January 4, 1917.—"We have been in our permanent camp about ten days now, and are quite likely to be here all the winter. After the winter come the floods, and before the floods have gone down the great heat comes on, so that October, November, and December are considered to be the only fighting months of the year. This year active operations did not start until December 14th, and the Cavalry withdrew to permanent billets on December 26th."

Captain W. H. Eve—January 17.—"I got your letter of December 3rd when we got back here to standing camp the day before yesterday. We are still all whole and flourishing, and I am very fit indeed. We left the base November 3rd. . . . The 'show' was to start on the 14th,[1] and we marched from here on the evening of the 13th. From the papers you will probably have gathered more or less what we did to start with.

"We were on the left or outer flank of the Infantry attack, away on the south. Marched all night, crossed the river Hai (nearly dry then) at dawn on the 14th; meeting with no opposition, then turned north up its western bank, and kept pace with the Infantry attack.

"Our part of the show was a success, a complete surprise to the enemy apparently, and in fact we didn't run into any of them for some time. We had a little sniping from Arabs and a few Turk Cavalry, but no real fighting. We had a longish trek though, about twenty-six hours for our horses under saddle, as we did not get back to doss down by the river until about 9 P.M., 14th.

"The next two days we spent in reconnaissance wide on the flank, had long days for the horses, no water from morning till night (luckily cool weather), but we had no fighting, only got shelled once or twice."

Sunday.—"I had a little scrap with the squadron. We got a sight of about 800 Arabs and Turks, and the Regiment, which was for duty that day, turned out after them. We could not catch them, but on going home had a rearguard fight, which devolved on my squadron, and wherein they pleased me.

"This rearguard business, the worst and most unsatisfactory to

[1] December.

do, is the Arabs' great game. We had two or three more days' reconnaissance. We accomplished nothing, and then were sent on a show to try and bridge the river Tigris higher up. This was a failure, and we rather got it in the neck, so you won't see anything of it in the papers. The Regiment that day was in reserve, so I saw very little of it, but of course I know what happened.

"On Christmas Eve we went off to deal with a hostile Arab chief down south. That meant another night march, and so on. But we found the bird had flown, and had to content ourselves with sacking his fort and village, and collecting what cattle and sheep we could.

"We got back here on Boxing Day, and it then rained for a week or more, and we lived in mud. But here, of course, we are in permanent camp and under canvas. I must say that so far we have been more than lucky with our weather when out.

"On the 7th we left here again, only got back on the 15th, during which time we were trekking about again, chasing after Arabs, and eventually were chased by them, as usual, on our way home. I had really no fighting. . . . I have told you all this about ourselves, though it is not in the least important, just to show you the sort of life we lead."

During this week the Regiment marched down to "Hai Town," a considerable place on the Hai stream, and collected some supplies and Arab arms. Private Massey gives a sad account of the soldiers' disappointment at the loss of a good meal in a deserted village.

"Here we seized several fowls and killed a few calves, taking only their livers, kidneys, and hearts, which we carried in our water-buckets. We camped close to the village for the night. After unsaddling our horses and watering and feeding were over, we lit a fire and commenced to cook the spoils of victory. But alas! when the water was nearly boiling for tea, and the livers and kidneys were frizzling in the mess-tins over a hot fire, the order was given, 'Fall in for line picket!' Of course that was enough to test the temper and patience of a parson, let alone a soldier, and a steady stream of expletives could be heard as we sloped arms and marched away. That night I had biscuits and bully, a poor substitute for fried liver and kidneys."

It was hard, but Private Massey and his "messing-in chum,"

RIVER FORT, HAI TOWN

ON THE TIGRIS

M'Nulty, had better luck a few days later at another village, and the diary triumphantly records how, "During the night M'Nulty managed to get several livers, kidneys, and hearts from the slaughtering place, and we had fried liver and kidneys served up hot before réveillé next morning, as well as a mess-tin full of hot cocoa, made from tablets I had sent from England."

Private Hugh H. Mortimer—January 18.—"Round about the back of beyond. Yours of the 5th ulto. duly to hand about five or six days ago. I say about, because one has no idea as to days and dates, &c., out here, Sundays included. Sometimes we get a volunteer Church parade when things are quiet, but that has been very seldom of late. The last one we had, last Sunday but one, I volunteered for one the night before, but what ho! the parade was for 11 A.M., and we had been on the trek reconnoitring and Arab-chasing five hours by then. We are quite seasoned hands at these quick turn-outs now; often we're all bivouacked down, giving it the bells in Snore Land, unless it happens to be your turn for guard, then somebody strolls round in a quiet hurry, kicks the sergeant in the ribs, and whispers, 'Turn out at once; parade two hundred yards west of camp midnight.' The kicks, &c., are passed on, and we are all saddled up, transport packed, &c., &c., and perhaps three or four miles away in less than an hour, and all that done in the dark, and no noise above a whisper, unless somebody gets a kick in the seat from a bobbery horse, and then there is some excuse for letting it rip.

"Yes, one day is much of a muchness with the next, unless it is that one has no time to think about what day it happens to be. Often I have asked five or six chaps the date to put in a letter, and then had to consult the orderly sergeant at the finish. Still, I don't suppose it would make any difference to one's destination if one happened to snuff it on a Friday or Sunday. We get in bags of warm spots now ever since we came right up, about two months ago. Sort of places that make the short hairs stand out on the back of the neck, and wonder if the next 12-pounder or so on has your number on it. I thought I knew a bit of what it feels to sit in a trench with them coming over in France, but galloping about in open country with shrapnel flying about licks it hollow. I saw one drop in a machine-gun section about fifteen yards on my right some fortnight ago, and that one bagged four men and three horses, and then several pieces hummed past me. . . . Then again the cod was galloping across a hail-storm for a mile from cover to cover. I took the Hotchkiss gun across—I am a nob at Hotchkiss guns now—and the remainder of the section followed some thirty yards behind at the canter. We got five shrapnel quick, and not more than thirty or forty yards away, and two of them in front of me and one behind the two parties. I looked round after the thud, having a good horse, and you talk about dust-flying—well, I thought 'There's only me in this section now,' and made a lightning spurt for a deep donga, dropped into it with a crash, and pulled up. About six seconds afterwards two more crashes, and there appeared the rest of the team, a bit pale and smiling rather sickly smiles, but not a scratch on either men or horses, so you may judge of the luck of the mob. The worst of this country is, you can't depend on the district for any supplies like you can in France; after we've had nothing but a couple of biscuits and one 12-oz. tin of bully from day to day for several days, when we get out of touch with the ration stands, that just seems to be the time when one could do with about four good sit-down knife-and-fork square meals per diem. When it's like that I go up two extra holes in my belt and try not to think of steak and chips or anything in that line. Never mind, I will make it all up when and if I get home. Roll on that time too. Have just received a letter from Fannie, and she says people at home say, 'Oh, he's safe in Mesopotoom,' but by the H.P. I never felt unsafer in France, Somme or no Somme. You may think that it's generally understood that the last round is for yourself, before being taken alive by Arabs, although the Turks seem to fight very fair, and there are no tanks to walk behind and no trenches to bob down in."

Lieutenant Chrystall—January 20.—"To-day we were within one and a half miles of Kut. 'So near and yet so far,' and we hear to-day some trenches have been taken, so things are looking up a bit. To-day we got a splendid ration of beef, a sirloin, probably the first one ever seen in Mesopotamia. . . . One has to be very drastic with these people, the women being just as bad as the men. If you do not take strong steps they will shoot you in the back, even if they have shown the white flag, which they usually do when we are advancing, but when retiring they shoot at you."

Captain Eve—January 20.—"When we approached our objective, the big native town [Hai Town], they came out with a white flag and pretended to be friendly, and we treated them as such, and *bought* supplies and things from them, and of course they made a fortune out of us. . . . Then Sunday, the 14th, we started back. As soon as we had left, our rearguard was heavily attacked (native regiments), and we had some casualties, and were very much delayed, and the Regiment sent to their support, but had very little to do."

January 26.—"I wish I could tell you *really* all about things out here, but it is quite impossible. Only I do wish now above all else that I could have gone to Infantry or gunners. I have slowly but surely come to the conviction that we are years out of date, and entirely or almost entirely useless, and will never take a real part again."

2nd Lieutenant J. O. P. Clarkson—February 2.—"Since my last letter we have done quite a bit. First we tried to go round the Pushtikuh Mountains,[1] so early one morning we started off in order to cross the bridge. It took us over two hours to get across, and then we were the leading unit of the 2nd Brigade. We went past a fort to the Wadi river for water, and then pushed on to try and cross the marsh. But we got hopelessly bogged, at least the guns and the transport did, although they had double teams in as it was. Meanwhile we were all anxiously looking at the sky, as there was a devil of a storm coming up. So those people who had crossed the marsh had to recross it again. We camped down just off the marsh, and had scarcely got the lines down when it started to pour with rain. We crawled into our valises (no tents), and slept, and woke up soaking, boots full of water and thoroughly wet. It was still raining. We got up at 5 A.M., and had to wait until they had got all the waggons out of the marsh, some having been left there the evening before. We waited about five hours. Luckily it then turned out fine, and we started back. Our things were more or less dry by the time we got into our camp by the fort

[1] "Pusht i Kuh," or Back of the Mountains, is the district so called by the Persians, their western district.

about twelve or sixteen miles from the Arab village. We got back to our permanent camp about 2 P.M. next day. The going after that rain was very heavy, and both men and horses came in very tired. We had a very short rest and now are out again."

The Regiment had returned on the 27th January to the Hai river, and from there during the following week made several reconnaissances to the westward, beyond Kut, where the Turks had a bridge across the Tigris, and some strong intrenched positions on the south bank. They had been dislodged from some points lower down, but still held on desperately to the westerly bends of the river in order to protect from attack the line of communications of their main force in Sannaiyat. Until the 4th February the Thirteenth, though at times under heavy fire from rifles and machine-guns and artillery, got off practically without loss. Then there was a sharp fight, in which the Cavalry was called upon to act dismounted in conjunction with the Infantry, and the Thirteenth had some casualties. An officer was killed and two wounded, with a few non-commissioned officers and men.

The officer killed, Lieutenant Munster, was much respected in the Regiment, "a very gallant, zealous, and capable officer," as his Colonel reported. All accounts agree in describing him as a man of exceptional character, who, though young, had already made his mark. Quiet and reserved, with strong religious views, he was spoken of in unusual language by many of those about him. "I am not good at expressing myself," one of them wrote, "but I may say his daily life was one which I shall try to follow: it was one of doing good to his fellow-men; my admiration for him was unbounded, and my grief worse and more intense than I ever felt in my life." His death seems to have been due to his unselfish devotion, for having led his men forward and seen all of them under cover preparing for another rush, he walked across the open to avoid exposing one of them in sending a message to another officer, and was shot through the heart. Such was the fire at the moment that his Colonel said: "I judged it inexpedient to attempt to recover his body, and to remove a brother officer, Mr Williams Taylor, and some men who were wounded, until after dark."

Nor were they the only two to distinguish themselves on this day. Captain Robinson, commanding "B" Squadron, had been

Lieut. D. A. Stirling

Lieut. J. A. Lord
(*Wounded at Imam Mahdi, 25th February* 1917)

2nd Lieut. J. F. Munster
(*Killed at Shumran Bend, 4th February* 1917)

Sergt. W. D. Tassie, D.C.M.

S.S.-M. J. Brearley, D.C.M.

wounded in leading the advance shortly before, and Sergeant Tassie of "D" Squadron received the D.C.M. for his coolness and courage in bringing up ammunition when the transport animals were shot down.

Other officers of the Regiment also showed great courage on this day, and it is a temptation to record what was afterwards written of them by an eye-witness—a Captain in one of the Indian regiments of the Brigade.

"Our Brigade was ordered to attack dismounted, Hussars on the left, ourselves in the centre, with Watson's Horse echeloned to the right rear. This meant that the Hussars had to advance across the open with no cover, whilst we had the cover afforded by the high banks of the nullah. After an advance of about 1400 yards the fire became so heavy that the Hussars decided to left-shoulder and make for the nullah, with the object of working down it and thus coming to assaulting distance. Their casualties had been pretty heavy. They eventually made the nullah in advance of our line, and cleared out some advanced patrols of Turks who were holding it there.

"Captain Willis and I were now sent down by my C.O. to gain touch with the Hussars by working down the nullah.

"I met Captain Newton and asked him who was commanding. He said he was for the moment, because Captain Eve and Captain Steele were outside, and he was reorganising the men preparatory to making a further advance.[1] Just as he spoke a renewed burst of machine-gun and rifle fire made me look over the top to see what was happening. I saw Eve and Steele helping in a wounded man who had been hit in the leg. The man had an arm round each of their necks. The Turks furiously opened up at them, and I don't know how they got away, for the ground all round was being thrown up by bullets.

"I don't think Eve realised he was doing anything exceptional, his only concern seemed to be getting the man in without giving him pain. What struck me most was the cool way he handed the man in, carefully caught up the thong of his crop, which he always

[1] Captain Eve, the senior Captain, was then second in command of the Regiment, owing to the absence of Major Twist, disabled, and on the particular day, as Colonel Richardson was on other duty, Eve had taken the Regiment into action. Captain Steele was next in seniority.

carried, and then jumped down into the nullah under a perfect hail of machine-gun bullets. He turned to me as if nothing in the world had happened, and we discussed the situation.[1]

"A few minutes later I went back and brought my squadron forward.

"By this time the C.O. of the Hussars and my own C.O. had arrived, and we all went down the nullah together.

"When we were talking to Eve, a man came up and said that a private of the Hussars was lying about forty yards outside the nullah and shouting for help. My C.O. turned to me and said, 'See if you can get him in,' but Eve said, 'Nonsense, he is a man in my squadron.' So he and Captain Jeffrey immediately left the cover and finally found the man. The man had only had his arm shattered, so after Eve had spoken to him he got up and they all came in safely. . . .

"Eve realised that it was impossible to leave these wounded men until dark, because we anticipated a counter-attack about dusk, and it was necessary to be hampered as little as possible then. . . . Also the presence of numerous Arab irregulars made the possibility of leaving our wounded out of the question. . . .

"It was one of the most gallant things I have ever seen, and was just typical of Eve. I know he did not give it another thought, and only considered he had done his duty, but that too is only typical of the man."

Captain Eve's own comment upon the day was short:—

"It went very well," he writes, "but we weren't allowed to hold on to what we had got, which rather took the gilt off the gingerbread. We had bad luck in officers, as you will have seen, Bob [Captain Robinson] and Williams Taylor being both hit, and poor Munster being killed. But the casualties among the men were very light, and in fact we were very lucky.

"This was undoubtedly from our own point of view the most satisfactory day we have had, though it was only a side-show."

[1] The man was Private T. O'Connor. In a letter from Mesopotamia to Mr Justice Eve he writes: "I was the last ammunition-carrier to get wounded that day, and your son and Captain Steele came out under heavy machine-gun fire and carried me in as we evacuated that position that night, thus saving me from being taken prisoner. There is no need to tell you he was loved in the squadron, as we have missed him as one of the finest soldiers and leaders of men any soldier could wish to follow."

LT. MUNSTER'S GRAVE

GRAVES OF LT. MUNSTER, PRIVATE KILLICK,
AND A CORPORAL OF THE 14TH HUSSARS

PRIVATE KILLICK'S GRAVE

BUSSOORIE

A letter of the 6th February gives a more detailed account.

Captain Eve—February 6.—"We came under pretty hot rifle and Maxim-gun fire, but not shell-fire. They shelled the guns and the horses in rear but not us. Well, the men are something to be proud of. They just advanced as they should. It was really just like a day on the sand-hills at Meerut. We had very few casualties in our advance, that is among the men, and the Turks did not wait for us in their advanced position, nor again in their first line, which was a big and deep dry canal. We got it fairly hot from enfilade Maxim-gun fire just before we got to this, but we got there all right. . . . All this time I had been with 'C' and 'D', the others on my left and behind. Then after we had got settled in our trench 'B' came in, and at last 'A', and I found I was commanding the lot. I also found poor old 'B' had dropped into it badly, Bob shot through the leg (slight), Williams Taylor badly in the thigh, and Munster killed. Barrett was therefore commanding, and only Hill left with them. The rest were all right, though Williams of 'A' had a bullet through his hat which made a furrow along the back of his head. I then found our flanks were in the air as neither of the other Regiments was up; but the Turks made no attempt to counter-attack, and we were pretty safe, except that one of their Maxim guns kept firing down the trench. So I couldn't go any farther without support, and reported by signal. Presently one of the others came up on our right, and our guns got very busy. Then the only bad thing that I saw happened. They tried to send our ammunition pack-horses up to us across the open. They got to within about 200 yards, and then got properly caught by Maxim-gun fire. The men got hit and the horses loose, and then the poor beasts stood, just being shot to death, one going after another. Sergeant Tassie on his own, and I think young Stirling did too, ran out to them across the open, and succeeded in getting one in, and he then went back and brought in ammunition off a horse that had been killed. It was a very fine thing, and I have pushed in his name for a D.C.M. . . .

"The Colonel was delighted with the Regiment, and so was the Brigade. . . . The men were very pleased with themselves, and I with them. They were just first-class. . . . Poor old Munster was shot right through the chest and killed instantly. He was such a nice man—very shy and reserved, but a real good sort, and every one is so sorry.

"My best bit of news of all is that Tassie has got the D.C.M. I know how pleased you will be. . . . He wears the ribbon, but I hope will have it presented in public some time. I am pleased and proud about it.

"Williams had another bullet clean through his helmet on Friday. He was trying to snipe a sniper, and was successful too—killed his man."

February 16.—"Yesterday, Thursday, we started at 5, breakfast at 4 A.M.—out to the same ground on the left flank. We kept the horses well back, and I was left in charge of them, and had a desperately dull day. However, I missed nothing, as the Division had very little to do except for the gunners. But it was a real good day—the Infantry did well, in fact it was quite a little victory. We completely cleared the south bank—took 1500 prisoners, including fifty officers, two battalion commanders among them, and killed a great number, including most of those trying to get back across the river, as we sank the whole of their pontoons and coracles while crossing. Also we got three of their aeroplanes. It was a good show altogether, and finishes this stage of the operations. The next, as you can see, must be to force the passage of the river and clear them out of Kut on the north.

"This will mean heavy fighting."

Private Massey, after describing how the enemy tried to get across "in pontoon boats and oracles," says: "They looked a dejected lot indeed, clothed in rags, no boots, and they had had very little food for some time. They eagerly took cigarettes offered them, and went limping off to the river dock for removal to the prisoners' camp, resembling as they went a tribe of lame and ragged beggars. But the Turk has a stout heart, is as brave as a lion, and will fight like the very devil on a handful of dates and a morsel of flour."

Pity that with all his fine qualities, which appeal so strongly to the British soldier, he has not yet learnt to treat his prisoners without brutality.

Captain Eve—February 16.—"I don't suppose we shall be wanted until we have the crossing secure, but after that we may have some fun. . . . We, of course, have had nothing to do with it really, but then we are only Cavalry."

February 19.—"We were in camp two days afterwards (after the action on the 4th), then were out on a foraging expedition on the 7th. On the 9th we were out on reconnaissance again all day and half the night, then had two or three days very bad weather, and on the 14th and 15th had two very long days and nights.

"But the last one was the final successful show which cleared out the Turks from this southern bank altogether. It was quite a good show, but we were sitting out wide on the flank all the time, and had really very little to do with it, and could see nothing. Still, it is quite good, and one feels something is accomplished at last. Now we have far the hardest part in front of us, but we shall do the job all right before long.

"Since the 15th we have been left in peace, and very glad of it we were. The horses begin to look a little better already.

"I have told you all this just to give you an idea how we are worked. Most days we are fifteen to twenty hours under saddle, and short and irregular water, besides long distances, and much night work, worst of all. Our unfortunate horses began to look like shadows, but are recovering a little now, and we have had one batch of very nice remounts, though we are still about one hundred short.

"All the men are very fit indeed, and so are we. It is desperately cold at times, but we are managing by degrees to get our tents out from the permanent camp.

"My squadron, of which I am now again in command, Twist having rejoined for duty, has sunk from six of us to four, as Pearson has left us to join probably the Australians, and Pardon is laid by with a badly sprained ankle.

"I expect we shall be busy again in a few days. We have been extraordinarily lucky in the weather, take it all round, and I should think Maude will take advantage of it while it lasts. Let's hope so."

February 21.—"I am starting my chat to-day because we are going out to-morrow, so I make certain of catching the mail. We shall have had a six days' easy—haven't been out since Thursday, the 15th, and it has been very nice. To-morrow I fancy the next stage of the proceedings is to start, probably crossing to the north. I don't imagine we shall have anything to do but sit on the flank at first, and until the Infantry as usual have done the work. . . .

"I started to write to mother after my chat to you last Friday evening, and then a most terrific thunderstorm with heavy hail and a gale of wind suddenly came on, and we had to turn out and stand to the horses, only just in time to stop a stampede. In fact, for a minute or two after we were with them, I thought they might go. I got to mine, Follow Me and Caprice tangled up together. We held them all right, but it was pretty miserable—pitch dark, blowing a hurricane, and hail you couldn't face. The men were top hole. When the storm passed, we got the lines down again (in water) and made all secure. We were, of course, wet through, and I then found our tent flooded out, beds in an inch of water, &c. . . . It blew and rained hard in the night, and the poor wretched men were wet through and in the open, all flooded or mud. Box and I kept dry in our tent, and it held all right against the wind. . . .

"The men, as a matter of fact, have got into the way of making themselves the most cosy little dug-outs. They dig down and cover up with their waterproof sheets. They are wonders—never have any of them sick—two rubbed heels is my total sick in the squadron, and yet in permanent camp there would probably be a sick-list of ten. . . .

"The river has risen tremendously the last few days, and is only a few feet from the top. Let's hope it does not rise any more or we shall be completely flooded out. . . . I have got thirteen extra men out from permanent camp, and with the new horses can turn out something like a respectable squadron. . . . I am still short of horses, could do with any amount more like the last lot. Do you remember how in peace time it was always too many horses and no men? And now it is the exact opposite, just when horses are wanted."[1]

February 23.—"Well, we didn't go out on Thursday, yesterday,

[1] The horse casualties from fire had been numerous, and many horses had been lost from fatigue and exposure.

CAPTAIN EVE'S CHARGER "FOLLOW ME"
(DIED AT SEA ON THE WAY HOME IN APRIL 1919)

the whole show being postponed owing to the river having risen too much. It has since fallen, and the show started early this morning, and so far is going very well indeed, and we are standing-to in readiness.

"Besides this show, we have taken a great part of the first and second lines of the Sannaiyat position and repulsed five or six counter-attacks, so all is very well. . . .

"Payne overheard the following between two of the men the other day. 'That there Major Fry, 'e's been made a Colonel.' 'What, 'im? *What's* wrong with him?' I think it good enough for 'Punch' if only one could draw a picture for it. . . ."

General Maude had certainly been lucky in the weather, the winter floods not having yet made the river impassable, and he was taking full advantage of his chances. The "quite good show" of the 15th had been the last of a series of carefully-planned and hard-fought Infantry attacks, which had gradually torn away the Turkish hold from point after point upon the south bank of the Tigris. The enemy was now thrown across the river, and in desperate anxiety for his long line of communications. They were still covered by a swollen river more than three hundred yards broad, but they were not safe from a resolute attack, and after some skilful feinting here and there the blow was about to be driven home.

Maude was not the man to lose an hour. On the 23rd of February, just before daybreak, some Infantry, English and Goorkha, were ferried across the Tigris at the Shamran bend, west of Kut, and after hard fighting succeeded in establishing themselves on the northern bank. Before dark the same afternoon a bridge had been completed. The end of the long suspense had come. The Turks had fought stubbornly for more than two months, but their line of communications had now been struck, and their main force must go or be captured.

CHAPTER XIII.

THE RECAPTURE OF KUT—RETREAT OF THE TURKS.

AT the same time that the Tigris was being crossed westward of Kut, the Turkish intrenchments at Sannaiyat were fiercely attacked in front, and the brave defenders, learning that the river defence had fallen behind them, at last gave way. On the 24th the maze of trenches which had resisted so many onslaughts was carried with a rush, and sweeping on rapidly the British troops replaced the British flag on the walls of Kut. On the same morning the British Cavalry poured across the bridge from the southern bank in pursuit of the retreating enemy, and the British gunboats pressed up the river again to join in his destruction. There was reason to hope that he had held on too long, and that his stubborn troops were doomed to be taken or destroyed.

This hope proved to be delusive. The Turks had to retreat very rapidly to extricate themselves in time, and they lost heavily in doing so; but they, or the bulk of them, did succeed in getting away. Many were slain, and some thousands remained in British hands, but by steady and determined rearguard fighting they checked their pursuers, and when they reached Azizieh, a hundred miles off, half way to Baghdad,[1] they were still a military force, if a defeated one.

The share of the Thirteenth in the victory and pursuit was not all that they hoped it might have been. Some eager spirits among them chafed at the caution with which the Cavalry was

[1] It may be well to note here that in the course of this campaign, where the Tigris was the main line of communication, the distances were usually calculated by river. The distances by road, when a road existed, were much shorter, roughly perhaps half the river distances.

handled and the pursuit carried on. But nevertheless the mounted men, although unable to storm entrenchments held by a resolute enemy, had throughout done much to cover and facilitate the operations of the Infantry and guns, and they now did all that mounted men could do to harass the retreat. If it did not mean converting the retreat into a complete rout, and annihilating a Turkish army, it was a material contribution towards that end.

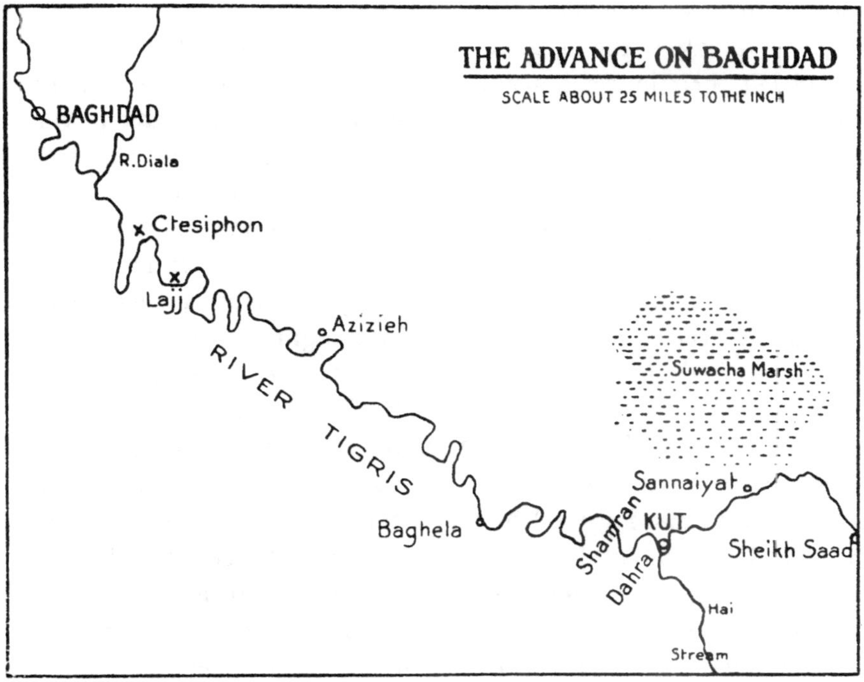

After three days of steady pressure, during which the Cavalry lost a good many men, they had pushed the Turks over a long stretch of road, capturing prisoners and guns and booty of all kinds; and if they were stopped there, half-way to Baghdad, it was only because they had got in advance of their supplies and of the supporting Infantry.

The following extracts from letters and journals tell the story in greater detail.

Lieut.-Colonel Richardson, D.S.O.—"On the 24th of February we left camp about an hour before daylight, reached Shumran, where a pontoon had been thrown across the Tigris the previous day by the Sappers and Infantry, crossed the river about 9 A.M., and gradually moved up to behind the Dahra Ridge, which at that time was the most advanced point held by our infantry. At noon we found what we thought was an opening in the Turkish line, went through it, and made a big left-handed sweep, until we were brought to a standstill by the Turkish rearguard, strongly intrenched. We were engaged with this till after dark, when we returned to bivouac, which we reached about 1 A.M. We were in the saddle again at daybreak on the 25th, and took up the pursuit of the Turks, who had left their intrenchments. At about midday we located their rearguard in position about Imam Mahdi, on the north bank of the Tigris, and the Regiment was ordered to attack dismounted, with the 14th Lancers on our left. We were closely engaged with the Turks until about 5 P.M., when orders were received to withdraw. Then we found that our casualties had been heavy, but mostly in wounded, the numbers being approximately one officer, Lieutenant Lord, and thirty-six men wounded. That night again we reached our bivouac after midnight, and were again on the move at daylight.

"On this day, 26th February, the Regiment was in reserve and not engaged. We bivouacked where we stood that night, and at daybreak took up the pursuit of the Turks, who were now in full retreat. Before midday we had taken some 300 prisoners, and had found six field-guns of large calibre, which had been left by the Turks. After that we replenished our nose-bags from a captured barge well stocked with barley, and followed the road taken by the retreating Turks, which was strewn with rifles, bayonets, small-arm ammunition, trench-mortars, to which the draft cattle were still harnessed, hundreds of 5.9 shells, carts full of implements, and even a motor-car. We bivouacked at dark some six miles from Azizieh."

In General Maude's despatch giving a summary of these operations is the following passage:—

"Our gunboats were in touch with and shelled the retreating enemy during most of the 27th, and his retirement was harassed by the Cavalry until after dark, when his troops were streaming through Azizieh in great confusion." General Maude adds that since crossing the Tigris his force had captured 4000 prisoners, of whom 188 were officers, and great quantities of guns and stores. But now the pursuit, in which the Navy and the Cavalry had co-operated in an unusual way, was broken off. The Turkish army had made good its hasty retreat, and was for the moment free to get away unmolested.

"We bivouacked at dark," says Colonel Richardson, "some six miles from Azizieh, and next day received orders to retire about ten miles, in order to allow supplies, which we had outrun, and the infantry, to come up. On our way back we collected all that was possible of the ammunition and stores that might be useful to us. On the 28th of February we spent a very welcome rest day in camp. On the 1st of March we marched to Azizieh, where a halt was ordered."

In fact, the Infantry having fallen behind, or rather, perhaps, having been kept back by the difficulty of bringing up sufficient supplies, there was a pause of five days during which the British force was closed up and got ready for a farther advance. Baghdad now lay immediately ahead, but as the Turks were not in such a condition of rout that they could be hunted through the great town *l'épée aux reins*, it was necessary to move with the force thoroughly together, and not to risk a check.

The following are further extracts from letters and diaries:—

Private Massey—February 24, 1917.—"On the 24th, réveillé was at 2 A.M., and we turned out on parade at 4.30 A.M. The previous night the Engineers had thrown a bridge across at Shamran bend, and early this morning, the 24th,[1] the Infantry had succeeded in getting across, and we followed soon after.

"The Turks, seeing their danger, had retreated the previous night, leaving only a rearguard to delay our advance, and cover the retreat of their main body. . . . After crossing the pontoon-bridge we formed troop, and went along in extended order, until we got close in on the enemy's rearguard. Here we dismounted and brought the Hotchkiss guns into action, and V Battery opened a heavy fire on the retreating enemy. The Infantry must have been heavily engaged, as many of their wounded were at the advanced dressing station, close to the bridge-head, and we passed several dead. During the afternoon we reached Dahra Barracks (Turkish), which were littered with Turkish dead. Still pressing on towards nightfall, we came in touch with an enemy flank patrol, and opened fire on them with rifles, Hotchkiss guns, and artillery. It

[1] As a fact the Infantry got across on the 23rd.

was now very late at night, so we returned back to Dahra Barracks for food and water for the horses. It was about 2 A.M. when we made down our beds and lay down for the night. When we had been in bed about half an hour, the order came for the 1st and 3rd Troops to saddle up at once, and I heard Mr Payne, the officer in charge, tell the troop sergeant that we were going out on the flank, on reconnaissance. On hearing the order, we jumped out of bed, saddled up our horses, and in a few minutes we were off in the early morning dusk. We were all tired and very sleepy, as we had only had half an hour's sleep, but we had a rattling good officer in charge of us, always giving commands in a fine soldierly manner, and a leader in whom we all had confidence.[1]

"As we neared the trenches we had fired on late the night before, we met a Turk, who gave himself up. Later on we came upon two Turkish officers, who gave up their revolvers and a valuable telephone instrument. On reaching the trenches, we found a dead Turk and one wounded man. We bound up his wound, laid him down, and gave him some cigarettes, and then put up his rifle to let the R.A.M.C. men know he was there, and when we came back later he was gone.

"We were now not many miles away from Kut and Sannaiyat, but on the north side of it this time. We had watched Kut from the south side so long that it seemed impossible we could be where we were then. We could see the 13th Division coming through Sannaiyat, and we afterwards passed through what had been a Turkish camping-ground, and kettles, dishes, and pans lay all over the place.

"We returned back to camp about midday, and heard that the officer in charge had been complimented for the reports he had sent in, and which had enabled the Division we left behind at Dahra Barracks to turn out at 5 A.M. that morning, and continue the pursuit of the retreating enemy. After watering and feeding our horses we commenced to feed ourselves. Porridge was made, and bacon and tea were made, and we had a hearty meal, after which we rested; but the place was beginning to smell, as the dead Turks had not yet been buried. . . ."

February 26.—"Next morning we moved on to where we knew

[1] This was Lieutenant Payne of "D" Squadron.

the Regiment had camped, but on arrival the Regiment had again gone on, but we went with the transport as escort. This was on the 26th, and on arriving in camp at night, we heard that the Regiment had been heavily engaged with the enemy's rearguard, and Sergt.-Major Goddard, S.Q.M.S. Edwards, Sergeants Bell and Hill, had been wounded, Private Davis of 'C' Squadron killed, and thirty-three wounded. . . . About 5 P.M. we again went on, intending to overtake and rejoin the Regiment, which we failed to do, until they were forced to call a halt as they were out of rations, and had eaten their iron ration. This will give any one who reads this an idea of the speed with which the Turkish rearguard was pursued and continually harassed and defeated.

"We reached camp at one o'clock the next morning, the 27th. . . . On the way up we had passed much abandoned booty, as the enemy, being hotly pursued, had to leave it to escape capture. There were guns of the 18-pounder class, as well as small camel guns which fired a shell not much bigger than a man's two fists. Thousands of rounds of artillery ammunition was lying strewn all along the track of the beaten enemy. Rifles and small-arm ammunition, equipment transport carts, and even a motor-car of German manufacture. . . .

"At night we rejoined the Regiment, who hadn't a biscuit left, but heard they had captured a Turkish barge, with a cargo of grain, flour, candles, dates, and supplies of all descriptions, so they had grain for the horses and food for the men. The transport arrived in camp, but they only had enough for one day's march; and although the paddle-steamers followed us up, we were now a day's march ahead of the dumping-ground, and two days' march in front of the Infantry."

February 28.—"So next day, the 28th February, we retired back a day's march, and came upon the food columns on the bend of the river. . . . On the 1st of March we remained in camp all day. . . . We got on board the captured barge that morning, bringing away whole-meal flour, dates, surgical bandages, and took off a lot of corn for the horses. M'Nulty and Moss made chupattees of the flour, baking corned-beef and dates and making pies, and we all had a good feed. . . ."

March 3.—"On the 3rd réveillé was at 5.30. After attending to our mounts we had breakfast, and then a big party of men, of which I was one, were ordered to stand-to for fatigue. We were marched down to the river, a distance of two or three hundred yards, and commenced to unload barges, which had arrived that morning with stores of all kinds, shells and bullets, and food for man and beast. We soon transformed the bleak ground on the bend of the river into a vast dumping-ground, and hundreds of tons of corn were soon stacked in heaps on shore. . . . At 8 P.M. that night we were again unloading a barge which we completely emptied of its cargo. We had no rest that day, but we knew we had done some good work."

March 4.—"Next day, the 4th, we remained in camp all day. During the afternoon we heard that the General was holding a parade to decorate men who were recommended for bravery on the field. We were afterwards to get ready to move on the morrow, as a report brought in by airmen reported a body of Turks holding trenches a day's march up the Tigris close to where they had a pontoon-bridge across."

2nd Lieutenant Guy Pedder — March 2. — "What do you think of it all? No doubt the papers have been shouting about old Mespot lately. When I wrote my last letter I little thought we should be quite so far on the next time I wrote. I think the rout of the Turks must have been a complete surprise for even our authorities. . . .

"We went through the Infantry and crossed the Tigris above Kut at 9 A.M., Saturday, 24th. Curiously enough, I met Masson, who used to be at Pat's at Repton with me. He had been building the bridge. We didn't do much that day, as we were held up when we tried to get round their flank; however, next day we went a long reconnaissance. I got a small batch of prisoners, and then we attacked the Turks' position. I was commanding the squadron, as Eve was commanding a wing. We advanced about 1½ mile across the open under very heavy rifle-fire and shrapnel. I lost both my sergeant-majors and two sergeants, besides others. The Regiment lost 35 and only 1 officer (Lord): how we got out of it I don't know, as hardly any one came out without a bullet through his coat, topee, or something. I tell you I lay pretty flat when we weren't advancing or retiring, but there was not a bit of

H.M.S. "TARANTULA"
"A TRIED AND TRUSTY FRIEND OF THE REGIMENT"

PRIVATE MASSEY'S "ORACLE," OTHERWISE CORACLE

cover! We camped down about 2 A.M., and were off again about 6 and went about 30 miles, but the Turks had cleared early next morning. We took about 300 prisoners and 6 heavy guns, and the road was littered with ammunition, kit, shells, carts, dead animals, &c. The prisoners were in a very bad way and could hardly walk. . . .

"Luckily the weather has been perfect, except it has frozen every night, and is quite cold in the day. I brought all my bedding on my spare horse, but of course we haven't got any tents. None of us got the chance of a wash or shave for four days, so you would have laughed if you had seen us. I am awfully fit. Hope the next letter I write to you will be Baghdad: it's under fifty miles from here. Best love to all.

"The monitors did excellent work and recaptured the *Firefly* and other boats."

2nd Lieutenant J. O. P. Clarkson—March 3.—"We started early in the morning[1] and trekked to the Tigris west of Kut. The Infantry had established themselves on the other bank, so we crossed in the early morning and followed them up. When they had cleared the trenches we went through to harass the enemy's retreat. . . . The next day we went in on what was supposed to be the retreating enemy, and found it was the main body, consisting of about 5000 men. . . . The next night we bivouacked by a water-hole in the desert. We had some armoured cars with us, and one of them got into difficulties a very short way off some Turkish guns, but the next morning was still intact. The next day we again went on and captured 400 prisoners and 12 guns, 6 of which were 5.9 howitzers, and one man found the breech-blocks, which had been buried in another place. By this time we were getting very short of food for the horses, but luckily we found a barge laden with grain that had been captured, so we filled up again. . . . We are now at a place about half-way between Kut and Baghdad. The Navy here—*i.e.*, the monitors—did great work. They rushed on ahead, and got right into the thick of it, firing 6-inch shrapnel at the Turk at a range of 400 to 600 yards. Every gun on board was worked, and in every direction. They captured several boats and barges, and what used to be the *Firefly*. The

[1] Of the 24th February.

Turks captured it at Kut, and have used it against us ever since, and in a few days it is going to come up and shell the Turk. The litter that was left behind by the Turks was awful—rags, clothes, camp furniture, ammunition of every description, and bombs, shells—in fact a bit of everything. We recaptured half a dozen of our fellows who had been captured the day before, and they had their boots and overcoats taken from them, as the Turks are very short of both. They even strip their own dead as well as our own for the clothes they can get off them.

"I also found a Constantinople paper which was written in French, and which said the Germans had gained a great defensive victory on the Somme against the English because—and here came long explanations. Also that Verdun was as good as lost to the French, &c. I was also shown some post-cards from Germany, but they were very badly written, and I could not decipher them. There was also an 'Illustratte *Zeitung*' there, with wonderful pictures showing how grateful the Pole was for the German liberation scheme."

(Lieutenant Clarkson read and spoke fluently both French and German.)

Captain W. H. Eve—March 2.—"Well, things are going *really* well, and we are now forty-five miles short of Baghdad, and resting a day or two. The difficulty is again supplies. . . .

"I wrote you Friday last, February 23rd, and we moved at 4.15 next morning, 24th. The fight at the river-crossing had gone very well, and we were sent over the bridge which our sappers had built about 9 A.M. The Infantry were still pretty busy further forward, and we waited about for three or four hours and watered. Then we really thought our chance was coming of getting them in the open, but their rearguard put up a very fine fight, and got into nullahs and ditches, and our show ended in a dismounted fight, and we weren't able to turn them out, and so it ended at dark. The squadron was out on flank guard and had next to nothing to do, and only one man wounded, and one of Payne's chargers, his best of course, killed. We camped north of the river, but weren't in till midnight, and I can tell you we were tired then. We had had nothing to eat, and there were no rations up, so we turned in straight off. It had been a very hot day.

"Well, the Turks' rearguard cleared off in the night and we had got Kut. For they had evacuated it directly we got the crossing of the river, to escape being bottled up there. Meanwhile the other Corps had taken Sannaiyat, but the remnants of the Turks got away from there too. The Turk rearguard really did well. I did hope and think our day might have come, but it was not to be. All the same there was nothing to grumble at, for we had got Sannaiyat and Kut, and the whole of the Turks were retreating as fast as ever they could.

"Our monitors came up the river as soon as Sannaiyat was taken, and passed through our bridge early the next day. Then—that is, Sunday 25th—the pursuit started. We managed to get some food early in the morning and left camp at 7. I was advance-guard, but we saw no Turks till the afternoon, except a few scattered prisoners which we rounded up. In the afternoon the other Brigade on our left found their rearguard and sent for us to help. They were in a very strong position, Infantry and Artillery, and we were sent in dismounted. The Regiment was on the left, 'B' and 'C' in front, and 'A' and 'D' in support, under me. We were very weak in numbers, only about 100 in the firing line from the Regiment, and they were much too strong for us. We went on, of course, but finally got held up, and we had a warm time. I took the support line up to within about 100 yards of the front (two squadrons). The Turks were shooting very well—rifle, Maxim guns, and their guns too—and people behind thought we should be about wiped out. But 'V' behind us helped us no end, and when orders came to retire we got out of it better than I expected, just before dark. 'B' and 'C' came back through me, and I then came, but by degrees. The men, all of them, were topping. We had heavy casualties, but only a very few killed.

"Only half of 'D' was there, as Payne with the other two troops had been out on a reconnaissance all the night before, and so had been left behind to come on with the supply column. Even so I lost both sergeant-majors, two sergeants, and a corporal wounded—no one killed. Sergt.-Major Edwards was only lightly hit in the leg and will soon be back at duty, but Sergt.-Major Goddard was very badly shot through the middle, and I am afraid was bad: I haven't been able to find out anything about him since, either, except that he was easier when he left the field ambulance. Sergeants Bell and Hill were both shot through the face, and Corporal Heathman through both legs. Lord was wounded, the only officer casualty. We were lucky to get out as lightly as we did. The monitors helped us too. I only had one horse very slightly hit, but shells fell right amongst them and they were very lucky indeed, too. It's an extraordinary thing: if you are going to be hit, you're going to be, and if you're not, you're not, and that's all about it. I was touched—my clothes—by bits of shrapnel, and I got a bullet through my revolver holster and then through my coat—near enough, but I wasn't touched. Corporal Watkins, my signaller, had two or three through his clothes, one of which was stopped in his pocket by his pocket-book.

"I rode Follow Me till danger threatened on the Saturday, and then Caprice, and on this day, the Sunday, I rode Follow Me. We didn't get back to camp till 11.30, but we got food all right then.

"We had to go back, no support ready to push these people out."

Monday 26.—"Left camp 7.30 and marched across the desert to strike the river much higher up—a long march, and wicked going for the guns in places. I rode Caprice. We found the Turkish rearguard in the afternoon—again very strongly posted. We were rear Regiment of the rear Brigade and did nothing, and the other Brigade found them too strong. But the Infantry were steadily coming up, and the monitors were topping, going ahead like anything and giving the Turks beans. They captured prisoners and destroyed Turkish ships, and did no end of good work. We had a day's food with us, and just lay down where we were for the night. It rained, but not much, luckily, but was bitterly cold.

"We advanced next morning, 27th (New College grind), at 7 A.M., but of course the Turk had gone by then. However, we rounded up about 240 prisoners, a hospital, and a battery of six 5.9-inch guns. We wasted a lot of time over this instead of pushing on, and in the end got no distance that day, and the infantry were only a few miles behind us. . . . It drove me mad. The Turks were obviously running like smoke—abandoning guns, ammunition-carts, and equipment of all sorts, and yet we didn't push on. It's beyond me altogether.

"A little farther on we found two large barges loaded with corn and all sorts of things. These were providential, for we filled all

TURKISH PRISONERS. FEBRUARY AND MARCH 1917

AEROPLANE ATTACHED TO 7TH CAVALRY BRIGADE

our nose-bags and corn-sacks and got a little loot ourselves as well. I took a Turkish trumpet, but I don't suppose I shall keep it. But we got useful things like chickens, ducks, candles, dates, sugar, &c. We then marched on about eight miles more and camped in the desert just before dark. I rode Caprice and Follow Me. Later in the evening Payne, with the other two troops, turned up with the welcome supplies at last, and also our regimental transport. The road of the retreat was a sight, every conceivable thing abandoned.

"Next day no supplies at all could be got forward, so we marched *back* the eight miles to where the captured barges were, and had a really welcome afternoon's rest there. *I* had the most glorious hot bath in a huge iron bath we got off one of the barges. I can tell you we were dirty. We left in the small hours of Saturday, and this was Wednesday, and you can't imagine how dirty we were."

"We marched again next morning, Thursday, March 1st, and came about seventeen miles to where we are now—a most delightful camp, right on the river bank.

"Two of the monitors are here. They *have* done well. They have recaptured the *Firefly* captured from us in the retreat last year, have taken a battery of guns here, and been up level with the Turkish infantry in their retreat and inflicted tremendous losses. They have sunk a lot of ships and taken over 2000 prisoners. And all this when you think they are unarmoured—only supposed to be proof against rifle-fire, and had all those heavy guns against them.

"I believe since the push started here we have taken 7000 prisoners, and their other casualties have been very heavy indeed, as we know from their dead. We believe we have only a remnant in front of us. Personally, I shouldn't be in the least surprised if we march into Baghdad almost straight, when we get going again. . . .

"It is lovely weather and we are all as fit as fleas, not a man sick. The horses, poor devils, of course, are showing it, and I have one or two tender backs and withers, but not many. They are just A 1. . . .

"It is a lovely day to-day, and the horses are spending the whole of it grazing, while we are cleaning up, &c. I have had a real useful morning at it, and have also mended my own breeches. In a few minutes I am going to have a hot bath. I hope we shall get letters away somehow by the supply people going back empty.

Whether this will catch this mail is a different thing, but you will understand, and if you get my wire you will know all is well, anyhow. I am most awfully fit—couldn't be better. . . .

"I shouldn't be the least surprised if we get there now with hardly any fighting at all; as it is, we have done little enough of it. The Infantry do nearly all of it, and the monitors have *really* done the pursuit, which we ought to have done. I do hope we go on to-morrow. I fret at every day's delay."

These passages have been given as they stand, for they show the generous impulse of a real soldier to award all due credit to other Services. And in truth it would be difficult to say too much for the resolute unwearied fighting of the Infantry during the two months' struggle on the Tigris, or for the dash and courage of the sailors in following up the pursuit. Nevertheless it is necessary to make allowances for the impatience of a fiery spirit always longing for a chance. Others had done well, very well, but the horsemen had done all that horsemen could do—and that was not little.

"We were bucked enough last night, and then we got the great news from France to crown all. It is only a telephone message as yet—no details—just saying the Germans have retired three miles on the Ancre. . . . We know nothing, not even of our own doings. I expect the papers at home have made the most of our success here, with very big headlines, &c., until the news from France comes to overshadow it. We do want news so badly."

That was the feeling at every part of the vast battlefield. In Europe and Asia and Africa and all over the seven seas millions of men were fighting for the Empire and doing their duty nobly, but "we do want news so badly" was the cry that went up from all of them, and they had to be content with very little, to find comfort and help in the smallest successes reported in a few vague words over the uncertain wires.

At the moment all was going well with the Thirteenth. They were not strong in numbers, for they had had casualties at times since the beginning of the advance, and had besides lost many horses from the wear and tear of the campaign. Captain Eve's squadron had been reduced to working as three troops instead of four, and had no senior non-commissioned officers left. But the men and horses were doing well, and the officers the same.

HALTS ON THE MARCH

"Lord is all right,—I mean going on all right. I helped to do him up. He had a very narrow shave, the bullet entering just above the collar-bone and coming out under his arm, only just missing the big artery there. Otherwise we are all most awfully flourishing. Twist seems quite all right now. . . . Sergeant Chipperfield is acting squadron sergeant-major, and very good indeed he is. . . . Payne I like better and better, and think I was more than lucky to get him. I am a better judge of horses than men, anyhow, first go off. . . . I am so sorry that Payne missed our show on Sunday, though I am glad he is still here safe, as he might not have been, of course, had he been there."

Another letter, written on the 3rd March, sums up the operations of the last week and the position as it then stood. At the risk of some repetition it seems as well to give it.

"Well, we had known pretty well for a week beforehand that a serious attempt was going to be made to put a pontoon across the Tigris above Kut as soon as the height of the river, which was then in flood from the recent rains, permitted, and this was done on the 23rd February; and during that night a Division was put over, and we followed on their heels next day, the 24th, and in the afternoon found a gap in the line of defence they were putting up to cover the retreat of the fighting troops from the Sannaiyat position, from which the big guns had been retired some days. The delay in getting the pontoon over the river, due to the flood, had, however, given the Turks just enough time to get the survivors from Sannaiyat—where the fighting has been very severe and bitter, and where the deep trenches were literally filled with dead—away, and when we got through it was only to bump up against their rearguard, who were well handled and fought so well that we had to force them back, after three days' fighting, from position to position, and until the Navy came along with their monitors and flyboats, and turned the retirement into a rout. The Turks then left so quickly that they abandoned guns, trench-mortars, doctors and hospital staff, a motor-car, shells, wounded, and many unwounded men who could not keep up, to fall into our hands or the hands of the Arabs, who kill, strip, and loot Turks and British alike, especially when wounded. Among the prisoners were a few German officers, and a few were also found among the dead. We were ordered to stop the pursuit three days ago by the Army Commander, as we had finished our supplies and were getting far from our base, so have been waiting by the river for the barges to come up; these began to arrive last night, and are coming up fast, so we may expect soon to be on the move again. We are now some forty-five miles from Baghdad, and our friend the Turk may put up a fight at Ctesiphon, although this is very doubtful unless he has been strongly reinforced, which seems unlikely. . . .

"The weather has on the whole been most lenient to us, for although the cold at night has been intense and the winds strong and bitter, we have only had a small shower of rain, and the days are now getting quite warm. The men are well and cheery, and the horses, which a few days ago were footsore and weary, are picking up and ready for more work. The Regiment has fought very well, and our casualty list—about forty in the last ten days—is comparatively small."

Such was the state of affairs when the Thirteenth received orders that the advance was to be resumed on the 5th March. The writer of that letter had longed for one gallop with his squadron. He was to get it at last. But before his confident words reached those for whom they were written his fiery spirit was at rest for ever.

Sergt. P. CHIPPERFIELD
(Died, 6th March 1917, of wound received at Lajj)

Sergt. H. KNAPMAN
(Killed at Lajj, 5th March 1917)

Capt. W. H. EVE
(Killed at Lajj, 5th March 1917)

Pte. A. WALLHEAD
(Killed at Lajj, 5th March 1917)

Pte. ALFRED JONES, D.C.M.
(Killed at Lajj, 5th March 1917

CHAPTER XIV.

THE FIGHT AT LAJJ, 5TH MARCH 1917.

"ON the fifth," writes Sir Stanley Maude in his despatch upon the advance to Baghdad, "the supply situation having been rapidly readjusted, Lieutenant-General Marshall marched to Zeur (eighteen miles), preceded by the Cavalry, which moved seven miles farther to Lajj. Here the Turkish rearguard was found in an intrenched position, very difficult to locate by reason of a dense dust-storm that was blowing and a network of nalas,[1] with which the country is intersected. The Cavalry was hotly engaged with the enemy in this locality throughout the day, and took some prisoners. A noticeable feature of the day's work was a brilliant charge made, mounted, by the Hussars straight into the Turkish trenches. The enemy retreated during the night."

Such is the brief official account of an action which was one of the most memorable in the history of the Thirteenth. In accordance with the custom of the times, when extreme reserve marked the despatches of the military commanders, no details are given, and the very number of the Regiment which made the "brilliant charge" is omitted. This reserve was probably necessary, yet the charge was a gallant feat of arms, and, if not a triumphant success, none the less creditable to the valour and discipline of the Regiment. News of it would have stirred the heart of the nation.

Extracts from several accounts by those who were present will be given hereafter; and it will be seen that they are not always in exact agreement. They reflect in some measure the rush and confusion which must always attend a Cavalry charge, and perhaps on

[1] Commonly written "nullas"—ravines.

that account they give a better picture of it than can be given by any attempt at a general narrative. The pace of galloping horses and the extent of ground covered make it impossible for any horseman sharing in a charge to note all the incidents that occurred, and it is equally impossible for him to remember in precise sequence all that he did see.

Nevertheless, it seems necessary to state briefly what happened on this day as far as one can judge from the varying and sometimes conflicting stories. Their discrepancies after all are mainly on points of detail, and a fairly accurate view of the whole affair can be formed by comparing them. The following facts appear to be free from any serious doubt.

On the 5th of March, shortly before daybreak, the Cavalry Division marched from Azizieh, the rest of General Marshall's force coming on behind them. The line of advance lay along the north bank of the Tigris, mostly over flat ground, with some low sand-hills on the right, away from the river. Somewhere between 9 and 10 o'clock the head of the force reached the village of Zeur,[1] and there the Cavalry watered their horses. Half an hour or so later they pushed on for Lajj, while the rest of the force halted for the night.

Lajj was the point on the river where General Townshend had left his flotilla and his hospitals before the battle of Ctesiphon in November 1915; and it was rumoured that the Turks intended to make another stand now upon the old field where General Townshend's advance had been checked.

At first there was no sign of an enemy, and the Cavalry moved forward unmolested. The 7th Brigade was leading in the usual march formation, the 13th Lancers, Watson's Horse, in front. Soon after the start, reports came in from some aeroplanes and armoured cars which had gone on to reconnoitre, to the effect that the country ahead was clear except for a Turkish convoy with a small escort of Infantry, which was retiring in the direction of Baghdad. After advancing another mile or two, the Brigade came in sight of some scattered parties of Turks, and the Thirteenth Hussars were ordered to come up abreast of Watson's Horse, which were under fire. They did so, "A" and "B" Squadrons advancing dis-

[1] The Regimental Diary says 9 o'clock, the Colonel of the Thirteenth says a little before 10.

Lieut. G. R. PEDDER
(*Wounded at Lajj, 5th March 1917*)

Lieut. E. F. PINNINGTON
(*Wounded and taken prisoner at Lajj, 5th March 1917*)

2nd Lieut. G. LYNCH-STAUNTON
(*Killed at Lajj, 5th March 1917*)

L/Cpl. A. W. WATKINS, D.C.M.
(*Wounded at Lajj, 5th March 1917*)

Sergt. F. SPANTON
(*Taken prisoner at Lajj, 5th March 1917*)

mounted, and remaining for a quarter of an hour or so in action. During that time they lost 2nd Lieutenant Welstead and several men, wounded by rifle-fire from concealed Turks.

It was then about 11.30 o'clock, and the day was a bad one for seeing. A strong east wind had risen at the backs of the Cavalry, and was sending along in front of them the "dense dust-storm" of which General Maude speaks in his despatch. Through this and the mirage the Thirteenth could make out, not very clearly, the supposed convoy, two miles or more away.

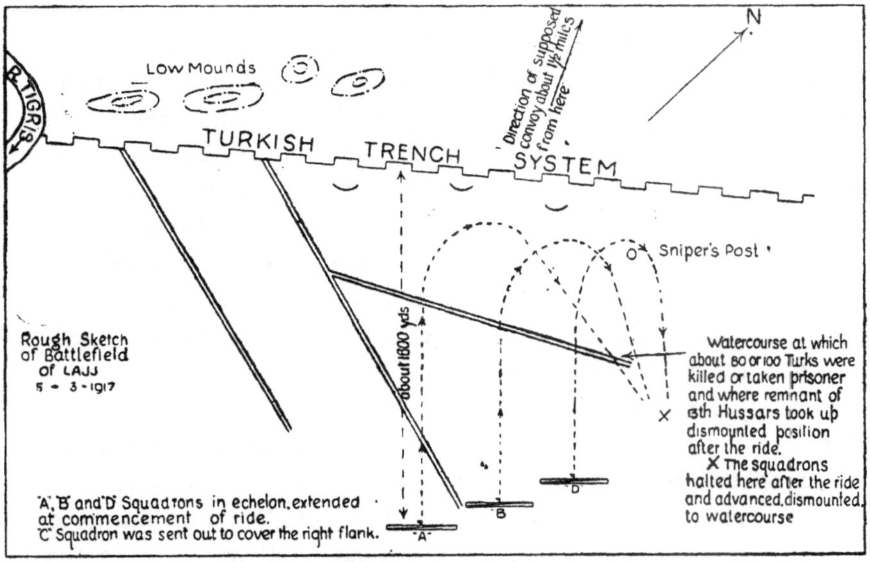

Their commanding officer now called up his four squadron leaders, and sending "C" Squadron under Captain Steele out to the right to cover his exposed flank, ordered the rest of the Regiment to advance in echelon of squadrons extended. The ground was flat and covered with low scrub. "D" Squadron, which led, was under Captain Eve, "B" under Lieutenant Dawson, "A" under Captain Newton.

The Regiment had hardly got into the formation order when a Turkish shell burst on its left, and rifle-fire opened from some hidden trenches on broken ground ahead of the leading squadron. No enemy guns or Infantry were then visible, nothing but the convoy and its escort in the distance; but they were evidently there, and it was impossible to say in what strength. As the aeroplanes had not made out any considerable body of them, they were presumably few, so it was decided that the Thirteenth should ride them down, and if possible gallop the convoy. The three squadrons accordingly drew swords and advanced at a trot, when it immediately became clear that they had in front of them not only scattered bands of Turks, but Infantry in concealed trenches with machine-guns and Artillery.

The order was then given to charge, and the eager horsemen swept forward with a cheer. A first low trench or water-cut was taken at a gallop, and some scores of Turks who held it threw up their hands, but as the squadron galloped on, some of these men seized their rifles again and fired into our men from behind, while in front the charging lines were met by a blaze of musketry which emptied many saddles, and broke the weight of the onset. Many brave men pressed on to the edge of the farther trenches, and even in places over them, but they found they could do little against Infantry in such numbers and on such ground. Their Colonel saw that the attempt to ride through was hopeless, and wheeling his three squadrons to the right at a gallop, and then to the right again, he withdrew them some distance until they were in comparative safety near the point from which they had started.

They were still under fire, but the remains of the squadrons rallied at once, and dismounted by order. The horses were sent away to the rear, and the troopers advanced on foot and occupied the trench which they had first taken. Leaving them there in the hope of another advance, or at all events of recovering their dead and wounded from the ground ahead of them, their Colonel went back to consult with the Brigadier and bring up support.

He returned soon afterwards, but only to be wounded by a Turkish bullet, and to find that the remnants of the three squadrons were too weak to push back the Turks and could hardly hold their own. They remained until dark in the Turkish trench. The rest of the 7th Brigade was similarly unable to advance, but held its ground in spite of a bombardment from about twenty concealed guns and a heavy rifle-fire, until the 6th Brigade worked round the enemy's left and threatened his line of retreat. Such of the wounded as could be found in the darkness were then brought in.

So ended this day. The Thirteenth had shown a fine example

Sergt.-Tptr. J. S. STYLES
(*Killed at Lajj, 5th March* 1917)

Sergt. G. ANDERSON
(*Killed at Lajj, 5th March* 1917)

2nd Lieut. E. V. ROLFE
(*Killed at Lajj, 5th March* 1917)

Sergt. W. GILBERT
(*Taken prisoner at Lajj, 5th March* 1917)

Sergt. A. HARRISON
(*Killed at Lajj, 5th March* 1917)

of combined dash and steadiness. They had charged boldly over broken ground, and immediately after the charge had re-formed and gone forward again on foot, as ready to fight as ever in spite of their heavy losses. For they had lost heavily. Of the officers who had led the Regiment into the charge, three had been killed, one disabled and taken prisoner, and five wounded. Of other ranks, twenty-five had been killed and fifty-two wounded. Among the wounded officers were the Colonel and the second in command. The senior squadron leader was dead, and another was badly wounded. One squadron leader only came out of the charge unhurt. In all, about one-half of the officers and men who rode at the Turkish trenches were killed or wounded.

Of the three officers who died in the charge, the most prominent was the Captain of the leading squadron, Eve, who fell sword in hand among the Turkish Infantry. Many of his brother officers have borne witness to his courage and character. No braver soldier ever rode in the Thirteenth, and none was more beloved. The other two were newly joined 2nd Lieutenants, Rolfe and Lynch Staunton, who had not seen a shot fired before that day.

The following are extracts from letters and diaries bearing on the story of the fight. As the narrative in the text has been taken mainly from Colonel Richardson's account and the Regimental Diary, others are quoted before these.

From Lance-Corporal Watkins, D.C.M., to the Hon. Mr Justice Eve.—"I was with your son, the (late) Captain W. H. Eve, on the 5th of March, when he was killed, and I was wounded almost directly afterwards.

"I was his signaller in most of the engagements in which we took part, and I should like to state, sir (although I am not in a proper position to do so), that he was the bravest man I have ever seen under fire, and under whom it was my good fortune to serve."

May 23, 1917.—"On the 5th of March 1917 we had been on the move since early morning after three days' rest, and about 3 or 4 o'clock in the afternoon we had word back from the advance-guard to say that they had got in touch with the Turks.

"We formed up, and your son, who (as I expect you know) was squadron leader of 'D' Squadron, received some orders from officer commanding Regiment, and we were ordered to advance.

"We went along at a canter, myself and Trumpeter Wren (he was afterwards killed) riding behind the officer commanding squadron.

"I noticed Captain Eve pulling his sword out of the scabbard, and shortly afterwards we all followed suit and we started to gallop.

"The squadron then had orders to wheel to the left, and Captain Eve, who had gone towards the right flank, dropped behind and I stayed with him.

"We followed behind the remainder, and I saw a Turk just in front of me put his rifle to his shoulder, and was going to fire at Captain Eve, but I just managed to get there in time to prevent him from doing so.

"Soon afterwards I saw a pal of mine (Private Jones, D.C.M.) lying on the ground badly wounded, so I dismounted and tried to get him out of it as best I could, and I then lost sight of Captain Eve and the remainder of the squadron.

"I did my best for my pal, but his wound was fatal, and I could not get him on my horse, so I looked about me to see if I could get help.

"I saw Captain Eve about twenty or thirty yards [off?] taking about twenty Turks prisoners at the point of his sword. He must have seen me, as he left the prisoners and came over towards me and told me to get mounted and draw my sword and go along with him.

"There were about fifteen or twenty Turks in a nullah a short distance from us, and they were firing at us and the other men lying about wounded and otherwise.

"We galloped towards them; Captain Eve jumped over to the other side of the nullah, and I remained this side.

"The odds were strongly against us, but I think we did a good deal of damage to the Turks, before your son fell from his horse, right into the centre of the Turks.

"I cannot say whether he was shot or if he had been bayoneted, but I did not see him move after that, so he must have been killed outright.

"His horse galloped away, and I was told later that it had seven bullet-wounds in its neck.[1] I was left by myself then, and I really

[1] Captain Eve's mare Caprice was lost for a year or more after the charge, but was then seen by his batman, Private Hogg, among the horses of an Indian Cavalry Regiment. She was recovered and brought to England.

did not know what to do, so I looked over my shoulder to see if there was any one near to give me a hand, when I saw a Turk about five yards away ready to fire at me, and before I had time to do anything I got a bullet through my left arm close to the shoulder; the same bullet broke the blade of my sword, which was in my right hand. My horse took me back out of it, as I was too dazed to do anything myself; and if it had not been for 'Chicko,' I should not have been alive to-day. I was told later that Captain Eve, when found, still had his sword tightly grasped in his hand, and I am sure it will be some consolation to you to know that he died a hero fighting to the last.

"I may also mention several occasions before that in which he proved himself one of the bravest.

"He rescued a wounded man from out in the open under very heavy fire and at close range, and on the 25th of February 1917, in a dismounted attack on the Turks, he bandaged several wounded men up, regardless of the danger to himself in being exposed to heavy fire in the open.

"A bullet [hit ?] him on that day, but luckily it hit his revolver-belt, and glanced off without doing him any harm.

"I was lying by the side of him at the time, and he said to me, 'Watkins, I think we are both marked men,' as I had been hit three times without causing any wound.

"One bullet passed through the sleeve of my tunic, and another stopped in my pocket, in a book that I carried, after passing through my pay-book; and I also got hit with a shell-case on my foot, which only caused a slight bruise, and I was able to carry on.

"I hope you will not think, sir, that I am trying to flatter myself in this letter, and making you think I am a hero, but I have only put down how things happened, and let you know as best I could how your son met his untimely death."

Major Twist—March 1917.—"I feel I must write and say how we shall miss dear old Billy. You will have heard how he met his death from those who were actually with him at the moment, and though I was quite close to him I suppose the dust and confusion of the charge, and also my being myself wounded, prevented my seeing him killed. He was, I know, most gallantly leading his squadron, and fell from his horse without a word. I believe if he had been asked to choose, this was the death he would have chosen, and I remember him often saying that if he had to die he would like to be killed on his horse. His men are heartbroken, and it was a sad day for the Regiment when we lost him."

Captain E. Wingfield Digby, V Battery, R.H.A. — "Well, here we are just outside Baghdad. . . . The saddest part of the whole show to me is poor old Eve's death—he was just one of the best fellows on earth—anyway he lost his life charging the Turk (intrenched). I believe he took his man, and was then shot through the head from below—the bullet entering under his chin. The Thirteenth lost pretty heavily that day; we got very heavily shelled too—lost all our layers casualties, and a direct hit on one gun and one waggon,—it was a very dusty day and high wind, and one couldn't see a yard. And when the wind dropped and mirage lifted in the evening, the Turks could easily see us in the open, and fired at us from three different directions."

2nd Lieutenant Guy Pedder—March 15.—"I am afraid the only comfort I can give you is that he was killed instantaneously, and also mounted, in a brilliant charge—a death which I know he would have preferred to all others.

"I remember him telling me at Makina that personally he didn't think a man could have a finer death than in jumping a large fence out hunting, but his was infinitely finer leading his squadron (I was directly behind him) straight at a small nullah full of Turks. He was probably shot by a Turk not more than two or three yards off. I did not see his body afterwards, as I was hit myself very shortly afterwards; but Mr Payne did, I believe, and no doubt he will have written to you.

"It is most awfully sad, and I shall feel it all the more when I get back to the squadron and he isn't there. I never saw a braver or more fearless person than he was: he almost made it impossible for any one under him to be afraid."

Lieutenant Fitzgibbon—March 21.—"I have seen by the way what no other Regiment has seen or done in this war—that is to say, the Regiment formed line and charged! It was a damned fine effort, but rather a mad one. We had been given

Capt. H. C. D. FitzGibbon, M.C.
(Wounded near Deli Abbas, 8th April 1917)

Lieut. A. M. Le Patourel

Capt. A. Vlasto, M.C., R.A.M.C.

Capt. H. G. T. Newton
(Accidentally drowned in Tigris, 25th April 1917)

Lieut. J. H. Hirsch

Pte. J. L. Roberts, D.C.M.
(Died, 10th April 1917, of wounds received near Deli Abbas)

wrong information by aeroplanes, and ran our heads against the most beautiful trenches full of Turks I have ever seen. The old Turk just sat back and waited for us, and, by God! he let us have it. The noise was something impossible to describe: how the devil we any of us ever got out of it I still do not know. The dust did us in, but also saved us. Well, we eventually got out of it, rallied, dismounted, and attacked on our flat feet, and had the satisfaction of seeing the Turk evacuate his position at nightfall. We rather got a dusting though.

"My squadron ('A') went into action very weak from previous casualties, and we went in sixty-six strong and came out thirty-three. Fourteen officers of the Regiment were knocked out and five killed. Newton and I scraped out of it safely.

"We had a good many shows besides this one, but none so good. It is something to be the only Regiment to have charged as a Regiment, and it did jolly well. We were the first Cavalry Regiment to enter Baghdad. We have lost our Colonel and also the second in command, Twist; the senior squadron leader was killed—such a nice fellow, Eve, shot through his head. I was the first to find his body, and took his things off him. It was very sad."[1]

Colour-Sergeant F. Spanton, "B" Squadron.—"After a few minutes we were galloping towards the enemy with drawn swords, at a fair pace. We were well received by the Turks, who blazed away at us as hard as they could, and when we got close some of their shots took effect, and one or two spare horses were galloping about. After we had crossed the first trench my horse was apparently hit and pitched over, causing me to be thrown. I landed between the lines of Turks, and remained quiet, waiting to see what would happen next. The Turks now had turned about, and were firing over me into the rear of the squadrons. The Regiment changed direction to the right, and passed out of my view; the Turks got out of the trenches and continued to send a hail of bullets after the disappearing squadrons. I watched this as I lay on the ground, weighing my chances of rejoining the Regiment. The

[1] Lieutenant Fitzgibbon crawled out into the open several times to look after the wounded. It was on one of these occasions that he found Captain Eve's body, and removed his watch, rings, and spurs.

Turks in front of me were now retiring, and moving to the left of the trench, and I thought if I remained quiet they would all probably pass me by unnoticed, and then I could get back to my squadron again. But no such luck: as the tail-end of the Turks passed by, one fellow stopped when he saw me, and raising his rifle to his shoulder he fired point-blank—but missed—the bullet not coming so near as the one that had cut the belt of my haversack in half a few minutes before. This man, evidently disgusted with his bad shooting, walked hurriedly away, so once again I thought my chances of getting back were good; but a group of three men coming along a little later came to see who I was, and lugged me off to a dug-out. I had hurt my knee rather badly when I was thrown, and couldn't get along over fast; this annoyed the Turks, as we were still under a heavy fire from the British machine-guns, so they jabbed me in the back with the butt of a rifle as a signal to hurry. . . . Sergeant Gilbert was brought up to where I was, and duly deprived of his possessions. After a few minutes Lieutenant Pennington[1] was brought in. He was holding his wrist: he had been badly hit in the forearm, the bone being broken. I tied him up with my field-dressing, and made a plug for him until we could find a dressing station. In the meantime, Private Morrison had been brought in, and he also was shot through the arm and was bleeding freely. We were now hurried off, and came to a dressing station, where the wounds of Lieutenant Pennington and Private Morrison were dressed. We were now handed over to an escort, who had instructions to take us to the Headquarters, 18th Turkish Division, for interrogation. The escort made a wide chukker to get to the Headquarters. After a while they called a halt, and motioned us to sit down. Now they thought was the best time to change clothes. They commenced with Lieutenant Pennington, taking his puttees and spurs, and tried very hard to get a gold ring off a finger on his broken arm, which must have caused him considerable pain; but being unsuccessful they let him alone, and turned their attention to Private Morrison, whom they robbed of almost everything; next they came to me and took what the others had left. . . . Now they turned to Sergeant Gilbert and served

[1] The name was Pinnington.

him in a like manner. After they had got all they could out of us they continued to walk to G.H.Q., where eventually we arrived. Here we were interviewed by an interpreter, who spoke excellent English, and who was wonderfully informed, but I am afraid the satisfaction he got from us was little. I spoke to him about my clothes and possessions being stolen by the soldiers, but he said nothing could be done, as they were allowed to do it. Two Turkish soldiers fighting for possession of my revolver were interrupted by an officer, who thrashed them both and took it himself. The Turks were still retiring and we were sent to travel with a Field Ambulance Section."

It will be seen that there was some measure of mercy shown by the fighting Turks towards the men who fell into their hands, but Sergeant Spanton's diary goes on to describe the lives and sufferings of the British and Indian prisoners during their captivity, and the story is painful reading.

2nd Lieutenant Pinnington, who was wounded and taken prisoner at Lajj, had joined the Regiment only a fortnight earlier with 2nd Lieutenant Rolfe, after a long voyage from England by way of the Cape and India. More fortunate than his companion, he lived to be exchanged to England within a year. His account of the fight at Lajj is as follows:—

"The 7th Brigade, to which the Thirteenth belonged, led the way, one of the Indian Regiments—Watson's Horse—furnishing the advanced and flank guards. The Division had been on the march several hours when the news filtered through that a Turkish convoy, escorted by a company of Infantry, had been sighted a mile or two ahead. The job of rounding up this convoy was assigned to the Thirteenth, and the Regiment drew ahead in the formation known as echelon of squadrons. We had not gone far when the whizz of bullets greeted our ears, and the order came to dismount for action—presumably owing to the impossibility, on account of the dust, of seeing what actually did lie in front of us. A lull in the dust-storm served to assure our leader that whatever opposition there was could be ridden down, and accordingly, after advancing a short distance dismounted, and snapping away with Hotchkiss gun and rifle, we were summarily recalled, ordered to mount, draw swords, and finally charge. Shells by this time were bursting overhead, and the storm of bullets through which we rode gave the lie to the report that, with the exception of the convoy and its escort, all was clear ahead. We were quickly on top of a line of Turks who had abandoned their trench and were scurrying back to join their comrades in the rear lines. The horrible screeching told its own weird tale of the fate they met. In the onrush I got 'winged' and was left, as I ultimately learned to my cost, a few yards from the Turkish main-line trench. As I stood there dazed amid the dust, I saw another squadron come galloping up. It was a sight I shall not readily forget. The leader to the fore with sword aloft, the line of panting horses, the grim eager faces of the men, the flashing swords—I thought of Lady Butler's painting 'Floreat Etona,' and marvelled at seeing the living parallel. Captain Eve was the leader, and as he approached 'with a swiftness not to be conceived,' I shook my right arm, which was hanging limply by my side, and shouted, 'This thing's shattered.' 'Hard lines, old boy. Never mind,' he called back, and was gone. The line had passed and disappeared in the dust.

"When once again the dust had settled, the Turks, observing that the attack was for the moment over, sallied forth from their trenches to pay attention to the casualties. An Arab relieved me of my possessions and led me to an officer in the trench. This officer, on seeing me approach, forsook the machine-gun he had been manipulating and shook my uninjured member warmly—a thing which no good Mahomedan should have done, as such contact with the 'infidel' is forbidden in terms of Koran regulations. Some one applied a tourniquet to my arm, which was still dripping, and I was taken to a bend in the trench where I came across Sergeants Gilbert and Spanton, and Trooper Morrison, the only other survivors amongst those who had fallen into Turkish hands. The horses of the two sergeants had been shot down, and Morrison was suffering from a wound in the forearm. We were led out of the trench and across the open country to the Headquarters of the Turkish forces (instead of rounding up a convoy the Thirteenth had ridden through a Division), and in the course of our journey to the rear had ample opportunity of realising how much nastier is the effect of British shrapnel than that of the Turk. Arrived at Headquarters, we were separated and interrogated

SCENE OF THE FIGHT AT LAJJ. 5TH MARCH 1917

in a very polite manner by one who spoke English fluently. In this interrogation, to my surprise and the credit of the enemy, there was not the slightest suggestion of coercion to extract information. After a few hours' rest, Morrison and I, with some wounded Turkish soldiery, were given seats in a rickety waggon, and with Sergeants Gilbert and Spanton marching behind, we set out. It was early the next morning that the palm groves of Baghdad, topped by the domes and minarets of the city's many mosques, came to sight. At Baghdad, Morrison and I were taken direct to a hospital, where a Jewish doctor, Alalou by name, dressed our wounds and gave us food and drink."

Lieutenant Pinnington was afterwards sent on to Mosul, and he writes: "It is perhaps worthy of mention that on many occasions during my imprisonment there I received compliments from Turkish officers on the bravery of the British Cavalry in making their ill-fated charge on the 4th [? 5th] March."

Captain Newton, the one squadron leader who came out unwounded from the "ill-fated" charge, evidently took much the same view of it as his subordinate. The entry in his diary for the 5th March is as follows:—

"Marched 4.45 past Zeur almost to Lajj, where enemy reported. Tried to take them on mounted, but found them intrenched and very strong, and had rather a bad time. Retired a bit, and came into action dismounted, and held on all day. Awful job getting wounded away at dusk. Eve, Lynch Staunton, and Rolfe killed. Pinnington missing. Colonel, Twist, Dawson, Pedder, Welstead wounded. Total casualties eighty-seven. Retired to biv. near Lajj bend, getting in soon after 10. An awful day, wind and dust terrible. A little rain at night. Took sixty-six men into action, aud brought out thirty-three."

In a letter of the 21st March 1917, Captain Newton gives a somewhat longer account.

"The poor old Regiment has been through some pretty troublous times, and has been pretty well knocked about, but it's done most awfully well, and got a great chit for itself. . . .

"I last wrote on Saturday the 3rd. They gave us a rest next day. We got orders to parade at 4.45 A.M. (on the 5th), and it was awfully cold and dark turning out. We had an uneventful march for some time, but quite unexpectedly came in touch with the enemy in the middle of the morning. We came into action against them, and there seemed to be very few, so the Colonel decided to ride them down. Then we found that we were wrong! There were a few scattered Turks in nullahs who ran away or held up their hands as we rode over them, but just behind were some beautifully concealed and very well-dug trenches, and we fairly caught it from rifles and machine-guns. We rode right up to them though, and even in places across them, but we couldn't do anything, and had to rally and go about, and eventually came into action dismounted. It was a pretty tough business though, as one did not know where any one was. Young Welstead was hit in the arm before we charged, and Twist during it. Billy,[1] Vavie,[2] Lynch Staunton, Rolfe, and Pinnington could not be accounted for at first, but later in the day we got forward a bit and found poor old Billy's body, and Vavie lying out wounded, and several other wounded men. Poor Billy was shot through the head, and death must have been instantaneous. He must have been shot in the very act of jumping a nullah, as his body was lying in it as he had fallen with his sword by his hand. Vavie was very badly hit in the knee, and must have had an awful day, as we couldn't move him till after dark, and the Turks had been at him and taken all his equipment off, and even the rings off his fingers, I believe. Rolfe's body we found after dark, and Lynch Staunton's next morning. Pinnington we have no news of, and he must be a prisoner, as his body was never found. When we had established ourselves in a nullah to hold on we continued to have a few casualties, as we had very little cover. Young Pedder got one through the shoulder, and soon after that Jimmy (the Colonel) got one through the arm. Then my troubles began, and I found myself in sole command of such as there were of three squadrons. 'C' were not with us, and it was not till much later that we got news that they were away on our right, though they had had a few casualties. We hung on all day, and I had a pretty anxious time, as there were an awful lot of Turks in front of us, and I couldn't make out what they were going to do. We had frightened them pretty badly, and they cleared off after dark. We collected all we could find and brought them

[1] Captain Eve. [2] Lieutenant Dawson.

SCENE OF THE CHARGE AT LAJJ. 5TH MARCH 1917

"CAPRICE" WITH PTE. C. HOGG, WHO FOUND HER IN MAY 1918
AFTER SHE HAD BEEN "MISSING" FOR 14 MONTHS

in; but we had no stretchers, and it was pretty hard work, and there were a lot of snipers at us all the time. Vavie was splendidly plucky. He was one of the worst cases, and we got him in on a ground sheet, and although I knew he was in agony he kept quite cheery, and would only apologise for the trouble he was giving us.

"I was simply dead-beat when we got back to the horses. I don't think I ever felt so done up in my life before. We reached bivouac about 10 P.M., and I was overjoyed to find Charles[1] safe and sound, and to find I was only second in command and not commanding officer."

2nd Lieutenant F. N. Payne to his Wife—March 13-16.—"At last I am able to write to you, and very lucky I am to be alive to do so, for we have been through a terrible time, and suffered heavy losses. . . .

"On March 5th the Cavalry, pushing on ahead of the army, ran into the enemy again at Lajj, thirty miles below Baghdad. The aeroplanes and armoured cars reported that they were about 300 strong, and protecting a convoy that was trying to get away safely, so we were given orders to round them up and bring them in, while the rest of the Division attacked the convoy. Our Regiment was sent round the right flank to get behind their trenches, the river being on the left. We started out in extended order, and gradually worked up to them, when about a hundred Turks seeing us coming, walked out of the nullah they had been firing at us from, and the Colonel gave 'Draw swords—charge.' The hundred dropped their arms and surrendered. We charged through them, and ran into their concealed trenches that were packed with a solid mass of Turks, and three thousand rifles from all directions, together with shrapnel and bombs, came pouring into us, and eighty-two of our good fellows went down. We had to get out the best way we could, and retire about a mile with what was left, and get away the remaining horses and advance on foot, for we had to recover our wounded at all costs: they were left all over the place as their horses reared and plunged till they toppled out of their saddles. Then came the worst part of all. I rallied all the men near me, after finding I was not killed or even wounded, and forty of us all told and cut off from any possible help kept the Turks in their trenches, and in about three hours made good the nullah, and held

[1] Captain Steele of "C" Squadron.

it till dark with our Hotchkiss gun and twenty rifles. We, of course, all expected to be wiped out, which would have been an easy matter, but they were satisfied to stay where they were except for a number of the boldest who walked out and stripped our wounded. I killed two of them myself, but we could not stop them, and that day the Turk seemed to return to his ancient savagery, for we found they had murdered several of our chaps that were lying helpless. Poor Eve was killed instantly, to the sorrow of every man in the Regiment: he always wanted to die in a charge, but we do want him so badly, and it seems such a shame that he isn't there to enjoy the victory in Baghdad. Dawson was very lucky to come off with his life—he fell right up against one of their sniper's holes and couldn't move with a shattered knee, and the Turks had taken everything off him, but couldn't get his ring off, so threatened to shoot him as they had done to seven others alongside him, together with the brave chap who tried to carry him in.[1] Eight of us worked our way up the nullah, and held them off till dark, when we got him out of the open. We had had nothing to eat or drink since 3 that morning, and had three miles of ploughed land to carry sixteen men in without stretchers, and fired on all the time. Some got hit several times on the way back. When we called the roll in the darkness, we found that thirty-two had been killed and twenty-six wounded. The Colonel, who is the bravest man next to Eve, was hit only at the very last: he was always to be seen in the thickest, carrying up ammunition and encouraging the men with a smile on his face, and you could hardly see him for the dust of bullets hitting the ground round him. The second in command was hit, Eve killed, and three young officers who had just joined killed. Pedder got a nice one through the shoulder, and two others wounded, so that Jeffrey and myself found ourselves alone in the squadron, absolutely staggered at being alive. I lost most of my men—all the best, of course—and have now only thirteen: we had to make two troops out of what was left of the four. When we moved on and saw the position, we stood aghast that it was possible, and the army commanders of the Infantry when they came on, sent messages of their deep admiration and gratitude for

[1] Private A. Wallhead, killed after putting Lieutenant Dawson in safety. Sergeant H. Knapman, too, was killed while helping the wounded.

the work we had done that would have delayed them considerably. I believe our Colonel is to be decorated later on, and we are having our reward now in that 'For Valour' our Regiment was chosen to garrison Baghdad when it fell.

"There were some very sad cases in these chaps that were killed. Out of five new officers that had joined and not been under fire before, three were killed and one wounded, and none of them more than twenty-one years of age. Only a few minutes before we went into the charge, and were halted awaiting orders, Eve called out one of my men[1] to present him with the D.C.M. for extraordinarily fearless bravery in the last show that had just been sanctioned that morning, and I promised to send a cable home for him if we got up alongside any of the gunboats. Another fellow, a great big, rough, hard-working fellow, remarked that if we could have one in the troop we could just as easy have two, and he was going to get it in the next show we got into or die in the attempt. I saw him try: he rode past me yelling and whooping and ran his sword through the foremost Turk, only to be bowled over the next second. When I went out later to find him, my sergeant, who had a shattered thigh, told me not to risk myself any further, as he had seen a Turk blow his brains out, as they had also done with the boy who had just won his medal. I have finished up after being in fifty shows without a scratch: they've chistled past my nose, through my coat and everywhere, killed four of my horses, one after another, and never touched me. You may call it providential or the hand of God. I don't know what it is other than miraculous good luck, for it's not right to think that I should be spared when such good fellows as Eve are struck down, and all three poor new chaps, who had never heard the terrible swish of a bullet pass their heads before, to meet death the first time. It doesn't seem right that I should be spared. . . . I had absolutely no thought whatever about my soul or the hereafter, in fact everybody seemed to be in a state of pleasant anticipation. When Eve galloped by me and I looked to him for orders he had a beautiful smile and just sailed along, and I waved my men on to follow him, and next second his soul had left his body. I don't believe he had a thought of death, or that it meant anything to him. I know only when we had

[1] Private Alfred Jones.

breathing-space in all these occasions did I worry about anything, and that was how you would get along, and whether I had left everything satisfactory. . . . Lynch Staunton, whose father was a Senator from Hamilton, was one of the six new subalterns who hadn't been under fire before. . . . Poor fellow, he rode right down into the Turk trench and walked his horse along it over the Turks, that were as thick as gooseberries, as much as to say you are my prisoners; and they certainly would have been, had not that murderous fire from their main position bowled him over. They took him prisoner, but we found his body next day: they had dressed his wound, but later decided he would be a hindrance to carry along, so murdered him and took everything off him."

2nd Lieutenant Guy Pedder, one of those wounded in the fight, writes to his mother from the hospital at Amara: "No doubt you will have heard by now how we ran into it on the 5th, but in case you haven't I will try and give you a rough idea of what happened. After camping at Azizieh for four or five days (I wrote to you from there), the Cavalry Division moved at 5 A.M. (we were centre Regiment of the leading Brigade) on March 5th with orders to camp at Lajj, which was about twenty-six miles away, the flying people having reported all clear. After we had gone about twenty miles we halted and watered, and fed on the river bank, three monitors steaming past us up-river, and an aeroplane over our heads—quite a dramatic picture. At 10.30 we moved off again, and almost immediately a sand-storm started, the strong wind taking all our dust out in front of us. We went about five miles, when one of our armoured cars brought in a Turkish officer, and the news that there were a few of the enemy scattered about, who were probably only waiting for us to collect them as prisoners (like the lot we took in the pursuit), so we got orders to move out to the right to verify this or collect prisoners. Eve told us this, and quickly changed on to his favourite mare. Three squadrons moved out to the right, the fourth ('C,' Stirling's and Ormrod's squadron) right out to the flank; 'A,' 'B,' and 'D' then advanced in line of troop column extended and almost immediately came under heavy rifle-fire. We walked about half a mile and then trotted; the dust was awful and you couldn't see more than one hundred to two hundred yards in front; all of a sudden, through the dust, I could see a lot of Turks in the open and in a nullah. Eve gave the order, 'Draw swords—form line—gallop,' and before we knew where we were we were into them. Some stood up and surrendered, others lay flat on their backs and shot at us at two or three yards' range; two fellows let off their rifles right in my face, but missed both Matches[1] and me. There was a small nullah at the part of the line where 'D' charged, and the Turk lying flat was pretty well safe, but we got a certain number; the noise was tremendous, bullets from revolvers, rifles, and machine-guns cracking all round. Beyond the first nullah, and running at right angles to it, was another much deeper, and we gave the Turks holding it a splendid target. I lost sight of Eve, who was just in front of me at the first nullah, but his orderly who was shot (wounded)[2] close by him, tells me he was shot from two yards' range, as he was bending down to charge a group of Turks. At any rate, he must have been killed instantaneously.

"There were a great many dead horses and empty saddles by now, and I tried to rally all the men who were near me; it was extremely difficult, as we were under very heavy fire at very short range; there was deafening noise and a hard wind blowing, and if you collected men in bunches, the bunches would very soon have a machine-gun on to them, so what men I got under hand I kept extended, and galloped a fair way back and handed over the horses and went up dismounted. The Turks' guns had now got our range and were very annoying, but their shells were bad, or I should not be writing to you now. I met Jeffrey, and we all went on by short rushes together, until we met the Colonel, who told us roughly where to go, and we got into the nullah we had first charged, and there we held on till dark, when we collected the wounded, who were lying out in front. I was hit clean through the shoulder about 2 P.M., having got through the charge and dismounted attack, and when I was sitting in the nullah talking to the Colonel, who got hit in the arm very shortly after me.

"We had nine officers casualties—two killed, Eve and Rolfe (from Aldershot, first time under fire), two missing, believed wounded and prisoners, Lynch Staunton and Pinnington (just joined from 'A'), and five wounded, Twist, Colonel, Dawson, Welstead, and self. 'A,'

[1] His charger.

[2] This was doubtless Watkins.

'B,' and 'D' lost very heavily. 'D' had thirteen wounded and thirteen killed, and I think we got off lighter than 'A' or 'B.' Dawson is the worst hit of us five; the Turks robbed him of everything, even taking his water-bottle and ring—we left him at Sheikh Saad, and I hope he won't have to lose a foot. Poor old Thirteenth, we fairly ran into it. My troop sergeant and best corporal were killed, new sergeant-major badly hit. And now I hear we have crossed the river again to the right bank, and are three miles west of Baghdad. I wish I was with the Regiment—it's sickening. . . .

"Later. News has just arrived Baghdad has fallen, only five days after I was hit: isn't it sickening I was not there? . . .

"I am awfully sorry about poor Billy Eve, but he died just the death he would have preferred, certainly a very noble one. He was one of the most fearless people under heavy fire I have so far seen, always rushing about looking after any one who was hit."

It need hardly be pointed out that 2nd Lieutenant Pedder, being when he wrote in hospital at a distance from the front, was not in a position to get the most accurate information about casualties or other matters connected with the Regiment. The Regimental Diary, written by the Adjutant, states the casualties as follows:—

"Killed, 3 officers; 22 British other ranks.
Wounded, 5 officers; 52 British other ranks.
Missing, 1 officer; 4 British other ranks.
Horses, 54 killed; 36 wounded."

This gives the total of casualties at nine officers and seventy-eight other ranks—of whom four officers and twenty-six other ranks were killed or missing. How many wounded died of their wounds does not appear, but none among the officers. The horse casualties were ninety, of which fifty-four were killed.

The following is an account of the day given by Private Massey of "D" Squadron in his diary of the war:—

"Next morning, the 5th of March (a day to be remembered in the annals of the Thirteenth Hussars), we moved off about 5 A.M. and proceeded up the Tigris, in the teeth of a terrible sand-storm, which blinded both horses and men, and through which we could only see 80 to 100 yards ahead of us. Suddenly, out to our front, we heard the sound of rifle-fire, and bullets whizzed amongst us. The report came through that a Turkish convoy was attempting to get away, and it was decided to make an attempt to capture it. Still under fire, the order was given to 'draw swords,' and we continued to trot on in extended order. As we got nearer, the fire became heavier, and the whizz-whizz of the bullets, as they flew in amongst us, told us we were in for 'hot stuff.' Then we got the order to canter, and when within a hundred yards of the Turks (whom we could now see through the sand-storm lining the parapet of the first-line trench) the order to 'charge' was given. Digging in our spurs, and with a loud shout, we charged the Turks in their trenches. Seeing us on top of them, the whole front line surrendered, holding up their arms, and some were waving white flags. The order was given to bring the left shoulder round, and immediately we commenced doing so, the enemy's guns opened on us with shrapnel; and he also had a second line 800 yards behind his first one, and we came under a heavy machine-gun and rifle fire, and our losses were rather heavy. On wheeling out of action, but still under fire, the Colonel rallied his men. The order was given 'dismounted action,' and No. 3's of sections led the section's horses to the rear, while Nos. 1, 2, and 4 of the sections formed line and charged with fixed bayonets. I went to the rear with the gun-section's horses, and many wounded men were being helped to the first-aid dressing station.

"I had gone all through the charge, riding my own horse, and leading the horse carrying the Hotchkiss gun. I saw M'Nulty at my side during the charge with his revolver in his hand. He was very excited and shouted across to me, 'Come on, Charlie'; a second later he fell shot in the head. My section-corporal was wounded in the elbow, and his horse had a nasty shrapnel wound in its neck. That I got through it all without a scratch seems hardly believable. Riderless horses, and horses carrying automatic rifles were galloping madly about, and I was told that I was the only one who had stuck to my led horse all through the charge.

"But the Turks played a dirty game on us, even those who had surrendered in the first-line trench fired on us when we had passed over them; and several men, whilst dressing our wounded where they had fallen, were shot at point-blank range in the head.

"But to return to my former subject: after the rally, and when a halt was called with the led horses well out of rifle-range, and knowing that the section-corporal was wounded, and that they would probably want the gun and ammunition up in the firing line, I decided to volunteer to return with it to the trenches. I asked the sergeant-major, and he gave me permission to take it up. I was about to start up with it, when I saw Private Galloway, who had caught a gun-horse which had got loose in the charge, leading it up towards the firing-line on foot. I thought that a much better way, and though one could not go so fast leading as riding one and leading, it had the advantage of offering a much smaller target for the enemy's machine-gunners. I told the sergeant-major my opinion, and he took my riding-horse from me, and I started to lead the gun-horse towards our fellows in the enemy's captured first line. On coming in view of the enemy, he opened on me with a machine-gun, and bullets were kicking up the dust at my feet, and buzzing round the horse and myself. I passed close to Corporal Taylor, who was lying wounded on the ground waiting for the Field Ambulance to fetch him in. He shouted to me to clear off to the right, as he had already been wounded once. I didn't think I was drawing the enemy's fire on him until he reminded me of the fact. After that I began to run, and seeing me coming up with the gun, two N.C.O.'s rushed out and hurriedly took off the ammunition. They didn't want the gun, and gave me a stern order to 'get away as quickly as you can.' I then started to run back with the gun-horse. When I had gone a dozen yards I remembered that I had left my rifle behind, and turned round to go back and fetch it, but the sergeant told me to go on and leave it. It was certainly a hot corner, and perhaps it was best I did.

"I then started to run back as fast as I could go, as the Turks knew I was bringing up ammunition as well as I did, and meant to make it hot for me, but I got back quite safely. On the way down I passed the Colonel talking to a wounded man.

"As soon as I had got back, however, Sergeant-Major Tassie called me, and ordered me to go up to the line again with more ammunition. I again started for the first-line trench. I arrived in safety, and having put it down was told to stay in the trench.

Here we hung on till nightfall, as the wounded on the field pinned us to the trench. Several were killed in the trench by rifle-fire, and the Turks heavily shelled V Battery, knocking out a gun and limber and killing several men. Whilst in the trenches the Colonel and Major Twist (second in command) were both wounded,[1] and the command fell on Captain Newton of 'A' Squadron.

"Towards night the Turks retired, so we collected the wounded, and slowly, very slowly we returned to camp, three miles to the rear, the end of a hard and terrible day. On arriving in, after taking over our horses we watered at the river-side."

The writer proceeds to describe how in trying to get a bucket of water, he was carried away by the stream and nearly drowned. He was rescued, and the day's entry finishes: "After getting into bed it started to rain, so I put on M'Nulty's cape and British warm, laying it all over the bed, as I knew poor Mac would never want his any more."

To close the story of the fight at Lajj, the Colonel's account of it is given in his own words. He had commanded the Regiment throughout the day, had led it in its charge, and after rallying it when the charge was over had taken it into action again dismounted. Conspicuous for his courage among many brave men, he had won the admiration of all; and the wound which disabled him towards the close of the action was happily not severe enough to prevent his commanding the Regiment again.

Colonel Richardson.—"The following extracts are taken from a letter written home about the middle of March, when events were fresh in my memory:—

"We started on the 5th of March, leaving camp an hour before dawn, and our orders were to march the Infantry to Zeur, twelve or fourteen miles on, and the Cavalry to Lajj, about six miles farther on still. Our Brigade, the Seventh, were leading, and we made Zeur, where we watered and fed the horses a little before 10, and we marched again at 10.30. Soon after we started armoured-car reports came in to say that there was a small convoy ahead of us proceeding in the direction of Baghdad with an escort of about 250 Infantry, and that otherwise the country was quite clear. The formation of our Brigade was

[1] Major Twist was apparently wounded earlier, when the three squadrons wheeled to the right.

the usual formation, with the 13th Lancers (Watson's Horse) leading and finding one squadron advance-guard, and one squadron each on right and left flanks; then the Machine Gun Squadron, Thirteenth Hussars and 14th Lancers, and V Battery R.H.A. We had gone about two miles when our Brigadier ordered me to take the Regiment up abreast of Watson's Horse with a squadron as advance-guard, and soon afterwards pointed out what he took to be the convoy on the march, and hinted that we might be able to get in on it with the sword and round it up. And so, having had a good look at this supposed convoy, which was about two and a half miles to our right front, I called up the squadron leaders, explained the situation to them, and sending 'C' Squadron well out to the right or exposed flank to work to the north of some small hillocks, I ordered 'D,' 'B,' and 'A' Squadrons in that order to advance in echelon of squadrons extended. (I should say here that the day was an exceptionally bad one for visibility, as a strong east wind had got up and was carrying along with us all the dust we raised and a good deal more, which filled the air and made it difficult to clearly distinguish things.) Of course a Regiment in this order covers a lot of ground and takes a certain amount of time to get its formation, and even before the formation was complete a shell bursting on our left warned me that we were seen, and that the enemy had at least one gun, and made me wonder whether the convoy was just a convoy, or whether it was not a gun or two with their ammunition-carts. At this time too, rifle-fire was opened on us from apparently a trench, for no Infantry were then visible, a few hundred yards to the front of the leading squadron; at the same time I heard rifle-fire well away to our left, showing that the advance-guard was coming into touch with the enemy. All this convinced me that we were up against a much bigger thing than the scattered bands of Turks reported by the aeroplanes and armoured cars, but already the leading squadron was on top of the first line of Turks, and the latter, about forty or fifty in number, were already running for all they were worth, or standing with their hands up. Some of these latter, I am told, picked up their rifles after we had passed and fired into our backs, but very few of that first lot reached their friends; they were either "stuck" or taken prisoners. In the meantime our impetus had carried us on two or three hundred yards, and into a very heavy rifle and machine-gun fire, from what I took to be a big nullah or water-cut some eight hundred yards to our front. Men and horses were dropping right and left, so I tried to wheel them right-handed and backwards until we had crossed the trench, or as it proved, small water-cut, from which the Turks had first opened fire on us. Having crossed that, I halted the remnants of the three squadrons ('C' Squadron was still operating by itself out on the right) some three or four hundred yards farther on, dismounted them, sent the horses back out of rifle-range, and advanced on foot until we had reached this trench originally held by the Turks, and which was still full of their equipment and rifles. I may say here that the test to which I put these squadrons—*viz.*, of halting and dismounting under fire, and when the squadrons were all mixed up and actually retiring at a gallop, albeit a slow one—was a very high one, and the way in which they responded was beyond all praise. It was only when we reached the Turkish trench that we were able to take a breather and take stock, and then only did I fully realise the cost of our ride. Major Twist, I heard, was hit, though slightly; Captain Eve was missing, and it was feared killed; Captain Dawson missing, Lieutenants Rolfe, Pinnington, and Lynch Staunton missing, and Welstead hit; my trumpet-major killed, and my orderly badly wounded, and a number of sergeants and men missing. It was patent to me that the men who were missing must be lying, if hit, between us and the big nullah in which the Turks were, and that our only chance of getting to them was to draw the Turks out of this nullah, and it was very soon apparent that we could not do this by a frontal advance without considerable reinforcements: so I went back to Brigade Headquarters to consult with the Brigadier, and pointed out to him that I thought the quickest and cheapest method was to outflank the Turkish position to the northward. He said he would consult with the Divisional Commander, and in the meantime sent up two squadrons of the 14th Lancers to fill the gap between us and the advance-guard of Watson's Horse. Soon afterwards I heard that the 6th Brigade was being sent to find the left flank of the Turks, and I believe they did find it, but late in the afternoon, showing that we could not be up against much less than a Turkish

Division, and that I believe was corroborated later. In the meantime we hung on to our line, but were quite unable to advance, and it was whilst sitting in this shallow water-cut that about 4.45 P.M. I was hit by a rifle bullet in the right forearm. At the clearing hospital I was able to ascertain from the books that night and next morning the approximate extent of our casualties from wounds, and since then I have had a few corrections and additions to make, with the result that I can append a pretty accurate list as under:—

Killed—Captain Eve, 2nd Lieutenants Rolfe and Lynch Staunton.
Wounded—Major Twist, Captain Dawson, Lieutenant Pedder, 2nd Lieutenant Welstead, and myself.
Missing—2nd Lieutenant Pinnington (reported a prisoner of war).

Killed—25 N.C.O.'s and men. Wounded—80.

"The wounded men are all most extraordinarily cheery, tremendously pleased with themselves, and keen as can be to get fit and go back to the Regiment; and I am glad to say that quite 50 per cent of the wounded cases are very slight, and if not already discharged from hospital, are likely to be there only a few days longer.

"At about 4.30 or 5 o'clock the enemy opened a heavy bombardment with about twenty guns, knocked out one if not two of the guns of V Battery by direct hits, and at the same time opened heavy rifle-fire, and it looked as if they intended to make a strong counter-attack. The event, however, proved otherwise, for as soon as darkness fell, about 6.30, they withdrew from their positions in haste; we were able to get all our wounded in, and eventually retired a few miles to the river. Next day the advance was resumed, and we were then able to judge the strength of the position, where the Turks evidently had intended to put up a big rearguard fight, and it became apparent that the shelling of 'S' Battery of the 6th Brigade from their left rear had caused the Turks to realise the danger in which they were of being cut off, and had consequently hastened their retreat. The Regiment received congratulations from many Divisional and Brigade Commanders for the fine show they had put up on this occasion, and it naturally was very pleasing to us to realise that we had effected good work, and that our heavy casualties had not been altogether in vain."

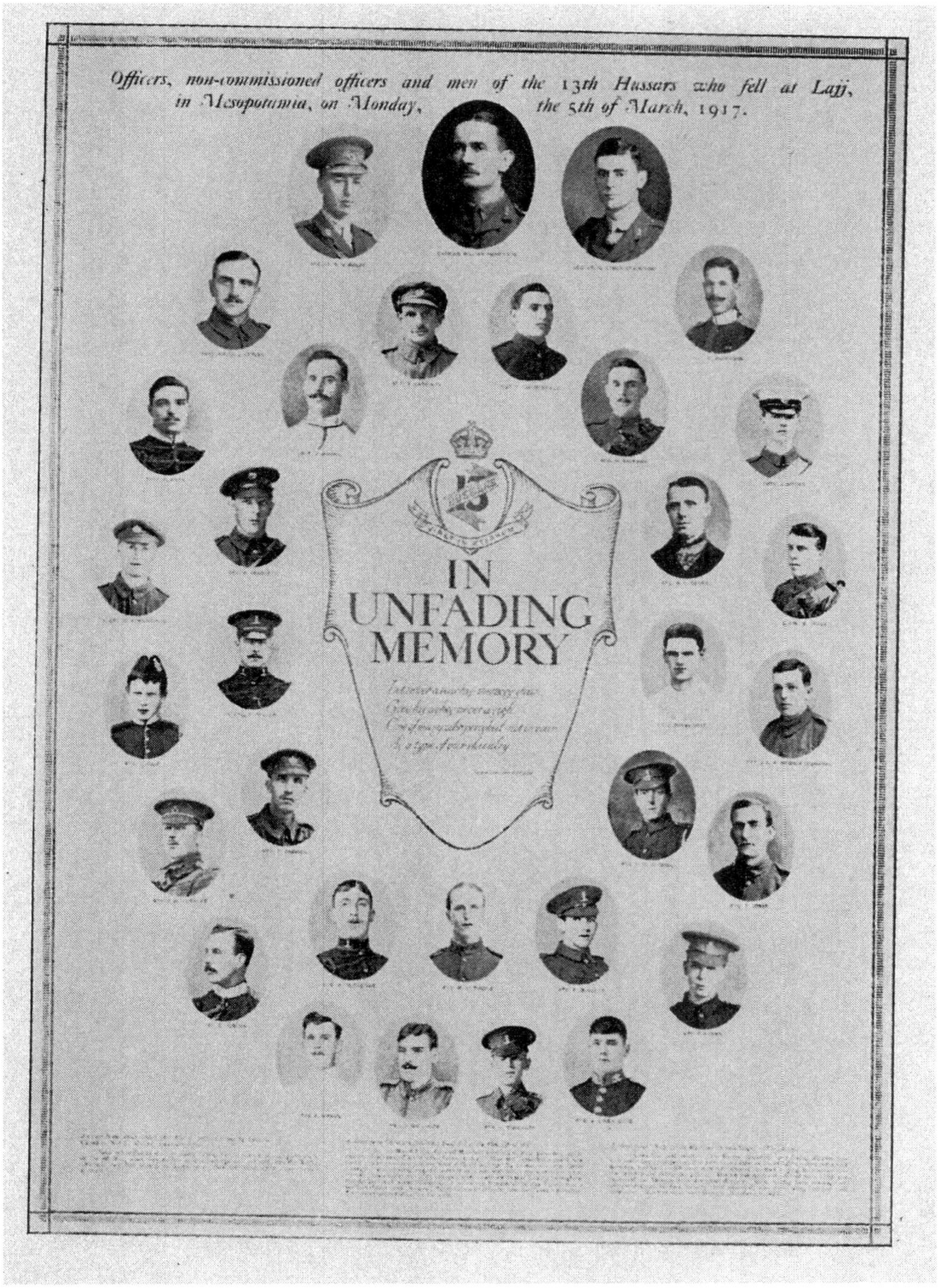
Officers, non-commissioned officers and men of the 13th Hussars who fell at Lajj, in Mesopotamia, on Monday, the 5th of March, 1917.
IN UNFADING MEMORY

General Maude's view of the action is given in a letter to Colonel Symons: "Your old Regiment did spendidly during the advance here, and particularly on the 5th inst., when they made a mounted charge and got well into the Turks, although they lost heavily in doing so, as they got hung up in some trench systems."

It may be noted that the loss in killed and wounded suffered by the Thirteenth at Lajj was greater than in the famous charge of the Light Brigade at Balaclava. The comparative nearness to England, the dramatic setting, and other circumstances, combined to make Balaclava a name of glory. The charge of the Light Brigade will always be for Englishmen one of the most splendid deeds in the history of the race. Lajj is, and is likely to remain for the nation in general, a name almost unknown. But to the Thirteenth it will always be one of the Regiment's great days, for the level sands by the Tigris saw that morning in the dust-storm a death ride just as brave and devoted as the one which has become immortal.

CHAPTER XV.

OCCUPATION OF BAGHDAD.

THE 5th of March, however glorious, had been a heavy day for the Thirteenth, and their ranks were sadly thinned.

But the Thirteenth formed only one unit of a pursuing army, and the Cavalry had to push on without delay to follow up the retreating Turks. At dawn on the 6th, therefore, the Regiment was again on the march.

One squadron, the one which had been commanded by Eve, remained for some hours with the transport in Lajj, to bury the dead and make further search for any wounded who might still be lying on the field. While they did so their Colonel, and all the wounded officers and men who had been brought in, were embarked and sent down the river to hospitals in rear of the army. The dead were buried together in a nullah on the field.

The rest of the Regiment, under command of Captain Steele, marched out weak in numbers, but still fit for further efforts, and proud of themselves. Their goal, Baghdad, now lay only thirty or forty miles ahead of them, and in common with the whole army they were eager to see the hard fighting of the past three months crowned by the capture of the famous city.

They had not long to wait, for only five days later the British flag was flying on the citadel of the Turks; but before that happened the brave enemy was to make one more stand, and take a further toll of British lives.

On the first day of the advance the Cavalry met with no serious opposition. They were tormented by another dust-storm and shelled by the retreating enemy, but they suffered little loss and made some prisoners. The line of march lay over the battlefield of

ONE OF THE ENTRANCES TO THE MOSQUE AT KHAZIMAIN

CTESIPHON

Ctesiphon. This position, says General Maude, though strongly intrenched, was found unoccupied. "There was evidence that the enemy had intended to hold it, but the rapidity of our advance

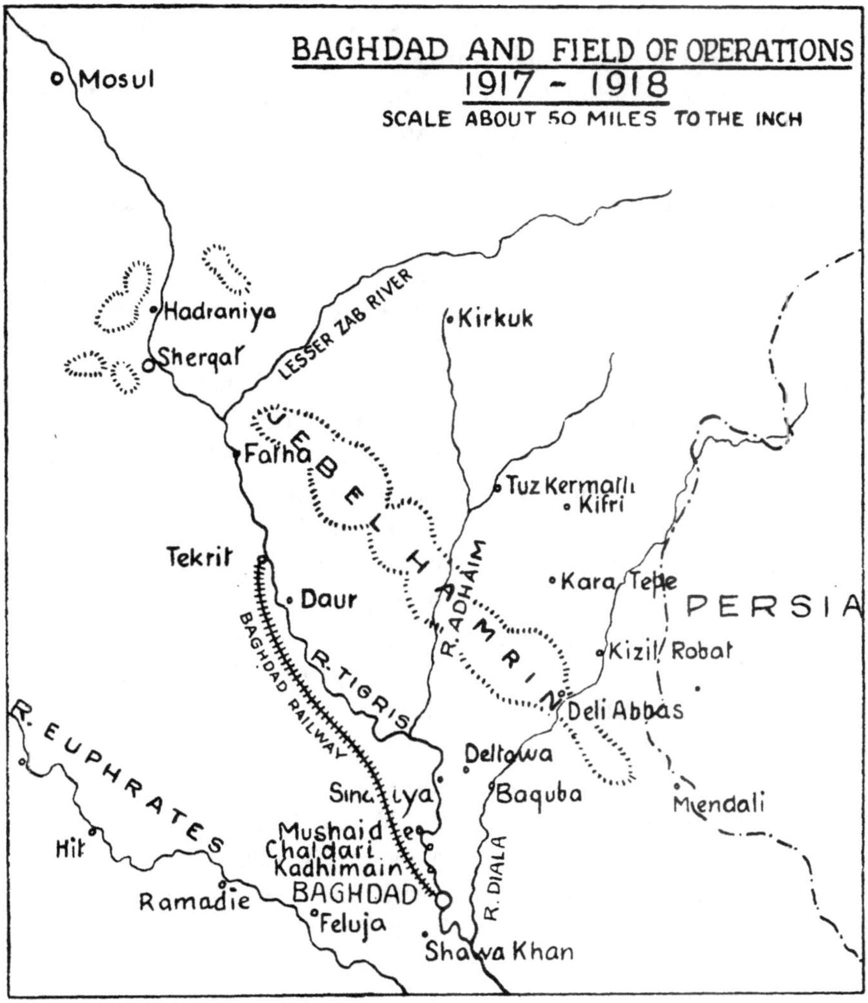

had evidently prevented him from doing so." The Cavalry believed that their swift and resolute attack at Lajj had accelerated the Turkish retreat and spared the army a possibly serious rearguard action there or at Ctesiphon, and it seems not unlikely that they were right. At all events Ctesiphon was not held, and the Cavalry bivouacked that night some miles beyond it, at Bawi.

But, however this may be, the next day showed that the Turks still meant to fight. About eight miles below Baghdad a considerable river, the Diala, runs into the Tigris from the north, and bars advance up the left bank, along which lies the road to Baghdad. The Turks had destroyed the bridge over the Diala, and as the river was at that time about 120 yards broad, with a strong current, it was a formidable obstacle. The enemy was not apparently in great strength, but he had more than once shown that he could conceal his troops with effect; and whether his Infantry was in strength or not, the farther bank was "defended by numerous guns and machine-guns, skilfully sited," while the ground in front of them was absolutely flat, with no cover. To force a passage was therefore no easy matter, and no attempt to do so was made that day.

The fighting that followed during the next three nights and days was desperate, and there is no incident in the whole campaign more creditable to the British Infantry than the repeated attempts to establish a footing on the right bank of the Diala. How attempt after attempt was foiled with heavy loss, the Turks destroying our pontoons and slaying the brave men who manned them, and how seventy of the Loyal North Lancashire got across in spite of all, and held their ground for twenty-two hours, and were at last relieved, has been told by others. It is a fine story. But not until the 10th of March was the Diala in British hands. The Turk had made a gallant stand in defence of the City of the Khalifs.

In this fighting the Cavalry had no direct share; but their work meanwhile had been hard and useful. When the advance along the left hand of the Tigris was held up by the Turks, it was decided to send a force across the Tigris with the view of turning the Diala position and getting at Baghdad from the south and west. Accordingly on the 8th of March a bridge was thrown across the Tigris, and the Cavalry, followed by an Infantry force, passed over to the right bank.

During that night the force marched in a north-westerly direction towards a place called Shawa Khan, which the Turks were reported to be holding. The march was much impeded by ravines and water-cuts, and was necessarily slow; but the Turks offered no serious opposition, and during the morning of the 9th Shawa Khan was

Capt. J. V. DAWSON
(Very severely wounded at Lajj, 5th March 1917)

Lieut. G. L. M. WELSTEAD
(Wounded at Lajj, 5th March 1917)

2nd Lieut. J. O. P. CLARKSON
(Killed near Jaffer's Tomb, 10th March 1917)

L.-Cpl. G. W. BOWIE, D.C.M.

The Rev. H. COOKE, C.F., M.C.

occupied, the enemy retiring to another position a mile or two farther back. This also was approached and attacked, but was still in Turkish hands at sunset.

The Cavalry during the earlier part of the day had been operating on the left flank of the force, away from the Tigris, but the horses being in distress for want of water, it was temporarily withdrawn to the river bank in the afternoon. In spite of some shelling and rifle-fire the Cavalry had suffered little loss, and the Thirteenth had till then had no casualties; but while watering their horses they were annoyed by some sniping from the opposite bank, and a promising young officer, 2nd Lieutenant Clarkson, was unfortunately killed — shot through the heart. He was the fifth officer of the Regiment who had been killed since the campaign began.

That night the enemy evacuated their position and the force pushed on.

"On the morning of the 10th," says General Maude, "our troops were again engaged with the Turkish rearguard within three miles of Baghdad, and our Cavalry patrols reached a point two miles west of Baghdad railway station, where they were checked by the enemy's fire. A gale and blinding dust-storm limited vision to a few yards, and under these conditions reconnaissance and co-ordination of movements became difficult. The dry wind and dust and the absence of water away from the river added greatly to the discomfort of the troops and animals. About midnight patrols reported the enemy to be retiring. The dust-storm was still raging, but following the Decanville Railway as a guide our troops occupied Baghdad railway station at 5.55 A.M., and it was ascertained that the enemy on the right bank had retired up-stream of Baghdad. Troops detailed in advance occupied the city, and the Cavalry moved on Kadhimain, some four miles west of Baghdad, where they secured some prisoners."

Meanwhile the force on the left bank of the Tigris, having forced the Diala, had also pushed on, and on this same morning, the 11th of March, they also entered Baghdad. The British flag was hoisted over the citadel; and the town, which was being looted and set on fire by Arabs and Kurds, was rapidly reduced to order. In the afternoon the gunboat flotilla with General Maude on board came up the river in line-ahead formation, and anchored off the British Residency. The capture of Baghdad was complete.

Thus ended, in triumphant success, General Maude's advance upon the capital of Mesopotamia. It had meant three months of severe and at times desperate fighting, during which the British force had lost heavily. But it had meant also the utter defeat of the Turks on their central front in Asia, and the restoration of British prestige in the East.

Indeed, it meant much more; and in writing this history of the Thirteenth it seems desirable to point out the full effect and significance of the victory in which they shared.

The recapture of Kut a fortnight earlier had already produced a striking effect. The 'Times History of the War' referred to this in strong terms:—

"Nor," it said, "was the effect confined to the Middle East. The Commander of the French armies telegraphed his warm and sincere congratulations on 'this splendid feat of arms,' and the defeat of the Turks made a great impression everywhere. Of its immediate practical result upon the course of the War in Asia there could be no question. Within a week the Turkish forces which had invaded Northern Persia were in full retreat for their own border, and the projected Turkish movements on the Euphrates were given up. In fact, the ambitious offensive of the enemy upon this central front in Asia had collapsed like a pricked bladder. The principle of a concentrated advance on the Tigris had already been justified."

Now Baghdad had followed Kut, and the immediate result of the second capture is thus described in the same 'History':—

"So fell Baghdad, the immediate base of Turkish warfare in Persia and Mesopotamia, and one of the most famous cities in all the East. If the recapture of Kut had produced a great effect, it need hardly be said that the fall of Baghdad made an impression vastly greater. In Germany it was described with unusual frankness as 'a deplorable event,' and on the Bosphorus the news of it was received with something like consternation; while among the Allies and all who sympathised with them it was hailed as a striking victory and an auspicious opening to the campaigns of 1917. Indeed, considering that the Asiatic theatre of war was

B SQUADRON ON THE WAY TO THE DIALA

CROSSING THE DIALA

WATERING IN THE DIALA

after all a secondary one, and that the capture of an Asiatic city could hardly have any material bearing on the issue of the European conflict, the weight attached to the British success seemed almost disproportionate. But it must be remembered that the conception of a great advance eastward by way of Turkey in Asia was a fundamental part of the German scheme of world-politics, and that the seizure by Great Britain of the eastern end of the Constantinople-Baghdad railway meant a heavy blow to this scheme."

These words were written in 1917, within a few months of General Maude's entry into Baghdad. The subsequent course of the war was to show that far from having over-estimated the value of that success, the writer might have said more. Now that the war has come to an end, there can be no doubt that the blow struck by the British in Mesopotamia, together with the almost concurrent victory of General Allenby in Palestine, had a material effect upon the whole issue of the war. By themselves they could not have brought about the complete triumph of the Allied cause. Nothing but the defeat of the Germans in the west of Europe could do that. But they meant, if duly followed up, the collapse of the Turkish military power, and the striking out of the Turkish Empire from the great confederacy. That meant the maiming of the German scheme of world-conquest. Without the aid of Turkey as a vassal, the Kaiser's dream of a vast Eastern dominion could never be fulfilled. And dominion in the East was no small part, some even think it was the more important part, of his scheme of world-conquest. This, however, is perhaps to say too much, and in any case, great as was the effect of the fall of Baghdad, much fighting had yet to be done, even in Mesopotamia, before the Turks finally recognised that they were not strong enough to stand alone against the British Empire. They had received a heavy blow, but they were not yet conquered.

To return to the Thirteenth Hussars and their share in the work of the Baghdad campaign, the following extracts from letters and diaries may be of interest.

The Regimental Diary shows that on the 6th March, the day after the fight at Lajj, the Thirteenth marched off at 6 A.M. with the Cavalry Division, and marched "without incident to Bawi (four miles north of Ctesiphon)," where they bivouacked. The next day the Cavalry "moved in direction of Diala river" and "came under shell-fire," but apparently suffered no casualties. "Retired and bivouacked near Bawi." On the 8th the Cavalry Division marched "with object of turning right flank of Turkish position." At 5 P.M. the Division was all across the Tigris and "march continued all throughout night." On the 9th, at 8.15, "shelled near Khair-el-Kharabeh," and at ten o'clock "took up position on left flank of Infantry." There the Regiment remained two hours, when they "withdrew and retired to water," on the banks of the Tigris at 1.50 P.M. "Sniping by enemy from opposite bank." In the afternoon the Cavalry returned to their former position and bivouacked. "Transport shelled during the day, but no damage done." On the 10th some of the Cavalry advanced towards Kadhimain, to the north-west of Baghdad; but there appears to have been no incident to report except "wind very high during the day." Finally, on the 11th, the Diary has the following entries:—

"4 A.M. Stood to.

"8 A.M. Marched with Division to Khair-el-Kharabeh, where the Regiment received orders to proceed to Baghdad.

"12.45. Entered Baghdad. Proceeded to bridge across Tigris, which had been destroyed by enemy prior to evacuation.[1]

"2 P.M. Returned to a palm bagh[2] on outskirts of town, where 'A' and 'C' Squadrons bivouacked. 'B' and 'D' Squadrons proceed to Kazimain."

"The Regiment was attached to the 35th Infantry Brigade for garrison duty."

In these curt and unemotional words the Diary describes the march from Lajj and the entry into Baghdad. It would not appear from this that the Thirteenth had any fighting during these six days, or was much excited by the capture of the city. But Regimental Diaries are not given to unnecessary eloquence. As a matter of fact we know that the Regiment lost an officer killed, and it believed that it had been specially selected, as a reward for its conduct at Lajj, to be the first Cavalry Regiment to enter Baghdad, and to form part of the garrison. The compliment was evidently appreciated.

[1] Baghdad lies on both sides of the river.

[2] Garden or grove.

THE GRAVES IN BAGHDAD CEMETERY

One young officer of the Regiment, 2nd Lieutenant Payne, speaks as if the work had been hard, and not free from fighting, even after Lajj. He had passed some years in Canada, and had perhaps learnt there to be more outspoken than the very reserved British officer generally is. "Since that date" (3rd March), he writes on the 13th, "we have been fighting day and night without any rest or sleep, till men and horses dropped with exhaustion, and had Baghdad not fallen the day before yesterday there would have been few of the Thirteenth left to tell the tale." His letter goes on to describe various incidents of the past week. The following are extracts:—

"Next morning, 4 A.M.,[1] the Division had to move on, and our squadron was left as escort to the transport, which didn't leave till the afternoon, so we took our men back to the awful battlefield to see if we could pick up any men alive, and to bury the dead. We buried sixteen men and three officers in the same grave, but got nothing in personal effects, as they had been stripped, some of them naked. All the time we were working we had to keep a Hotchkiss gun going to keep off the Arabs, who were coming in fast to pick up loot, of which there was tons in the way of arms and saddlery.[2] . . .

"That same night we passed through another strong point at Ctesiphon that they had not stopped to defend, and in the moonlight I rode over to see the great arch of Ctesiphon that is supposed to have been built in the time of Nebuchadnezzar or one of those birds. It has a great façade of about 200 feet high that opens into a courtyard, and alongside is this wonderful arched roof that looks as though it had been a banqueting-hall."

The writer is a little out in his dates, but the majority of his countrymen knew not much more about the matter than he did.

"The enemy had gone back to their last and strongest position above Diala, where there is a fork in the river. We weren't successful in smashing that place, so our Division and half the Infantry crossed the river with the intention of striking the railway north of the city, and we were out three days and nights without any rest,

[1] 6th March.

[2] The bodies of these men and officers were afterwards removed to Baghdad, and buried in the British Cemetery, under the superintendence of the Rev. F. H. Cooke, M.C.

trying to force a way round, and the farther we went the farther we had to come for water, as it was death for men and horses to be cut off from the river. In forcing our way down to water on the morning of the 12th[1] we lost an officer and man, being sniped from the other side of the river. We were all done up, no sleep, little food, and the horses dropping under us after three weeks' fierce fighting, and slowly making our way to carry on the attack on the flank of the Infantry, which always necessitated the enemy widening his front and weakening his strength, when news came that the Black Watch entered Baghdad at 5.30 A.M.

"It was too good to believe, but then orders were passed down, 'The Thirteenth Hussars will report at once to G.H.Q. at railway station for orders *re* garrisoning town.' There was new life in all of us at once; we hadn't had a sight of Baghdad and didn't think we were likely to for an age, but the next minute we were hoofing it at a trot along the highroad that had been used for hundreds of centuries, over trenches that the Turks had held the night before, and over dead bodies of both attackers and attacked. But we got into the city. One seething mass of Arabs greeted us; the same scum that the day before would have delivered up any of our wounded to the Turks now brazenly brought out wounded Turks to us that were not able to get away in the rush, dirty devils always ready to join in with the winning side, but always with an eye to scrupper the unfortunate of either if they can catch him unprotected.

"After standing by for an hour or so Jeffrey was told to take two squadrons and one squadron of machine-gunners to protect Kazimain, three miles up to (?) the river. So I found myself in command of 'D' Squadron, and here we are in the lap of luxury (Eastern). Kazimain is where all the pilgrims come to the great mosque, and is entirely composed of caravanserai sort of hotel quarters: there are 5000 of these buildings. We are quartered in the biggest, and even the men are able to have a wooden bedstead to lie on each. It is a big square building with a courtyard in the middle, in which we have picketed over 250 horses comfortably. Jeffrey and I and the Intelligence Officer have a large stone-floored room with four bedsteads in it; there are no windows, of course, but you let down great heavy wooden shutters when the sun is too hot. The Sheikh

[1] Afternoon of the 9th?

TURKISH BARRACKS

ARMY COMMANDER'S HOUSE

BRITISH, AMERICAN, AND FRENCH CONSULATES

BAGHDAD

(elder of the town) brought in beautiful carpets and bedding for us, and the women and children in hordes followed us in the street, all making a tinkly sort of yodelling song. They are in terror of the Turks coming in, and sacking the place, and taking their women. We weren't looking our best by any means: no change of clothes, unshaven, and unwashed for five days, the dust from the storms begrimed into our skins, and our clothes in rags. But to-day we are all happy and clean. The men, who haven't had a bath or change since January 24th,[1] can now be seen having their boots polished by Arab children, and strutting about like lords; they do so love to have something to order about. You would laugh to see them when we take prisoners: they make them hold their stirrup-irons while they mount their horses, and hand them their coats to put on—anything for the sake of exercising power. . . .

"However, it was all too good to be true—to go to sleep and not get up till daylight, and have your sleep out, was too good to last more than two days, and we have orders to turn the job over to an Infantry battalion, and rejoin the remainder of the Regiment in Baghdad, where we go into the Cavalry barracks, which no doubt will be just as nice: but the heavenly paradise of it all here, with the river alongside, and all the water you like, fresh good green stuff, and milk. This is buffalo milk, but quite good, and we get lettuces, onions, and mutton—in fact, we don't have to open a tin of any sort. Our bread is *chupatti*, the round flat things which the Jews make. . . ."

War certainly teaches a man to be content with the essential things of life.

"The Mosque here is most beautiful—that is, the outside, for no infidel is allowed inside the gate.[2] It is simply enormous, and faced with silver till it reaches the roof and four towers, which are gold.

"I went down the Bazaar yesterday afternoon, and bought a pair of saddle-bags for five rupees. They are made of the very finest carpet of true colours, bound with leather; it's a shame to use them on a horse, but my other ones are in rags from carrying so many tinned foods. . . . I have never carried anything in the way of loot about with me, nor let my men do so, because it is such

[1] ? 24th February, when the pursuit began.

[2] This from the officer of a conquering army.

an incentive for the Turk to put a bullet through you if you are wounded, and he finds them on you; but now the Mesopotamia War is over so far as we are concerned, and we shall live like human beings again under cover from the sun and cold nights, and be able to think of such things. The rest of the Cavalry Division are getting three days' rest before pursuing the enemy farther up. We, as far as we know, are to be in Baghdad for a while at least. . . .

"Late evening. I have just been taking a stroll round the town in case we don't see it again. The officers are allowed to go in pairs armed, but we haven't dared to let the men out yet; they have been in trouble already, climbing their way on the roof to the quarters of the Sheikh and pinching bedding; also there are intermittent shots fired now and again from mysterious spots, and there has to be a house-to-house search for any stray Turks in hiding. We went through endless streets with the houses almost touching across to each other. Every window is glassless, but securely barred, and great wooden shutters slide down to shut out the hot sun. Down by the river the Eastern picturesqueness is very novel,—to see the Arabs lifting water by appliances that have been used as long as the world was peopled; the coracles (round boats) spinning round and round as they in time get across the stream: they are allowed to spin, as it lessens the resistance to the current. To get alongside the water and amongst the palms with their green crop underneath is wonderful after weeks of the desert. I don't know of anything more depressing and hopeless than to—as we often and often had to—be going all day in the broiling sun and dust and wind, and find it hopeless to try and get to the river, and must camp where we are, the horses unwatered, and every one finished their water-bottles, and knowing that you've got to be on the move at 4 A.M. again, with the expectation of meeting the enemy."

Surely the soldier on service earns his pay.

March 14.—"We turned out of our delightful quarters at Kazimain this morning, and have now taken over the Cavalry barracks in Baghdad—a horrible, flyey, and hot place that is nothing more than bare mud walls and filth, since the outside Arabs have stripped it of everything. These brutes got in and looted the

A TYPICAL WATER LIFT

Bazaar clean; the civil population always expect that to happen on such occasions, and just sit calmly by and watch them wreck their shops. . . . I am writing this in agony from the flies. There are five other chaps in this bare stone-floored room with me who have their heads buried under their blankets, having given up any attempt at reading or writing. . . .

"We found absolutely nothing in the town. The Bazaar, as the multitude of dusty and smelly arcades are called, are rows and rows of little cubby holes with the meanest supplies of merchandise, and as they were all ransacked when the Turks left, we could buy nothing at all. The whole town is most disappointing, even were it not half in ruins: everywhere is dilapidation and dirt. . . .

"Gowan, whose father was a member for Vancouver, is in charge of the armoured cars we have attached to the Cavalry; he is an extraordinary daring chap, and when in France with the 7th Battalion had the back of his skull blown away—which he carries in his pocket—had a bayonet through his stomach, and was discharged from the Canadian Army as no further use—is now the admiration of every one for his daring exploits!"

Lance-Corporal Bowie's diary, from which quotations have already been made in earlier chapters, tells the story of the advance from Lajj to Baghdad in very concise terms.

"On the following day, the 6th,[1] we carried the pursuit of the Turks past Ctesiphon. A very bad sand-storm raged during this day, making our progress very difficult. We bivouacked that night near Bawi, and the next day was spent in a well-earned rest, while a pontoon bridge was being thrown across the Tigris, which, during the day, Turkish airmen tried very hard to destroy by bombing, but only succeeded in hitting one of our ammunition waggons, killing the entire team, drivers, and everything within a radius of 300 yards. The following day we crossed the Tigris without opposition, whilst the Infantry fought their way across the Diala river some miles farther up. We at once made a long night-march to get at the enemy's left flank, which we attacked, forcing him back a distance of some two miles, bivouacking that night on the ground from which we had driven the enemy. During the same [day?] one of our officers was killed while we were watering our

[1] March.

horses in the face of the enemy's fire. The following morning we started to blow up a railway bridge, but had to abandon this enterprise owing to a terrific sand-storm which overtook us. After a few more hours' rest we proceeded on our way to Baghdad, where we encountered no opposition, our Infantry having driven out the last of the enemy's rearguard during the earlier part of the morning, the Thirteenth Hussars being the first of the Cavalry to enter the city, which we did about 11 A.M. on the 11th March. The following morning the pursuing Cavalry took Khasmain, whilst the Thirteenth Hussars remained bivouacked in a large orchard. Having been selected to garrison the city, we crossed the Tigris on the morning of the 14th, and took up our quarters in the old Turkish Cavalry barracks, which were situated just outside the citadel. These we found to be in a filthy condition, and infested with vermin. Happily our stay here was very brief. . . ."

Captain Newton's diary is also very concise, but as an example of a day's work during the advance, the following entry of the 10th March may be quoted. After noting on the 9th "horses and men done," the diary goes on:—

Saturday, 10th.—"Saddled up at 5.30. Brigade off to water Jaffer's Tomb. Left with 'A' and 'B' (right wing) to escort Divisional troops. Two troops sent off on search for some missing R.E.'s and wireless. Sent with remainder to escort 6th Brigade second line down river to refill. Awful dust. Found dead R.E.'s in pontoon, and punished Arabs. Went back to where we had left Division leaving 4.30. Found Division 9.15. Awful trek in dust. Then moved to Hilawiyni in gale and dust-storm. Reached bivouac 12.30. To bed 2.30. Orders to be saddled up by 6."

Sunday, 11th.—"Réveillé, 4. Gale still blowing. . . ."

In a letter written from Baghdad the same writer gives a summary of the whole march from Lajj.

"We had a longish trek next day (the 6th), but saw nothing of the enemy, and we bivouacked about 6, but had no transport, so not much food. Transport came in about midnight, but I'd got my blankets with me and was fast asleep by 9 o'clock. We marched at 6 A.M., a bitter cold morning, but our mess-boxes had turned up with the transport, so we started with our tummies good and full.

TURKISH CAVALRY BARRACKS OUTSIDE BAGHDAD USED AS A HOSPITAL

VIEW OF BAGHDAD FROM ROOF

"We spent an idle day, sitting about a good deal while reconnaissances went out; but it was a long day, and we didn't get in till 9.30 in the evening, to find a grand dinner waiting for us. It was midnight again before I got to bed, as there were a lot of things to do. Next morning we were up at 4.30 and ready to move at 6, but then heard we shouldn't move till about 1, and then cross to the other side of the river again. We lunched at 11.30, but didn't move off till 2.30, and it was about 6 before we were across the bridge. We trekked about all next day, and in the evening heard we were to go on, and we did, but came up against the Turk positions and couldn't get on, so came back to bivouac, thank goodness, as men and horses were nearly done in. It was while we were watering in the afternoon that poor young Clarkson was shot dead by a sniper, the only casualty we had that day, and most awfully hard luck."

Such is too often the lot of the soldier on service—long, dreary, uneventful days with no excitement, nothing but discomfort and fatigue; and then, suddenly, the call to show what his training has done for him, and perhaps to give up his life for his country. That is what sets apart and ennobles the profession of the soldier and sailor—the constant readiness to face death. Women with their quick perceptions understand and are grateful, nor do they alone understand. In spite of all prejudices and jealousies, men too know in their hearts that the first honour is due to those among them whose choice and pride it is to guard the nation with their lives.

"We were up at 4 A.M. and ready to move at 5.30. We didn't move till about 8, and then I was sent with two squadrons to escort transport to refilling-point. It was not a particularly hard day's work, but it was another long day. It was 9 P.M. before I found the Division again with my transport, and then we trekked on in an awful dust-storm, and eventually reached our bivouac about 12.30. I got to bed at 2.30, was called at 4.30, and we trekked at 8. We crawled along, every one, men and horses, dead to the world, did a bit of Arab strafing, and at 10.30 received the news that the Infantry were going into Baghdad, and that the Regiment was to march in to form part of the garrison. My word, we were proud and pleased, and every one just brightened up, and we were as cheery as anything. We entered the town on the side of the river, but couldn't cross till a bridge had been built. However, 'B' and 'D' were sent off to occupy Khazimain, a town about six miles away and a great religious centre, and 'A' and 'C' were told to camp in a delightful garden, all green and shady with date-palms and orange-trees. It was just like coming into fairyland after the time we have been in the desert, and we felt so peaceful and happy. Directly after we had fed I went to bed, and slept and slept and slept till 7.30! It was grand."

Private Massey, of 'D' Squadron, has also left an account of the march from Lajj from the point of view of the trooper. He describes how the Regiment marched nearly to the Diala river, and how on the 8th March the Cavalry crossed to the right bank of the Tigris.

"It was now beginning to get dusk. All that night, and during the early hours of the 9th, we continued marching, going in a north-westerly direction. We crossed many big nullahs, which we found great difficulty in crossing. When daylight broke we found ourselves within a few miles of Baghdad, and according to rumour we were to attempt to cut the railway above Baghdad, to prevent the Turks getting their rolling-stock away. We were heavily shelled during the day, but our Regiment escaped without casualties. We proceeded to water late in the afternoon, and it took us a long time to reach it as the river was a long way off. We were sniped at on our way, and on reaching the river we had to go down to get water in buckets under the fire of a sniper on the left bank of the river. Lieutenant Clarkson, of the M.G. Brigade, brought a machine-gun into position on the bank, and stood up trying to find the sniper with his glasses. He was instantly shot through the heart and fell down dead.

"When we had finished watering, we moved into a big garden with a large wind-wall round it. Here we had something to eat, a piece of biscuit and a bit of bully. The horses were very badly done up, but we soon moved on again. Where we camped that night is more than I can say, as for the next few days I lost all sense of our position, as we seemed to be all over the show, here, there, and everywhere. Anyway, we camped somewhere in the desert that night, close up to the firing line.

THE REGIMENT ENTERING INTO OCCUPATION OF THE TURKISH CAVALRY BARRACKS AT BAGHDAD. 13TH MARCH 1917

"The transport was shelled coming in, and suffered several casualties amongst native drivers. We had six hours in bed."

March 10.—"Réveillé on the 10th was at 4 A.M. We fed the horses and had breakfast, and afterwards saddled up and went to water. Two troops were put to guard a signalling-post all day which was on the top of a high mound, close to where the Turks had had trenches the previous day, and from where we had been fired on. It was pitch dark when we moved off, very late at night, and the worst sand-storm in which it has ever been my luck to be in came on. It was a terrible experience, and I never want to experience such a storm again. How we reached camp, or who led us in, I couldn't say, but it ceased when we reached camp. Got in after midnight, with nothing to eat and drink, and réveillé ordered at 3.30 A.M. next morning. Transport late at arriving in. Lay down in our clothes and one blanket for a few hours, and slept like the war-worn soldiers we undoubtedly were."

March 11.—"On the 11th we started off again early in the direction of Baghdad, and met several parties of Arabs who were fleeing from the city on entry of the British that morning. . . . After proceeding a little farther, word was passed down the line that we were going to be the first Cavalry Regiment to enter Baghdad. As we neared the entrance other regiments stood aside, and we passed on and at last met a battalion of the Black Watch, who had been the first Regiment to enter Baghdad. After proceeding a bit farther, we came to Baghdad station, and here we halted for half an hour. We then mounted again, and proceeded round the end of the railway. . . . We were now entering the city, but with the exception of a few snipers all was quiet. White flags were everywhere, and after riding through several streets we called a halt close to some Cavalry barracks, which we entered soon after. Only two squadrons went in, however, as there was not enough room for the Regiment. The barrack rooms above the stables were roomy and airy, but filthy. . . . On the 14th the Regiment moved into fresh barracks on the left bank of the river."

There for the present we may leave the Thirteenth to get a little rest, and to regain as far as possible the cleanness and smartness which had suffered during the rough work of the march. Their goal was reached, and they were part of the force which had occupied the enemy's famous citadel.

CHAPTER XVI.

OPERATIONS IMMEDIATELY AFTER CAPTURE OF BAGHDAD.

BAGHDAD had been captured, and a heavy blow struck at the Turks—a heavier blow perhaps than any one understood at the time; but much remained to be done before the British Commander in Mesopotamia could give his troops the full repose they needed.

In the first place, the Turkish force which he had beaten at Kut and driven through Baghdad must be pursued in its retirement northwards until it was destroyed or hunted away out of reach. General Maude was too good a soldier not to know that the annihilation of the enemy's armed force, rather than the capture of any city, was the real object to be kept before him. And the necessity of a rapid pursuit to the north, up the line of the Tigris, was all the more urgent because of the near approach of the flood season. Control of the river "bunds," or dams, and irrigation works up-stream of the city must be secured, or it would be possible for the Turks to cause extensive inundations, not only crippling the movement of the British army, but doing serious mischief in other ways.

Then it was important for the security of the military position that the surrounding country on all sides, not only to the north, should be cleared of Turkish troops for a considerable distance, so as to allow of the pacification of the neighbouring Arabs, and the free collection of supplies. To the westward there were Turkish troops on the Upper Euphrates, which threatened mischief down the course of that river on the British flank and rear. To the eastward, on the Diala, it was believed that there were also some Turkish troops, dominating a tract rich in crops and forage.

BAGHDAD—CLOCK TOWER IN TURKISH INFANTRY BARRACKS SQUARE

A STREET IN BAGHDAD

FRAMEWORK OF HANGAR IN BAGHDAD

Moreover, General Maude had to keep in mind the great strategical considerations involved — the exertion of continued pressure on the Turkish base in Asia Minor, and the possibility of co-operating not only with the Russians on the right, who were driving or following the Turkish invading force out of Persia, but in a measure with General Allenby in Palestine. The Russians, though weak, were comparatively near, and it might be feasible actually to join hands with them. Palestine was far away, out of reach, behind the great Arabian desert; but the harder the Turks were pressed on their central front, the less strength they would be able to spare for their right.

General Maude's plan of operations therefore, after the capture of Baghdad, was to follow up that blow without delay by pushing out three strong columns. One was to strike at the Euphrates westward, and occupy Feluja on that river, thirty-five miles from Baghdad; another was to press the pursuit to the northward, up the line of the Tigris and of the Baghdad Railway; the third was to move out north-east towards the Persian frontier, so as to secure the rich Diala district, and if possible, with the help of the Russians, to crush the Turkish force retreating from Persia.

During the short campaign which followed, all these objects were in a measure attained. By the 19th March the western column had established itself on the Euphrates, and thus commanded the whole course of the river down to the sea, clearing Maude's left and rear from Turkish pressure. The northern column had a hard fight at Mushaidi, twenty miles up the railway line, where the retreating Tigris force turned to bay, with the result that it was, on the 14th and 15th March, driven from its intrenchments and completely beaten, streaming away to the north in confusion. The eastern column at the same time surprised and occupied Baquba, a point on the Diala thirty miles from Baghdad.

It was with this eastern column that the 7th Cavalry Brigade was connected; and indeed during the remainder of the war, which was to last for more than a year and a half yet, the Brigade was chiefly employed in the country to the east of the Tigris.

The Thirteenth Hussars were left at the close of the last chapter in occupation of the Turkish Cavalry barracks in Baghdad, resting after their severe exertions. But their rest was short, for on the

23rd of March, by which time they were becoming tired of the heat and the dirt and the flies, they were warned for active service again, and on the following day they rejoined the Brigade at Baquba. The Regiment was now in good condition, men and horses restored and fit for service, and the prospect of further work in the open was hailed with pleasure.

General Maude's despatches of 1917 show what the work was. The Turks, heavily defeated as they had been, were by no means inclined as yet to give up the game as lost, and the British forces east of the Tigris soon found that the enemy was still capable not only of resistance, but of determined attempts at offensive action. His troops consisted of two main sections, the 13th Corps, which was retiring from Persia, and the 18th Corps lately beaten on the Tigris, but rallied and reinforced in a few days from the Turkish reserves in the North. These two Corps began to act in co-operation, and to threaten attacks upon the British eastern and northern columns. It was an excellent example of the tenacity of the Turks, and of their quick recovery after defeat.

"About the 26th of March," writes General Maude, "there were indications that a converging movement was being made by part of the 13th Corps down the right bank of the Diala from Deli Abbas, and by the 18th Corps along the left bank of the Tigris . . . apparently with the intention of assisting the withdrawal of the 13th Corps from before the Russians." These advances were resisted by the British northern and eastern columns, the former driving back the 18th Corps northwards, and the latter dealing with the 13th Corps. "On the 27th the enemy made a determined attempt to move down the right bank of the Diala . . . but our Cavalry, skilfully handled in some difficult ground, resisted the enemy's advance from successive positions, inflicting severe losses, and finally checked the forward movement that evening. Next day the enemy fell back towards Deli Abbas, followed up by our Cavalry."

On the 2nd of April the British and Russians joined hands. From that time until the end of the month there was some stiff fighting all over the country, and the troops suffered severely from the heat, the constant dust-storms, and the difficulty at times of getting water. The work of the Cavalry was incessant and very

INSIDE TURKISH CAVALRY BARRACKS, BAGHDAD

arduous, now in trying to get round into the enemy's line of retreat, now in falling back before him and luring him on into a position where the Infantry could close on him, and always in covering the front and flanks of the columns. There was no conflict of the nature of that at Lajj, no mounted charge into the enemy's troops, but there was steady hard work, of great value to the army; and of this the Thirteenth had to do its full share. General Maude, who refers repeatedly to the Cavalry in his despatches, sums up in the following words the effects of the operations in this part of the country:—

"As a result of the fighting during the month of April the enemy's 13th and 18th Corps had been driven back on divergent lines. . . . The 13th Corps had twice taken the offensive, with results disastrous to itself, and the 18th Corps had been defeated and driven from its selected positions on four occasions. Our total captures for the month amounted to some 3000 prisoners and 17 guns, besides a considerable quantity of rolling-stock and booty of all kinds. The objectives which we had set out to reach had been secured, and the spirit of the enemy's troops had been broken."

Coming so soon after the winter advance and the capture of Baghdad, this was a good piece of work, and creditable to all the soldiery concerned, whose spirit never faltered. Indeed, according to their General, "as conditions became more trying, the spirit of the troops seemed to rise," and to the end their discipline, gallantry in action, and endurance were as conspicuous as ever. But their exertions had of course imposed a heavy strain upon them, and now that the enemy was for the time powerless for further trouble, General Maude determined to give them the rest they needed.

"The increasing heat," he writes, "now rendered it necessary that the troops should be redistributed for the hot weather, and that every provision possible under existing conditions should be made with a view to guarding against the trying period which was rapidly approaching. Whilst it was necessary to hold the positions which had been so bravely won, and to strengthen them defensively, the bulk of the troops were withdrawn into reserve and distributed in suitable camps along the river banks, where they could obtain the benefit of such breezes as were available, and where a liberal supply of water for drinking, bathing, and washing was obtainable."

The Thirteenth, among other regiments, were allotted one of these standing camps, a shady grove of palm-trees on the eastern bank of the Tigris at Chaldari, about nine miles above Baghdad, and very thankful men and officers were to settle down in it.

Nevertheless, it may be observed, the Regiment did not begin to enjoy its comparative peace and comfort until nearer the end of May than the end of April. They marched into Chaldari on the 14th of May, but on the 16th they marched out again for another week's hard work under the blazing sun. Though the two Turkish Army Corps had ceased fighting, the Arab tribes on the upper reaches of the rivers above Baghdad had been giving trouble, and before settling down for the summer it was found necessary to punish them for some of their misdeeds. Columns were accordingly sent against them, and to one of these, which operated on the Tigris, the Thirteenth were attached. Some tribes were duly brought to book, and the troops did not get back to camp until the 24th May, when the heat had become tremendous. But the tribesmen had learnt that British troops could move against them whatever the temperature might be, and the lesson was a very salutary one.

The following extracts give the story of these two months, March 24th to May 24th, from a more personal point of view.

Lance-Corporal Bowie—Diary.—Lance-Corporal Bowie's remarks about the Cavalry barracks in Baghdad have been quoted in the preceding chapter. His diary goes on :—

"Happily our stay here was very brief, every one being very thankful to leave these so-called barracks, which we did on the morning of the 23rd, rejoining our Brigade. The next morning we recommenced our pursuit of the enemy, who were located by Captain Robinson (with 'C' and 'D' Squadrons as a patrol) on the banks of the Diala river the following evening. We attacked them at dawn the following morning, and for the next four days our two Brigades were hotly engaged with the enemy, who were finally driven back on to Deli Abbas at the foot of the Jebel Hamrin Range. We bivouacked some ten miles back, and there formed a line of outposts. Meanwhile the enemy had apparently retired right up into the hills. On April 8th, our Brigade moved out from behind the huge mound where we had been under cover and

G.H.Q.

HOSPITAL SHIP

BAGHDAD

marched down to the river Diala, which was some ten miles distant, to refill our water-carts and let the men do their washing and have a much-needed dip in the river while the horses were grazing. After a pleasant day here, during which our patrols saw no signs of the enemy, we went back to the mound. On the following morning, the 9th, the Regiment left the Brigade to make a reconnaissance of Deli Abbas. We had not travelled very far before we encountered no less than seven columns of the enemy, who had marched down from the hills during the night. This force proved to be the 13th Army Corps, some 28,000 strong. The Regiment immediately took up a position with all Hotchkiss guns, on a long low mound situated a few miles S.W. of Deli Abbas, and from here we were able to inflict heavy losses on the enemy, owing to their eagerness to surround and annihilate what they took to be an isolated Regiment. In this action Lieutenant Fitz-Gibbon, although wounded himself, succeeded in getting all our Hotchkiss guns safely out of action, when we were in danger of envelopment, and were forced to retire amidst a perfect hail of lead and shell. Meanwhile the remainder of our Brigade and the 6th Brigade had deployed for action. Now commenced what proved to be one of the hardest fights in which the Cavalry had yet been engaged, the enemy trying their hardest to effect an enveloping movement by forced marches. This object we were only able to defeat by a most stubborn 'rearguard action,' which we had to maintain until the arrival of our main force. Our two Brigades engaged them all day, and by nightfall our outposts were on the Serri, some eight miles from the mound where we had first come into contact with the enemy. Early the following morning they renewed the attack with increased violence, and slowly forced us back the whole day to the Tomb of   , some few miles north of Deltawa, where they succeeded in getting well round our left flank, and were slowly enveloping us, when we received the most welcome news that our Infantry had arrived, and had succeeded by forced marches in getting well round the enemy's right flank, which considerably relieved the strain from us. Our outposts on this night, the 11th, were about one mile south of the Tomb. Early the following morning we, co-operating with our Infantry, made a very strong attack and succeeded in driving the enemy from their position at Chaliyeh, where they left 315 dead on the field after a very fierce engagement. Renewing the attack at dawn the following morning, we forced the enemy back again to a small village called Serai Lik, where they again left the field full of their dead and wounded: some 200 dead and 900 wounded were counted. The following morning we again attacked the enemy, whom we completely routed, and after a very short stand they speedily retired on to Deli Abbas, a village situated right at the foot of the Jabel Hamrin range of hills. For two more days we continued to harass the enemy's flanks until their main force had retired right up into the hills again. Our Infantry now formed a line of outposts some few miles from Deli Abbas, whilst on the 17th inst. the Cavalry marched back to the village of Deltawa, where we went into bivouac. Here it was decided to form the Regiment into two composite squadrons, owing to the great number of casualties we had sustained amongst both men and horses through wounds, sickness, &c. This was done on the morning of the 19th inst., and the remaining men and horses were sent into bivouac at a small village called Sindiyeh, some few miles farther back. In this district the Arab rifle thieves were very active, and on several occasions succeeded in getting clear away with rifles, in spite of the utmost vigilance, during the time we had a bivouac near that village. During the next fortnight the Regiment remained in bivouac at Deltawa, at the same time making wide reconnaissances between the line held by our Infantry, who had 'dug in' near the Serri, and the foot of the Jebel Hamrin Range. During these we occasionally came into contact with enemy patrols, with whom we had several sharp skirmishes. It was during one of these that Private Keany of 'C' Squadron was captured while carrying despatches.

"On May 4th we received orders to march down to Chaldari and there pitch our summer camp. We accordingly marched to Sa Salekh and from here to Chaldari, where we proceeded to erect our summer camp on the left bank of the Tigris some nine miles north of Baghdad. On the camp being completed, we went into it on the 13th inst., and on the same day the following awards appeared in our Regimental Orders for gallantry in our action of 9th April: Lieutenant Fitz-Gibbon the M.C. and Private Roberts the D.C.M.

BAGHDAD—STORKS

BAGHDAD—THE NORTH GATE FROM INSIDE

THE TRAMWAY—BAGHDAD TO KHAZIMAIN

Private Roberts died from the wounds he received the day following the action.[1]

"Three days after this, the 16th inst., the Regiment received sudden orders to turn out at 4.30 P.M. on an Arab strafe. It was made known to us later that these Arabs had literally cut to pieces the whole of a British survey party, consisting of three British officers and their orderlies, and some sixteen native soldiers who were their escort.

"We marched at 4.30 and crossed the river at Baghdad. We bivouacked that night some ten miles east of it. Marching again at daybreak, we encountered a terrific sand-storm which made the going very difficult, nevertheless we succeeded in covering some thirty-nine miles over the waterless desert before bivouacking for the night.

"Continuing again early the following morning, we trekked another eighteen miles or so in the direction of Mushada, when a halt was made. Reconnoitring from here, we located the Arabs some distance from the village. We rested until dawn the 19th inst., when we attacked them and inflicted heavy casualties amongst them, at the same time driving them before us on to General Cobbe's column, which had marched down from Samarrah. This column exacted a terrible revenge for those murders, and literally mowed them down with machine-gun fire. We inflicted some 3000 casualties in all amongst this huge band of Bedouin Arabs. This incident had a wonderful effect on the Arabs throughout the country, and no doubt prevented them from committing many similar outrages.

"On the conclusion of this we marched back to our camp at Chaldari."

Private C. T. Massey, 'D' Squadron—March 24.—"On the 24th we turned out to graze, and while the horses were grazing Private Cox came up with the order to 'turn in and stand to.' We moved off in a north-easterly direction about midday, after buying oranges and limes from Arabs on the square. We rejoined the portion of the Regiment that moved off earlier in the day, at a camp next morning, the 25th.

"We were now on the right bank of the river Diala, and I kept

[1] The Regimental Diary of 14th May says: "Moved into standing camp at Chaldari." . . . "The tents were on the edge of a palm grove, but the horses were picketed down in the open."

no notes of daily events, but we had many severe engagements with the enemy. It was on Easter Sunday or Monday that we met the Turkish army coming out of the Jebel Hamrin Range. We were then at Deli Abbas. It was the 1st and 3rd troops of 'D' Squadron which bore the first shock of the Turkish attack, but we were greatly outnumbered, and retired only in the nick of time under a hail of bullets. A few horses came down; one man was killed and two wounded. My riding-horse got away, so I jumped on top of the pack-horse, but the gun-case had not been properly strapped up and the Hotchkiss gun fell out; but No. 5 on the gun, a man named Cartwright, who was riding behind, dismounted and picked it up, and I halted and it was put in the case again. We continued to retreat for several days, but it was part of a clever scheme; the Infantry meant to get in between them and the hills from which they came. The Infantry, however, came up too close, and the Turks retired again, before our Infantry could get behind them, but they suffered heavy losses in the retreat back to the hills. It was during these twenty-five days on the Diala that our horses suffered more than in the whole fighting up from Kut. We could only water once a day, and many a time I have felt faint from thirst and chewed grass for moisture, as well as wild oats and barley."

A few days later the writer was sent on leave to India, and his diary comes to an end.

Letter from 2nd Lieutenant F. Norman Payne—March 25, 1917.—"After the Regiment getting nicely settled down to their garrison work they were suddenly ordered away, and are now seventy miles beyond Baghdad.

"Poor chaps may be out weeks and weeks, and already have been five weeks without any kit, it all still being at Bassouia Camp.[1] The boats coming up river are fully laden and won't stop to pick up anything like that."

From Private Hugh H. Mortimer to his Mother, April 3, 1917.

ON THE TREK, 3. 4. 17.

"MY DEAREST MOTHER,—No doubt you all thought my last letter a bit of a hash, which it verily was, but if you only knew the conditions under which I wrote it you'd think I was lucky to

[1] The old camp on the Hai stream. The kit was not received till the middle of May.

DEPARTURE FROM BAGHDAD

DEAD TURKS

CAPTURED ARABS AWAITING EXECUTION

THE DIALA AND KHALIS CANAL CAMPAIGN

get it off as it was. I wrote part of it in Baghdad, as I was in the City Military Police for a week, and could not manage to post it then, so I had to take it with me when the order came to get on the move again. Do you think they could do without the Thirteenth Hussars here when there's any dirty work to be done? Not likely. We got the job as Cavalry garrison in the capital because we'd distinguished ourselves on the drive-up, and incidentally lost more than half, but when reinforcements came up and not many of them, out we had to go again: the Cavalry Division could not manage without us.

"We've been living mainly on fresh air, biscuits, dates, and water, with an occasional bully or fresh-meat stew when we can buy or pinch a sheep from Mr Arab. Lor' knows where we are now, but we can't be far from the Russians; in fact, I believe one column of ours is in communication with them. Our aeroplanes have been over to 'em several times.

"Well, that's that. We have had some very exciting little times since we left B. [Baghdad?], and I thought it was all up when ten of us, all that's left of the 4th troop, 'C' squadron, had to go out twelve miles in advance on reconnaissance, as we got cut off by their Cavalry twice, and had to gallop miles for life and ford a canal, known as Kelly's Canal,[1] about eight feet deep. Still we got back none much the worse.

"Have you ever experienced the thrill one gets when something happens to lift one out of the blackest depths of depression, &c.? Well, it was like this, I hadn't got a fag or shred of baccy, and hadn't had a smoke for days; was browned off to the eyebrows on bully and biscuit diet with occasional spoonful of jam thrown in here and there—more then than now—and was trying to snatch forty winks—we were having a rest day—under my saddle with flies and mosquitoes buzzing around, sweating like a bull with the heat of the aft'noon even with one thin shirt on, when a bloke kicked my feet and shouted, 'Cheero, Morty, mail's up.'

"Please keep the newspaper cuttings of all our doings out here. Now don't forget that, as I shall love to read 'em over afterwards, and we don't get much news of our own doings out here.

"I think I shall manage to wangle the ten-shilling note as soon

[1] Khalis Canal?

as we get settled somewhere, sometime. Please thank —— and —— and all other well-wishers and contributors for me separately and individually.

"Yes, I received a very nice little parcel from St Silas's, but have not had any time at all to acknowledge it, as you can guess by your mail, which always takes precedence with me. . . . I expect we shall be slackening off out here shortly as 'tis getting very hot again, especially from about 10 A.M. to 4 P.M., when every effort is an effort and one sweats rivers, lying down—in fact, it's really worse lying down than knocking slowly, one doesn't seem to notice the heat so much then. We have to make bivouacs with our blankets thrown across swords and twigs stuck in the ground to get a little shelter from the sun, as there is next to no shadow here, and the sun is nearly overhead again. Roll on, we've just about got J. Turk beat to the wide now, so it's about time we had a long rest. Then we are made to clean and polish everything up, and it's some job that, as all steelwork is eaten in with rust, and everything more or less the same since we came on service.

"Well, my candle has just given me the wink and I'm finishing this by moonlight, so cheerio to all of you. Hope you are all in the pink. . . . Take great care of yourself, mother, love, and all the others as well. Am fit as a fiddle myself when I can get enough to eat."

Private Mortimer was decorated soon afterwards with the Military Medal for bravery in action, and after reading his letter one is not surprised to hear it.

The Regiment, it may be observed, though weak in men, and still weaker in horses, was now strong in officers again. Colonel Richardson and others wounded at Lajj had returned, and though one or two more had since been hit, they were doing well. But one, unhappily, had lost his life in a very sad way: Captain Newton, who had commanded a squadron at Lajj, and had afterwards gone through the April fighting, was drowned in the Tigris on the night of the 25th.

He stepped overboard in the darkness from a river steamer, and though life-belts were thrown out at once he was swept away by the current, which was very strong. A boat was lowered, but could find no trace of him. Captain Newton's death was much deplored. An officer on the staff of the Division, in a private letter to his own people, writes of him: "Poor Tiger Newton is dead. He walked overboard and fell into the Tigris on his way to India on leave. . . . He is a great loss to the Thirteenth Hussars, where he was immensely popular. He was a splendid character, was greatly esteemed by all who came in contact with him, and exercised great influence in his Regiment." His Colonel wrote: "Quite apart from my personal sorrow, I know how grieved the whole Regiment will be, for both officers and men were very fond of him." And General Jones, who had commanded the Brigade at Lajj, wrote of him as follows: "At the fighting at Lajj, where the Thirteenth Hussars had such heavy losses, he was the Senior Officer left of the squadrons in the most advanced line, and was in constant communication with me by telephone after the charge. It was an anxious time for most of us, and perhaps especially for me, as I was commanding the Brigade. Nothing cheered me so much as his cheery voice. We were momentarily expecting a counter-attack in greatly superior strength, and our Battery was stricken silent; but he never lost heart, and I am sure his example went far in putting heart into his men, if such was necessary. It did me good anyhow. It was also greatly due to him that at dusk we were able to carry away the officers and men who were lying wounded between the lines. Had he been alive now, his name would have gone in for a Mention in Despatches. He died as many others have died, but he has not lived in vain."

In other letters mention is made of "his outspoken Christian life," and he is described as "one whom everybody loved."

It may be noted that in the meantime one of the Thirteenth subalterns, 2nd Lieutenant J. H. Hirsch, had passed through a very tragic experience, being on board the *Cameronian* transport, on the way to join, with several men of the Regiment, when she was torpedoed in the Mediterranean. His letter describing the affair shows that "there was no panic," all the men going quietly to their posts on deck, and setting to work to lower the boats. The two destroyers which formed the escort "were simply wonderful—the rate they came alongside! When we were hit they dashed round, making a smoke-screen. Then they dashed round the boats and rafts and swimmers, like two old hens guarding their young, picking up men when it was possible, who were shouting for help.

A lot we passed were drowned, as they had put their life-belts on wrong. . . ."

Then the submarine was sighted. The writer was by that time on board one of the destroyers. "The other destroyer had signalled there was a submarine near us, so the guns were manned, and we saw the wake of the beast, so we fired and only just missed." Luckily the shot was enough to drive away the submarine, otherwise the slaughter might have been terrible, for, according to Lieutenant Hirsch, "there were about 1200 men on the destroyer and about 50 officers," and she was, of course, exposing herself to great danger in trying to save life. However, no second torpedo was fired, and after circling round the wreckage for some hours after the ship had gone down, the destroyers made for Malta, and in spite of a "very, very rough" sea, succeeded in getting there safely next day. But 140 lives were lost, among them the lives of eight privates of the Thirteenth.

Lieut. E. Bristol

Lieut. A. E. Annett

Lieut. C. A. F. Wingfield

Major and Q.-M. A. Cooke

Lieut. W. Madgin

Lieut. A. Williams

Lieut. L. A. Ormrod

CHAPTER XVII.

THE SUMMER OF 1917.

THE summer of 1917 was comparatively quiet; General Maude writes of it in the following words:—

"For the remaining five months of the period under review the heat was considerable, and during the latter part of June, July, and beginning of August it was intense. Consequently, movements could not be undertaken by either side without grave risk of incurring substantial casualties from heat-stroke and heat exhaustion. The troops enjoyed a well-earned respite from the continuous fighting in which they had been engaged during the five months terminating in April, but they were by no means idle. Our hold over the area was made secure, defensive positions and pivots were prepared, and training was carried on in the early mornings and late evenings as the temperature permitted. Manly sports, too, which are so essential to the wellbeing of the soldier, especially when temporarily inactive in the military sense, were freely indulged in, with beneficial results to the health and future fitness of the army for service in the field. Arrangements had been made for a proportion of the troops to proceed to India on leave during the summer months, and those who had been on service for a considerable time derived much benefit from the change and rest thus obtained."

India during the hot-weather months is not usually regarded by Englishmen as a pleasant country, but no doubt its settled cantonments and civilised cities were preferable to camp-life in Mesopotamia, and its cool "hill stations" were a godsend to men worn out with exposure to the fierce heat of the Arab plains.

General Maude's remark that the troops who remained in Mesopotamia were "by no means idle" is well within the mark, for not only were a considerable part of them engaged during the month of May in such punitive expeditions against the Arabs as the one mentioned in Chapter XVI., but at some points these expeditions extended into June and even the beginning of July, when, as he says, the heat was intense. Nor were such minor movements the only operations which the troops were called upon to undertake. In June the Russians reported that in consequence of the increasing heat they had found it necessary to evacuate the line of the Diala river, and they withdrew across the Persian border from which they had advanced. This necessitated the occupation of the district by a British force, and in August this force had to drive back a body of Turkish troops. Moreover, in July it was thought necessary to push a column forward on the line of the Euphrates, and there was some fighting with the Turks in this direction, fighting which was stopped only by "a blinding dust-storm" and "the commencement of an abnormal heat-wave." On both flanks of the army, therefore, the troops had active work to do in the summer months.

Nevertheless the bulk of the British troops about Baghdad did no doubt enjoy a peaceful summer in their standing camps along the banks of the Tigris; and this was the case with the Thirteenth, who were left undisturbed among the palm groves at Chaldari. There they rested and recruited their strength for any further campaigning that might occur when the summer was over. Before that time arrived the regiment was thoroughly fit for service again—its numbers full, in officers, men, and horses, and its health excellent.

The following brief extracts from letters and diaries seem to be all that is available for this peaceful interlude in their Mesopotamian war record.

Captain S. O. Robinson—May 13, 1917.—"It is too hot to fight out here now, and we are sitting down for the hot weather after a very successful winter campaign. I expect people at home look upon this as a side-show, which of course it is compared with Europe, but all the same there has been plenty of hard fighting, under most difficult and trying conditions.

CHALDARI CAMP. 1917

"Those at the top, responsible for running the show, have run it very well indeed.

"I have not had any letters from any of you for weeks, but expect letters will turn up eventually; the advance was so rapid that they are bound to have gone wrong."

2nd Lieutenant Guy Pedder—June 7.—"Just been told I may go on the next leave to India; if so, I shall sail from Basra on the 18th, and will probably go up to Mussoori or some hill station. . . .

"Our routine at present is: 5.30-7, training remounts; 7-8, stables, and nothing else all day but lie in a tent and get bored, so I shall be very pleased to get away. It's not really hot yet, but quite hot enough to stay in a tent all day. Absolutely no news. . . . . Well, cheerio, GUY. No letter from you for three weeks."

2nd Lieutenant Pedder had returned shortly before from hospital, on recovery from his wound received at Lajj.

2nd Lieutenant F. N. Payne—July.—"The Cavalry have a splendid camp all along the river. They have an ice-barge, cinema, ice-cream shop with many coloured drinks, polo, bathing, and fishing, so that it is ten per cent cooler than in Baghdad. They can dress as they please, and are pretty comfortable.

"Most of the officers have had mud walls built round their roomy tents."

A pleasant picture, and a contrast to the earlier phases of the Mesopotamian War, when our ill-equipped and out-numbered troops were struggling with all the initial difficulties and hardships, or vainly throwing away their lives in desperate assaults on the Turkish trenches to relieve Kut.

2nd Lieutenant Pedder—August 28.—"Here I am back again and very fit. Most of the people here are very fed up. They had one very hot spell, otherwise the heat hasn't been too bad, and they have got plenty of ice and soda, &c. It is still pretty warm, but much pleasanter than Basra or being on the river coming up. . . .

"Robinson lent me a very good book the other day, 'In Mesopotamia,' by Martin Swayne.

"We are giving a concert to-night, and each mess has got at least three guests.

"In a week's time the Division is holding a race-meeting. I am not riding, as I have had no time to train my horses. . . .

"A week after our meeting there's to be a huge regatta at Baghdad, and then I suppose war, at least I certainly hope not before then."

September 4.—"Many thanks for your letter, dated July 8th, which arrived a week ago. We won't get another mail for another week, when we expect the Colonel, who arrived Basra on the 1st.

"Hurrah, it's getting cooler. However, when it gets cool we'll move, and I don't know which is preferable—heat and peace, or pleasant climate and fighting: I wouldn't vote for either.

"I got back just in time for the Cavalry Division Race Meeting, but not early enough to train 'Matches' for the Lajj Cup. The meeting was a great success. The Regiment did awfully well—won three events out of six. Ormrod won the big event, Baghdad Cup given by Maude; Hartigan won the Kut Cup, and was second in the Lajj Cup; and Peat, who is now my servant and was Eve's, won the Br. Ranks open, a very fine performance, as there were forty-two starters, and it was a flat six furlongs. Our Brigade showed its superiority by winning every event.

"They are starting a great Sporting Club at Baghdad with its grounds 1½ miles outside, where there will be golf, tennis, cricket, racing (?), polo, and boating. It's a great scheme, and will be a godsend if we are stuck here after cessation of hostilities.

"Did you see the Colonel has got D.S.O. now as well as Croix de Guerre, also FitzGibbon has got M.C. as well as Italian Medal?

"I go out shooting occasionally, but there's not much. . . . It's still too hot to go out or do anything before 5; however, it's not as unbearable as it was."

Captain S. O. Robinson—September 10.—"It is still very hot here in the middle of the day, and when you do get any wind it is generally accompanied with dust.

"We are expecting a large draft of men this week, and shall be very glad to get them, as we are very short at present."

September 18.—"We have had a dust-storm for the last three days, which is very unpleasant, but the wind usually drops at night. I am writing this in a mud hut, so am out of it for the time being.

"The British army is more or less settling down in the occupied portion of this country. I only hope we shan't get a rude shock.

CHALDARI SUMMER CAMP, 1917

THE REGIMENTAL MESS

"We have quite transformed the ancient capital of the Caliphs. It is now a modern Eastern town, quite a different place to what it was six months ago."

The following extract is from the diary of Lance-Corporal Bowie, continued from the extract in Chapter XVI.

"On the conclusion of this[1] we marched back to our camp at Chaldari, and here commenced our summer routine and the training of remounts, and drafts of men from England for the next operations.

"On the 31st May the Cavalry Brigades were inspected by Lieut.-General Sir Stanley Maude, who, after making us a very stirring speech congratulating us on the splendid work we had done during the advance on Baghdad, presented the awards which had been won by men of our machine-gun squadron at Baghela in our action of the 25th February, S.Q.M.S. Luff, Sergeant Bush, and Lance-Corporal Lonsdale all receiving the Military Medal for gallantry there.

"On 16th August we held a boxing competition at Chaldari for the Cavalry, the most notable event being the fight between Private Wells, Thirteenth Hussars, *versus* Private Harriss, Fourteenth Hussars, when Wells added greatly to his reputation by gaining the verdict over Harriss, who was at one time the Bantam Champion of India.

"Some little time after this, the Regiment was invited by the Sheikh of a neighbouring Arab village to beat up his estate for wild boar. This we accepted, and all ranks had a most enjoyable day's sport in consequence.

"It was during our stay here that the first Baghdad Racing Meeting was held. It was on September 1st, and this proved a great day for us, some of our own horses winning some most exciting races. Six races there were, numerous entries, big fields, and a band to enliven the proceedings. There were twenty runners for the Kut Cup, which was won by 'Whiskers,' belonging to Lieutenant Ormrod, and ridden by Lieutenant Hartigan, both of the Thirteenth Hussars, and almost as many for the Baghdad Cup, a two-mile steeple-chase, which was won by 'Venus,' owned and ridden by Lieutenant Ormrod. For the Lajj Cup another big field

[1] The expedition against the Arabs in May 1917.

turned out; this was won by 'Flying Heckle,' owned by Major Daniels, and there were more than a score of runners for the Deli Abbas Cup. This was won by 'Percival,' a horse owned and ridden by Lieutenant Denning. Dearly, indeed, are these cups prized by their fortunate possessors, who were presented with them by Lieutenant-General Sir Stanley Maude at the conclusion of the day's sport.

"Shortly after this, on September 10th, we held our first Baghdad Regatta. This was a very small affair, but contained many popular events, such as pontoon races, launch races, gufa races, dinghy races, and tilting competitions. One race in particular which we were very proud to carry off was the gufa race. In this we gained both first and second place, beating all other competitors by several hundred yards. It should be here explained that a gufa is nothing more or less than a large circular basket about 3 feet in depth and 8 feet in diameter, covered with skin. This race was really the tit-bit of the day's sport, much amusement being caused by our frantic efforts to propel such an unwieldy craft on such a river as the Tigris."

So ends Lance-Corporal Bowie's diary for the summer. It must be admitted that if the Thirteenth had shown themselves forward in action they were not behindhand in sport. The two things do not perhaps always go together, but it is remarkable how often they do.

Lieut. W. G. Newman

S.S.-M. (afterwards Lieut.)
C. M. Douthwaite

Lieut. G. F. Earle

Lieut. J. W. Biggar

Lieut. C. W. Jemmett

R.Q.-M.S. H. J. Edwards
(*Wounded at Iman Mahdi, 25th February* 1917)

Lieut. A. H. Wood

CHAPTER XVIII.

AUTUMN OF 1917—RAMADIE, MENDALI, TEKRIT.

WHEN the summer of 1917 drew towards an end, and the weather began to grow more suitable for action in the field, General Maude had worked out his scheme for further operations, and was ready to move.

His force was now facing due north, towards the distant Turkish base in Asia Minor, and it was evident that this must be the main line of warfare during the ensuing winter—the line of the Tigris still, but the Upper Tigris as it comes down from the Northern Highlands. For the moment, however, the Turks on this line were giving no trouble. Hard hit by the fighting in April, they had remained for some months content to hold the river down to Tekrit, more than a hundred miles above Baghdad; and before committing himself to an advance upon this position, which was strong and well prepared for defence, General Maude decided to secure both his flanks so as to be quite free from trouble when delivering his main blow.

Facing his left, on the Euphrates, the enemy had a considerable body of troops at Ramadie, a few miles beyond the British advanced posts; and on his extreme right the Turkish 13th Corps, falling back before the Russians from Persia, but not closely pressed, seemed inclined to assume a threatening attitude. In September he struck simultaneously in both directions.

On the Euphrates a considerable force of Infantry with one of his Cavalry Brigades, including the Fourteenth Hussars, advanced on Ramadie and carried out a very successful attack. The end of it was that the Turkish force was not only beaten but completely surrounded, the Cavalry getting round into their rear and cutting their line of retreat. In these circumstances they were forced to surrender, and more than three thousand prisoners fell into General Maude's hands.

At the same time another force, with the Cavalry Brigade to which the Thirteenth belonged, pushed out eastward and seized Mendali, close to the Persian border. They made no large capture there, for the enemy did not attempt to hold it, but they drove him off and established themselves right up to Persian territory.

It will be seen from the map on page 201 that in this part of the country, to the north-east of Baghdad, a range of hills, the Jebel Hamrin, cuts across the district lying between the Tigris and the Diala rivers, and even extends beyond the Diala southward. This range, a mass of difficult highlands, was held by the Turks, and was a formidable obstacle to any advance against them, while from its lower spurs issued a great system of canals upon which depended the fertile tracts below. To complete the expulsion of the Turks from these tracts, rich in supplies, and to prepare for a further advance in the future, General Maude's right-hand force now pushed on and seized the head works of the canals, penetrating some way into the Jebel Hamrin. There they established themselves, opening up the country with great labour by bridging the canals and making roads. General Marshall, who commanded the force, writes that "seventy-five bridges of various sizes were built in this area alone, and the Jebel Hamrin, which prior to our occupation was a roadless tangle of hills, was gradually pierced by a very complete and convenient number of roads suitable for wheeled traffic."[1]

Thus both on the Euphrates and on the Diala, to left and right, General Maude had freed himself from immediate enemy pressure. It remained to deal with the Turks to the north on the Tigris.

These, consisting of his old enemy the 18th Corps, greatly reinforced, had not been passive spectators of the British movements on the two flanks. They did not move in time to prevent or hamper these movements, but in the middle of October 1917 they pushed down the Tigris for a considerable distance below their strong intrenched position at Tekrit, and assumed a threatening

[1] The Jebel Hamrin is described by the Colonel of the Thirteenth as not unlike the Cheviot Hills.

attitude. General Maude at once struck at them in front, and bringing in the Cavalry from his wings, tried at the same time to work round their flanks, the result being that they retired again to Tekrit, and there gave battle. It was a strong position—"an intricate system of trenches, many of which had been well and deeply dug, with numerous machine-gun emplacements"; and, as every one knows, the Turks in such a position are among the most tenacious soldiers in the world.

Nevertheless, on the 5th of November they were attacked and completely defeated, losing 2000 men, of whom 300 were killed, and though the brunt of the trench-fighting fell of course upon the Infantry, the Cavalry did their share. General Marshall, who was in immediate command, records that in the afternoon "a spirited attack by the Infantry coupled with a charge by the Cavalry Brigade on the flank, threw back the Turks in disorder, who fled panic-stricken." Most of them got away under cover of strong rearguard positions, prepared beforehand, which held up the British pursuit. Still, it was a very successful action, and during the night the Turks, after burning a large quantity of ammunition and stores, withdrew the whole of their forces to a point forty miles up the river.

At Tekrit, for the second time in the Mesopotamian campaign, the Thirteenth, or two squadrons of them, got an opportunity of attacking on horseback, sword in hand, which is always the chance the Cavalry soldier longs to get. They lost an excellent officer, Captain Robinson, who was commanding the two squadrons concerned, and was killed in the charge. Another officer, Lieutenant Williams-Taylor, was wounded and taken prisoner. Their total loss was six killed and twenty-two wounded—only a fourth of the loss sustained at Lajj eight months before, but still a sensible one for two weak squadrons. It was a fine affair, and the Regiment was proud of itself.

After the action, and the rapid retreat of the Turks, the main portion of the troops on this line were brought back to a better position, but Tekrit was held and was never again in Turkish hands.

Unhappily, this was the last battle fought by the army of Mesopotamia while under the command of General Maude. On the 18th of November, at the height of his success and reputation, he died of cholera in Baghdad, and not only for the troops he had led to victory, but for the nation, his death was a real misfortune. Few men have been more beloved by those who served under him, or more trusted as a bold and yet careful commander. He was no doubt more fortunate than his predecessors, in that he was given a large and thoroughly equipped force to do the work demanded of him, while they had been called upon to achieve victory with troops insufficient in number and very badly equipped, but this in no way detracts from his merits. He was not like them, asked to do impossibilities, but what he was asked to do was not easy, and he did it thoroughly well.

The following extracts are from papers bearing upon the fight at Tekrit from the point of view of the Thirteenth Hussars.

By Lieutenant T. Williams-Taylor.

"TEKRIT, *November 5th*, 1917.

"In the afternoon we advanced under fairly heavy shell-fire, but with few casualties, and dismounted in a depression of the ground about half a mile from the Turkish position. There was plenty of firing, but we were under good cover. The Infantry were attacking on our right nearer the river and town of Tekrit. At first we were ordered to attack dismounted, and had just prepared to do so when the order was countermanded.

"We sat about smoking and watching the progress of the fight to our right and half right. Then 'B' Squadron was ordered to mount. Robinson told me he had orders to go in with the sword. We formed up under cover, the composition of the squadron being roughly as follows:—

"Three troops (2nd, 3rd, and 4th) less the Hotchkiss guns and gunners not armed with the sword.

Captain Robinson	Squadron leader.
Mr Hartigan	Second troop.
Mr Jones	Third troop.
and myself	Second in command.

"The total strength could not have been more than forty swords (this total I have arrived at after discussing it with Sergeant-Major Potter, who was then S.S.M. of 'B' Squadron).

"We trotted out in squadron column, and soon came under heavy but inaccurate rifle and M.G. fire. We then formed line and extended. It was impossible to see what was going on in front owing to the smoke and dust. Captain Robinson brought the squadron to a walk, while Alston and I went out to see what was in front of us. We had only to gallop a short way to see the enemy position. It looked formidable with well-made trenches and machine-guns, and more Turks than we could count. We also saw our own Infantry to our right. I reported to Robinson, who galloped off half left to see what was going on there. He soon disappeared into the dust. Our position was now unpleasant, advancing at a walk over perfectly flat ground under heavy and increasing fire from rifles, machine-guns, and guns, so I brought the right shoulders up until we were following Robinson's line, and gave the order to charge. This was the last I saw of the squadron, as I was a considerable distance in front trying to keep in touch with the squadron leader.

"I crossed the front-line trench and found that the Turks were very demoralised, and I could not get near any of them I then went on and crossed another trench (a communication trench) and had a go at a Turk who was half in and half out of the trench. He was very low down, and my horse jumped, so though I got him I don't suppose it killed him. I found Robinson surrounded by some thirty or forty Turks. They apparently had been paralysed with fear, which was now working off when they saw only two of us. Robinson turned his horse and galloped at them, but was shot down point-blank. I followed him when he started to gallop, but my horse was shot and came down (just by Robinson, who died almost instantly), pinning me down underneath him. We exchanged shots with each other for some time at a range of about five yards, and I got a bullet through the shoulder, and shortly afterwards a clout on the back of the head from a rifle-butt, and some minor injuries.

"I was unconscious for a time, and came to, to find the Turks looting me. They got my pistol, glasses, &c., when a Turkish officer came running up firing his revolver off in the air and shouting at them. They departed into the trench under a shower of what I suppose was abuse from the officer, who was now firing at them, but did not seem to hit anybody. The Turk then came up to me and asked in bad French if I was an officer. He then made me a long speech, and told me he was a captain most distinguished of the M.G. Corps, and that he was pleased to meet me, and that he was a Socialist from Constantinople, &c., &c. I thanked him and told him I was glad to meet him, and that I was also very distinguished, but would be able to chat better if he took my dead horse off me, which he did under a very heavy fire from our people. He was very good indeed, and had my various damages seen to there and then in the trench. He subsequently took me back to the Turkish C.O. of the Regiment, where I had coffee and cigarettes before going on to the Field Ambulance. I found out that there were some four thousand Turks at the point where we charged."

Lance-Corporal Bowie's Diary.

"On September 25th we marched out to commence operations again, our first march being to Conningham Post, some 12 miles distant. The next day we marched on to Baquba. Here we took the opportunity of giving our horses a good swim in the Diali river, which runs by the side of the village. Early the following morning we moved on to Belad-Rus and after resting here for a few hours we made a night march at about 8.30 P.M. on to Mandali, a village situated right at the foot of the Jabel Hamrin hills. This place we attacked at dawn, and speedily discovered that it was held by a very small force of Turkish Cavalry, about one squadron, who, needless to say, did not waste any time with us, but took to their heels and succeeded in escaping through the pass into Persia, where, as we discovered later, they fell into the hands of the Kurds, those very fierce tribesmen who never lose an opportunity of revenging themselves for former outrages they have suffered at the hands of the Turks. Our chief capture here was some 300 camels, which are the enemy's chief means of transport in this country.

"We went into bivouac near this village and remained there for about 14 days, while at the same time we made daily reconnaissances well into the surrounding country. During this period we were forced to live on the country, our transport failing to

F.S.M. A. Bald

S.S.M. A. Potter

R.S.M. S. F. Seekins, M.C.

W. H. Tetheridge, D.C.M.

Cpl. G. Rayner, D.C.M.

Cpl. J. Stevens, D.C.M.

arrive with our much-needed rations for some unaccountable reason, and our fare consisted mainly of dates and oranges, which, fortunately for us, were very plentiful in that district. Even our horses had to be rationed with these dates.

"On October 18th we left Mandali and made a forced march on to Kizil-Robart, making a short halt at the Salt Water Lake in the Jabel Hamrin hills, some 23 miles distant, to refresh our horses; we marched again at dawn and covered another 17 miles at a very fast pace. We came into contact with the enemy a few miles south of the village. We at once deployed for action and engaged them, and at the conclusion of a very severe day's fighting we had driven them back on to the village. Renewing our attack early the following morning, we succeeded in driving them out of it, and finally captured the village at about 11.45 A.M. on the 20th instant. Leaving our Infantry to occupy this place, the following three days we devoted to reconnoitring the surrounding country.

"On the 23rd instant the following names appeared in our Regimental Orders for conspicuous gallantry in our action at Lajj on March 5th: S.S.M. Brearly, Cpl. Bowie, L.-Cpl. Stevens, L.-Cpl. Watkins, and S.S. Couch. All were awarded the D.C.M., and a very proud Regiment we were on this day.

"The following morning we left bivouac and marched through the pass, leading our horses owing to the very rough nature of the ground. Mounting at the farther end we marched on to Sharaban, where we bivouacked for the night. The following day we moved off to Baquba, where we arrived at about 4 P.M. It was whilst here that we celebrated the anniversary of the Balaclava Charge, in which the Regiment took part. Making an early start the next morning, we marched on to Sadiyah. Crossing the river Tigris at midday, we bivouacked at a spot which afterwards proved to be the site of our winter camp. We were not destined to rest here very long, however, as two days later we were ordered out to take part in some operations on this side of the river. We marched out on the 29th instant, and bivouacked at Sinajah some 15 miles distant. From this place all our movements had to be made under cover of darkness. Marching again on the night of the 30th, we rested at Istabulat, from which we made another night march on to Samarrah, arriving there at about 10 A.M. on November 1st.

Here we rested until 8 P.M., when we marched again to a point some 12 miles west of Daur. We waited here until dawn, when we commenced an elaborate demonstration, whilst our Infantry were attacking Daur from the south. During this day we came under a considerable amount of shell-fire, and here the Regiment was subjected to a very severe test, while our Colonel displayed the utmost coolness and courage in manœuvring us at the walk in the very face of the enemy's Artillery, in order to draw their fire and thereby enable our observers to locate their batteries.

"Our casualties during the demonstration amounted to 1 killed[1] and 5 wounded, whilst later on the same evening an enemy airman tried hard to inflict further casualties amongst us by bombing, but fortunately without success.

"On reaching bivouac that night we received the stimulating news that our Infantry, under General Cobb, had succeeded in capturing Daur; after attacking a strong trench system of a 9-mile perimetre encircling Tekrit, they had defeated the Turks after heavy fighting, capturing some 319 prisoners and the river boat *Julnar*, which was aground near by. This steamer was captured from us by the Turks in our attempt to reach Kut with provisions for General Townshend and his heroic force in April 1915.

"The following day, November 3rd, whilst resting in bivouac, we were again bombed by the enemy airmen, but, luckily for us, the majority of our horses were being watered at the river at the time, and this undoubtedly saved us from a stampede, our casualties this time being 10 of our transport mules which were killed outright, and one man slightly wounded.

"On the night of the 4th we again moved up, marching on to Tekrit, which is on the river, some 5 miles north of Daur. Here the Turks had installed themselves in a very strong position. Whilst 'standing to' at dawn awaiting the order to attack, we were once more bombed by enemy airmen, but luckily without result. A short time after this we commenced our attack against the enemy's right flank, whilst our Infantry engaged them frontally. We fought our way by a succession of bounds up to a large hollow, which afforded us excellent cover, and was situated some 1500 yards from the enemy's main position. Having a short breather here to

[1] Private J. Tunnicliffe.

refresh both man and horse, we suddenly received the order to mount, and galloping over the top we, under a splendid moving curtain of fire from our Artillery, drew our swords and charged straight into the enemy's trenches, passing through our own Infantry on the way, who cheered us madly. We took their first and penetrated into their second line of defences, where we dispersed a considerable number of their reinforcements who were being massed for a counter-attack, inflicting on them heavy losses. An incident which enraged us beyond measure was the tactics which the Turks resorted to on this and previous occasions when we had attacked them in a similar manner. Immediately on us charging their front line, they would throw down their arms and cry for quarter, which on us lifting our swords and granting, thus leaving them to surrender to our supports while we charged on to their second line, they would deliberately regain their rifles and fire into our backs. On realising this treachery again, we wheeled about and literally cut our way through this treacherous mass, and it is thought that any who may have escaped our avenging swords were brought down by our Hotchkiss gunners, who maintained a steady fire on our left flank throughout the action.

"On regaining the deep nullah from which we commenced our charge, the roll was called and it was found that our total casualties were 6 killed and 22 wounded. Amongst the killed was one of our most popular officers, Captain S. O. Robinson, the leader of 'B' Squadron, whilst Lt. Williams-Taylor, being wounded, was captured, together with Pte. Kasper of 'C' Squadron.

"Having secured sufficient horses from the native Regiments to replace our heavy casualties, we rejoined the 6th Brigade and bivouacked some 10 miles back on the right bank of the river, which we were more than glad to sight again, men and horses having fought and worked for over 70 hours on the desert without any fresh water. We had a well-earned rest here the following day, and we received many congratulations from various Corps and Divisional commanders for the work we had done and its moral effect on the enemy. We commenced our march back to Sadiyah on the morning of the 7th instant, doing easy marches owing to the exhausted condition of our horses. We arrived at Sadiyah at noon on the 12th instant, whilst on the 18th the following names appeared in our Regimental Orders for gallantry in our last engagement: Lts. Norton and Jones both winning the M.C., Sgt. Tetheridge the D.C.M., whilst Sgt. Matthews, Pte. Corbyn, and Pte. Mortimer all won the M.M."

From Lieutenant Chrystall, October 7, 1917.

"We are out on operations again now, as I expect you will have seen; a good show has been managed, and Johnny Turk is not feeling too well just now. We have practically wiped out one of their divisions, and how he is going to do anything more in this part I don't know, except perhaps beyond raids on communications, &c.

"We are living on the country now, and very rough fare it is too—sand-grouse, dates, and rough grass, with a little bhoosa, sort of chaff barley."

Operations leading up to the battle and capture of Tekrit, November 1917. By Colonel J. J. Richardson.

"After continuous trekking since the 17th October, including the enveloping manœuvres, which resulted in the evacuation by the Turks of the Jeb-el-Hamrin on the left bank of the Diala river, and of Kizil Robart, we reached Sadiyah on the Tigris on Friday, October 26th, and encamped on the right bank about a mile from the remainder of the Division. Owing to smallpox—contracted by men who had been assisting the L.P.O. at Mandali—having broken out in the Brigade, we were segregated, and we were looking forward to a well-earned and much-needed rest, to making up our equipment and to getting our winter clothing.

"But on Sunday the 28th we received orders to proceed by night marches to Samarra. We were joined by the Division at Istabulat on the 31st, and marched that night for Samarra and bivouacked for the day in the old Izakhi Canal.

"On the night of the 1st-2nd November we marched to a position west of Daur, co-operating with the 7th Division in driving the enemy out of this place, and bivouacked that night at Daur.

"Daur is a very strong natural position, being protected by the broken and undulating ground rising to the plateau some 50 feet high to the westward, and having, owing to the same ground, a covered way of retreat to Tekrit, some ten miles to the north.

"At 7 A.M. on the morning of the 3rd the Division marched out to reconnoitre Aujah-Tekrit, and came under artillery fire at the former place; the 7th Brigade continued its march to carry out its task of testing the strength in which the trenches to the west and north-west were held by the enemy, and drew sufficient fire to enable the G.O.C. to report them held in strength. The Regiment suffered some seven casualties, of which one proved fatal, from shell-fire. We returned to bivouac at Daur that evening, reaching camp about 7.30 P.M.

"On the night of the 4th-5th the Division made a night march to a position west of Tekrit to co-operate on the 5th with the Infantry in their assault at that place.

"The Brigade was detailed to work close on the left flank of the Infantry, and the Regiment was held in reserve all day. This gave us an opportunity of watching from a good point of vantage the systematic advances of the Infantry behind an intense artillery barrage, and it was good to see them (in this case Punjabi Infantry) advancing shoulder to shoulder, slowly and deliberately, to the assault. What most struck me as a spectator was the apparent immunity from casualties of this system, although next day we learnt that they had had their fair share of casualties.

"The two native Cavalry Regiments of the Brigade—with the exception of two squadrons—had been operating on foot against the trenches to the west of Tekrit, and the 6th Brigade had been acting as a flank-guard some way to the north and north-west, and by its presence there had prevented an enemy Division, which had been marching presumably from Khanal Kharninah, from joining hands with the force holding the Tekrit position.

"About 4.45 P.M. General Norton decided that the time had come when we could materially assist the Infantry, and ordered me to send a squadron mounted to find the enemy first-line trench at a point about 1½ miles to the north-west of Imam Arbain, which was almost the only conspicuous landmark, and from there to sweep right-handed up the trench towards the left flank of our Infantry, who appeared to be assaulting the trenches between them and the Imam, from a south-westerly direction. I accordingly ordered

Captain Robinson to take his Squadron ('B') for this duty, and Lieutenant Norton to hold his Squadron ('C') in readiness to support 'B.' When 'B' was well under way, 'C' Squadron was ordered out, and soon afterwards the Brigadier ordered me to take out the two remaining Squadrons in support. This I did, 'D' Squadron leading, and as we approached the trench line swords were drawn. I had meantime lost sight in the dust, and in the failing light, of the two leading squadrons, but saw about 100 or 150 men on foot on some rising ground, whom I took to be Turks who had held up their hands to the two leading squadrons, and I turned towards them, intending to round them up and prevent them from taking up their rifles again, as our experience at Lajj taught me they were apt to do. As I drew nearer, however, I saw they were our own Infantry, and seeing a British officer I rode up to him and asked him the situation.

"He told me that the two squadrons had gone on to our left front, and that there was a party of about 50 or 100 Turks some 200 yards straight ahead. I decided to go for these, and we passed a squadron of Indian Cavalry (13th Lancers) on our left as we advanced at a slow gallop with our left on a communication trench running much in the same direction I was pursuing. A good deal of rifle fire was being poured into us, and, from the sound of it at very close range, probably from the communication trench, but personally I saw no body of Turks nearer than 1000 yards.

"Meanwhile we had crossed some three lines of fire trenches in which a few wounded Turks were lying, and on my left I saw the two leading squadrons returning from their ride. They were then about 400 yards away, and the light had become so bad that I distinctly saw the flashes from the enemy artillery some 2000 yards ahead of us. I had already gone far beyond the objective given us by the Brigadier, and now that I had found the two leading squadrons, I was chiefly concerned in getting the Regiment out of action whilst the light remained. I accordingly had 'Rally to the Right' sounded on the trumpet, in the hope of drawing 'B' and 'C' Squadrons to me, but they were too far off to hear the sound in the din of the firing, so I wheeled to the right again and returned to the neighbourhood of the first-line trench.

Lieut. T. Williams-Taylor
(Wounded at Shumran Bend, 4th February 1917; wounded and taken prisoner at Tekrit, 5th November 1917)

Lieut. M. G. Hartigan, M.C.

Capt. S. O. Robinson
(Wounded at Shumran Bend, 4th February 1917. Killed at Tekrit, 5th November 1917)

S.S.-M. F. J. Tegg
(Killed at Tekrit, 5th November 1917)

Pte. F. G. Jasper
(Killed at Tekrit, 5th November 1917)

I was then in rear of the squadrons, and seeing a few men struggling back with wounded comrades and wounded horses, I sent out a message to Major Twist to retire the horses out of range, then to dismount the men and bring them up dismounted to cover the retirement of any stragglers who had been wounded or unhorsed. I myself stayed with the R.S.M. (Seekins), the Trumpet-Major, and a few men to give a hand to a few men who had got left with wounded comrades and horses.

"After allowing about half an hour for any stragglers to rejoin, I retired the men on to their horses and we rejoined the Brigade. Here I learned that Captain Robinson had been killed, Lieutenant Williams-Taylor and some seven N.C.O.'s and men were missing. 'C' Squadron had 22 horses killed and 'B' 9; in addition, we had about 19 men wounded; my orderly was hit in the chin, and his horse through the neck. One man, Pte. Thomson, died of wounds *en route* to bivouac at Aujah Nullah, which we reached about midnight.

"Here I will interpose an account of the doings of 'B' and 'C' Squadrons, as gathered from reports by officers and men.

"'B' Squadron rode straight to the point to which they were directed without opposition, and Captain Robinson, finding our Infantry in position, there halted and waited until 'C' Squadron came up; then he explained the situation to Lieutenant Norton, and his decision to lead the two squadrons against a mass of Turks, 300 to 400 in number (estimated), who could be seen some 1000 or 1200 yards away to his left and left front,—and I may say here in parenthesis, that his decision to go beyond the objective given by the Brigadier was eminently the right one, if not the only one: our mission was to assist the Infantry, and that was the only way in which he could do it under the altered circumstances, so the two squadrons wheeled to the left and rode to the north-west, one on either side of a long—and I think the first-line—trench. They went through the Turks, and in this advance Captain Robinson was killed, shot through the head by one of the four Turks whom he was trying to take on with his sword. His Trumpeter, Maguire, claims to have shot this particular Turk with his revolver. Having gone through the mass of the Turks, they turned about 1200 or 1500 yards from where they started, and rode through the Turks again, and they claim to have accounted for at least a man per man. It was soon after this that I saw them returning, and they pulled up near our Infantry and came under Major Twist's orders. The men were much elated with their 'ride,' and our losses were, comparatively speaking, extremely light, as the rifle and machine - gun fire was very heavy, and as they returned, they came under fire from two or three batteries—at a low estimate. At one time when I was looking, I saw the sky thick with bursting sharpnel, but it was all very high, and the gunners must have been firing wildly. It must be remembered, too, that by this time the dusk was becoming the dark.

"Next day, the 6th November, we learnt that the Turks had completely evacuated Tekrit during the night, so I sent an officer's patrol, under Lieutenant Stirling, to go over the battlefield and to gather what information they could about our dead and wounded. On its return I learnt that our Infantry had buried six dead, presumably belonging to the Regiment, although they could not be identified, as the Turks had stripped them completely, even to their identity discs; so on the following day, the 7th, I obtained permission to take a party for the purpose of burying all our men together, and succeeded in getting the Divisional Padre—Ryan—to accompany it. We only succeeded in finding four bodies—viz., those of Captain Robinson, Sergeants Gray and Newman, and Private Francomb, and these we collected and buried, and the Padre read the burial service over them. We also found Lieutenant Williams - Taylor's horse, dead, where it fell, and from its position, and from the report of the man who saw him lying under it, pinned by its weight to the ground, I felt confident in assuming that he had been taken prisoner by the Turks.

Note added on 6th February 1918.—"This assumption has been confirmed by wire received to-day from London, and I have every hope of hearing soon that S.S.M. Tegg is also alive, but a prisoner of war.

"I will conclude this short account of an incident which deserves a prominent place in the annals of the Regiment, by giving a list of the 'immediate rewards' which were bestowed by the Com-

13th HUSSARS.

The Charge of Squadrons B and C at Tekrit in Mesopotamia on Monday, the 5th of November, 1917.

Capt. S. O. ROBINSON.

S.S.M. F. J. TEGG

Sergt. A. S. NEWMAN.

Sergt. JOHN GRAY.

Pte. F. G. JASPER.

Pte. WM. THOMSON.
(1914 Star.)

13224 Pte.
MICHAEL MacDONNELL.
Served with the 6th Dragoon Guards in S. Africa,
1900–1902
(No portrait obtainable.)

Pte. WM. FRANCOMB.

"Then the cavalry came in on the extreme left. They went in with the point, jumped the trenches and scattered the Turks who were fleeing in confusion after being driven from their trenches in the assaults by the infantry. This charge was carried through 1,000 yards beyond the trench held by our infantry when machine-gun and field-gun fire was encountered at close range. The cavalry wheeled about and covered their withdrawal by dismounted action."

The Times, 15th Nov., 1917.

Extract from a Letter of one who took part in the charge:

"The most pleasant part was to hear our infantry cheering us as we cleared the trenches and galloped between them and the Turks."

mander-in-Chief on individuals for various acts of gallantry performed on the afternoon of the 5th November. These were:—

Lieutenant D. J. E. Norton " H. R. Jones	*The Military Cross.*
Sergt. Tetheridge	*Distinguished Conduct Medal.*
Sergt. Matthews Pte. Corbyn " Mortimer	*The Military Medal.*

LIST OF CASUALTIES FOR THE PERIOD UNDER REVIEW.

Killed.

Captain S. O. Robinson	November 5, 1917.
Pte. Tunnicliffe, J.	November 3, 1917.
Sergt. Newman, A. S.	November 5, 1917.
" Gray, J.	November 5, 1917.
Pte. Francomb, A. W.	November 5, 1917.

Died of Wounds.

Pte. Thomson, W.	November 5, 1917.

Missing.

Lieut. T. Williams-Taylor (Who has since been reported Prisoner of War in Turkey.)	November 5, 1917.
S.Q.M.S. (Atg. S.S.M.) Tegg, F. J.	November 5, 1917.
Pte. Jasper, F. G.	November 5, 1917.
" MacDonnell, M.	November 5, 1917.

Wounded.

November 3, 1917.

L.-Cpl. O'Connor, J.
Pte. Liddle, A.
" Miller.
" Barker, J. H.
" Beales, H. J.

November 4, 1917.

Pte. Bell.
" Bennett, W.

November 5, 1917.

L.-Cpl. Reaves. A. E.
" Lindsey, H.
Cpl. Hudson, H.
Pte. Stubbs, R.
" Jeffrey, J.
" Halstead, C.
" Manby, E.
" Byrne, J.
" Williams, D. G.
Cpl. Leutchford, W..
Tpr. Burder, H. A.
Pte. Viney, A. V.
" Taylor, C. E.
" Delaney, C.
" Ward, A. J. F.
" Andrews, F.
" Hibbett, A.
" Wilson.
" Stewart, H. W."

Extract from letter from Lieut.-Colonel J. J. Richardson to General Symons, C.M.G., dated 7th Cavalry Brigade, M.E.F., November 9, 1917.

"DEAR SYMONS,—Altho' this letter may seem to you to be rather belated you must make allowances, for we have been on trek 'on operation scale' continually since the 17th of last month. . . . By great good fortune we bivouacked at Baghdad on Balaclava Day, and I was able to get (from the E.F. canteen) a bottle of beer a man and ½ a bottle of 'pop' per officer.

"We have just completed, and are returning to our base from, the operations against Tekrit, where the Rgt. again distinguished itself in a mounted charge against the Turks entrenched, and enabled the complete capture of the strong position to be consummated before nightfall. Our losses, altho' they include the loss of Bob Robinson (killed), were extraordinarily—and to me unaccountably—light, being 1 officer and 5 O.R. killed, 1 officer and 2 O.R. missing, and 16 O.R. wounded. The missing include Williams-Taylor and S.S.M. Tegg, but I have very great hopes that the former and possibly the latter will prove to be prisoners of war. Two days later I was able to collect and bury together and hold a service over the corpses of Robinson, Sgts. Gray and Newman, and Pte. Francomb. We claim to have accounted for at least 100 Turks, and Sgt. Gray's sword was thick with blood up to the hilt.

"Please let General B.P. and as many present and past 13ths know this, and that I feel confident they will hear nothing but good of the Rgt. The men were simply splendid, and *ipso facto* the officers also.—Yours sincerely,

(Signed) J. J. RICHARDSON."

From Colonel Richardson to Sir Frederick Williams-Taylor.

"7TH CAVALRY BRIGADE, M.E.F.,
12th November 1917.

"DEAR SIR FREDERICK,—Many thanks for your kind letter of August 27th, which has been forwarded to me from home, and reached me last night; many thanks also for your kind offer to help financially in the publication of the record of the Regiment in this

War; I have noted this, although I do not—at present—anticipate that any appeal with this object will be made.

"We arrived back from the operations which resulted in the capture of Tekrit yesterday, and I was on the point of writing you an account of our share in them and of the charge in which your son Travers was wounded. The Army Commander has interested himself personally in the case of your son, and will, I have no doubt, have cabled you fully and given you the reasons why we think we are justified in thinking his fate to be no worse than that of a prisoner of war.

"Well, about 4.45 P.M. on the 5th November, the Regiment was being held in reserve, so that should an opportunity arise to assist the Infantry who had been assaulting the Turkish trenches all day, we might be put in without delay, and our Brigadier evidently thought the moment had arrived, so he ordered me to send out a squadron to find the first line of trenches and to sweep up it towards the Imam, which was the most conspicuous mark; 'B' Squadron was the first for duty, so I gave Captain Robinson his orders, and the squadron went out. No sooner were they started than I received orders to send a second squadron, and 'C' Squadron, under Lieutenant Norton, was sent out. I at once received orders to take out the remaining two squadrons in support; this I did, but the first two squadrons had got such a good start of me that what followed I had to learn from the survivors of the charge. It seems that on approaching that point of the first-line trench which had been given him as an objective, Robinson realised that it had been evacuated by the enemy, who had retired in a N.W. direction, and that the left flank of our Infantry were also approaching that point; he therefore halted his squadron, and when 'C' Squadron had overtaken him, he explained to Norton the situation, and his decision—which incidentally was the true soldier's instinct, and from a military point of view quite correct—to attack the mass of the enemy, a few hundred in number, who could be seen retiring in the open about 1200 yards to his left flank. Accordingly the two squadrons wheeled to the left, and almost in line, galloped along, one on either side of a long trench, which had several strong points, at one of which Robinson was killed, shot through the back of the head; and your son was last seen pinned to the ground by his horse, which had been shot, and trying to load his automatic pistol, and surrounded by a number of Turks. The man who saw this said he also thought that Travers was wounded, but there is only his evidence on this point. When the squadrons had got back to the neighbourhood of our Infantry line they were dismounted and advanced about 200 yards, with a view to covering the escape of any of our men who had been unhorsed or slightly wounded; but it was now almost quite dark, and after allowing a reasonable time for any stragglers to rejoin, I ordered a retirement on the Brigade.

"The Turks retired during the night, and our Infantry next morning went over the ground, and buried six bodies, including Robinson's; they were unable to identify these as the Turks had stripped them of everything, including identity discs; I however sent a party that day, Tuesday, to ascertain what was possible about our missing, and they identified the bodies of Captain Robinson, Sergts. Newman and Gray, and Pte. Francomb. Two bodies, buried by our Infantry, they were unable to find, and so remained unidentified; but I think we are more than justified in presuming, under the circumstances, that Travers was made prisoner of war by the Turks, wounded perhaps but not necessarily so, and personally I have every hope of hearing in course of time through our Red Cross and the Red Crescent Associations that he is safe and well in the hands of the Turks.

"I will say no more at present beyond expressing our universal regret at losing so popular and so capable an officer, and our great sympathy with you in the painful suspense which the uncertainty of Travers' fate must impose upon you.—Believe me, yours sincerely,

(Signed) J. J. RICHARDSON."

Extracts from letter from Lieutenant Guy Pedder to his Sister, 15th November 1917.

"At last we are back at rest, so I will try and give you a sort of account of our doings of late. We have been on the go now for just on 8 weeks, so you can imagine how topping it is to get back into a decent camp with a tent to yourself, &c. This morning I had a cup of tea and a biscuit when I was called, followed by a boiling-hot bath and a change of clothes—what luxury! . . . The mail goes out to-night. . . .

"Our Brigade left Childari about Sept. 25th, a week after the other Brigade had gone out on the Ramadie stunt, which was a great success, the Fourteenth doing very well, but losing their C.O., and Medd Bridges is now commanding them.

"After four long marches during very hot weather (Robinson went down with heat-stroke, but rejoined a fortnight later), we took Mandali, some 100 miles N.E. of Childari up on the Persian boundary, after very little opposition and no casualties.

"We stayed there almost three weeks, during which time we had a certain amount of patrol-work, &c., to do. Then, after two long marches N.W., we took part in an affair in the right flank to kick the Turks out of a pass over the Jubal Hamrin—it was quite a relief to get among some hills, and away from miles and miles of flat desert. We had very few casualties, and kicked the Turks out more by surprise and rapidity of movement than by fighting. The last morning the Brigade covered 15 miles in 1¼ hours!!—a pretty hot and uncomfortable pace in full marching order over rough ground. We then denied Qusil Robart to the Turks for three days, and got shelled a bit; however, it was all very chusie until we were suddenly hustled back to the Tigris, *viâ* Bacuba, to a more central camp, so we don't know what happened on the right flank since we left.

"We took about five days to get here, and were then told that we should be here for 21 days, the Brigade having covered some 400 miles in under five weeks. Our three weeks' rest developed into two days, when the Division which had reassembled were hustled up N. on the right bank; we reached Samara after three very cold and unpleasant night marches, and were very weary, so we lay hidden in a nullah all the fourth day without any shade. We moved again at 7 P.M. for our fourth night march, and after covering some 27 miles found ourselves right out on the left flank [of the Infantry], who captured Daur that day, whilst we demonstrated only on the flank and got a bit shelled; however, we underwent a new experience to us, but a very old one to Londoners, as Fritz came over and dropped some bombs—very unpleasant being bombed by an aeroplane; you feel so helpless standing in an open desert holding your horse's head. We camped just S. of Daur, and next day had to go right out on the Infantry's left flank whilst they captured Aujah, and there we had to reconnoitre the enemy's position at Tekrit—a long march there and back; we were shot at a good deal, but were very lucky, and had under a dozen casualties in the Rgt., and only one man actually killed. We got back late that night, and were woken up at 7 A.M. next morning by Fritz dropping bombs on our Div. camp, which he repeated again that evening, when luckily nearly all the horses were down at the river watering. The Rgt. had only three or four casualties from these bombs, but it was a horrid sensation hearing the bomb come tearing through the air, and not knowing where to expect the crash. That night we moved at 10 P.M. (another night march) for the battle of Tekrit.

"We were sent well out on the left flank, where we stayed all the morning, and from where we saw our Infantry take two lines of trenches under cover of a heavy artillery barrage, and also where we were bombed again twice. This time I actually could see the bombs leave the aeroplane, and watch them drop right up to the moment of exploding on striking the ground. About 3 P.M. we were moved right up on the Infantry's flank, and stayed in some hidden ground. At 4.39 the Infantry attacked, and a ¼ hour later we were galloping through our Infantry, who cheered us like mad, towards the Turks. Robinson's squadron led, followed by 'C'; then about 1000 yards behind 'C' the Colonel took us —'D' followed by 'A.' 'B' and 'C' got right into the Turks and beyond, or as much into them as you can expect, considering the place was full of deep trenches (we who were some way behind 'B' and 'C' crossed four lines of trenches), and Bob was killed at the farthermost point reached by the charge—instantaneously—shot through the head. We got his body and buried it the next day after the Turks had cleared off. Williams-Taylor, also in 'B,' was last seen lying wounded under his dead horse. However, as his body was nowhere to be found next day, we hope, and have reason to, that he is alive and a prisoner. 'D' was the third squadron in the charge, and though we crossed four trenches we never actually reached the Turks, whom we saw just in front; and when the Colonel, who was leading us, saw 'B' and 'C'

THE GRAVE AT TEKRIT

galloping back, he gave the order to retire and took us some way back where we got into dismounted action. My horse (not Matches, whom I was resting) was badly cut through the stifle, just as we got the order to retire—however, he just managed to get me back and out of it, altho' he had to jump four trenches; he was a very good young horse, and I was sorry to lose him. The Rgt. was again very fortunate, as the rifle-fire, altho' of course (why, I don't know) not at all on the same scale as at Lajj, was quite unpleasant enough. We only had some 28 casualties and 60 horses. Personally, I don't think the game was worth the candle, especially when you think of Bob, the best fellow in the Rgt.; but other people don't all seem to agree with me, and the Corps Commander gave the Division great praise for our assistance all that night, and the Rgt. got it doubly from the Brigade Division. The most pleasant part was to hear our Infantry cheering us as we cleared the trenches, and galloped between them and the Turk.

"Now we are back in that central position, for a few weeks' rest, I hope, I mentioned before on the Tigris.

"We charged at 4.45 P.M. on 5th Nov., exactly three years since I left for the 11th, and didn't get back into a camp, having collected wounded, &c., until 2 in the morning. We took six days to get back here, and the last night I had a tragedy, as the Arabs came into our camp and stole my saddle-bags, which were full of kit.

"Robinson was far and away the best fellow in the Rgt., and it's a tremendous blow to me. . . .

"Hind has just joined us—also two new officers with a draft."

From Colonel Richardson, 7th Brigade, M.E.F.,
13th November 1917.

"I must just write you a short line to let you know how your brother, Captain S. O. Robinson, or Bob as we all loved to call him, died. I shall hope later on to send you a more detailed account of the action, but in the meantime an abbreviated account must suffice.

"On November 5th the Cavalry Division was co-operating with the Infantry in the attack on Tekrit, and at 4.45 or thereabouts in the afternoon, I received orders to send a squadron to charge the Turkish trenches on the left of our Infantry. Bob's squadron, 'B,' was the first for duty, and I sent it, and soon afterwards sent 'C' Squadron in support; these two squadrons rode up to the Turkish front-line trench, found it evacuated, found the left flank of our Infantry, with a mass of Turks estimated at 3 or 4 hundred retiring in the open to the N.W. or left front. Bob at once, with the instinct of a true soldier, decided to charge these Turks, and proceeded to do so with the two squadrons.

"When they reached the big mass of Turks, a mêlée naturally occurred. Bob was seen taking on four Turks with his sword, and was shot through the head by one of them; he was killed instantaneously (incidentally his trumpeter, Maguire, from whom we gather this description, claims to have shot this particular Turk with his revolver). The squadron went on a couple of hundred yards or so, then turned, and came through the Turks again, and eventually reached our lines in safety. By this time it was dusk, and after dark the Turks retired; next day our Infantry came up, and found several bodies stripped, even of their identity discs, and buried them lightly in the trenches. We had gone back that night to bivouac some miles, but on Tuesday (6th) I sent a small party to go over the ground to ascertain what they could; this party found and identified Bob's body and that of three others of our men, and next day, the 7th, I went myself with the C. of E. Padre, Lt. Stirling, and 5 men, collected all the bodies, buried them deep, and the Padre read the burial service over them. We could not mark the spot, but we took records of the exact position where they were buried.

"Major Twist is writing, and will see to his effects in the Regiment.

"In offering you all my deepest sympathy in your loss, I should like to add that his loss will be most keenly felt in the Regiment, and by many outside of it. If his brother officers loved him for his kindness and simplicity of character, his men adored him for the same qualities and for his justness, and I personally mourn the loss of a true friend; but he died the death of a gallant soldier fighting sword in hand against superior odds, a death that few Cavalry officers can ever hope to meet—to so few is it given,—an end that

Sergt. JOHN GRAY
(*Killed at Tekrit, 5th November* 1917)

Lieut. D. J. E. NORTON, M.C.

Capt. F. NORMAN PAYNE
(*Died of cholera in Baghdad, 14th November* 1917)

Sergt. A. S. NEWMAN
(*Killed at Tekrit, 5th November* 1917)

Pte. WM. THOMSON
(*Killed at Tekrit, 5th November* 1917)

will live long in the history of the Regiment, if not in the history of the Army. R.I.P."

From Captain Chrystall, 14th November 1917.

"I herewith send you a short account of the part of the battle of Tekrit, Nov. 5th, 1917, which deals with the Cavalry, and showing how co-operation with the Infantry can be carried out even in a flat country.

"Our troops consisted of—

1 Cav. Div. of 2 Brgds.
1 Inf. Div.
1 Inf. Brgde.
The usual division of corps artillery.

"The enemy had—

2 Inf. Div. in trenches round Tekrit.
250 sabres.
1 Inf. Div. in reserve, some 12 miles to the north.
The usual artillery and heavy guns, and 5.9's and heavy howitzers.

"The Cav. Div. had the task of protecting the W. flank, and early in the morning of the 5th had gained touch with the whole of the enemy's entrenched flank, a distance of some 6 miles. The Infantry, meanwhile, were marching into position for an assault on the line *Z*-X. Owing to the heat, and telephone wires being continually cut, this attack had not reached farther than the first line of trenches.

"However, at 4.30 P.M., Nth Div., Nth Bde. leading, attacked, under a heavy barrage by our guns, the trenches *Z*-X, 2nd and 3rd lines. Immediately, on Infantry's left, was the Cav. Div.—2 Regiments of Nth Bde. at point B in position of readiness (Point B was a deep water-course, with a bluff on the N. side), and 1 Regt. slightly N.W. on the left, with the Nth Bde. to the N.

"On the right of these two Regiments was a subsection of M.G.'s, 'R,' and on the left one section 'L.' These M.G.'s were, until the Infantry attack came off, firing on enemy's trench-line X-Y, and covering the advance of dismounted Cavalry patrols, enabling the latter to approach to within 800 yds.—1000 yds. of enemy's trench. During the whole day the Turkish gunners were busy ranging on the Cavalry with heavy field-guns, and a number of casualties resulted both in men and horses.

"As the Infantry attacked the 'R' section, (sub) 'R,' was able to bring heavy fire in enfilade on trench at point *Z*, and this had the effect of completely protecting the flank of our attacking Infantry, and it was significant that during the whole advance to X-*Z*, no enemy fire was opened from *Z* flank. The Infantry remarked that this co-operation undoubtedly saved many casualties. A few moments later the Infantry had carried the 2nd line of trenches, the Hussars and a squadron of Lancers (Ind. Cav.) having been collected at Point B, emerged from cover, and proceeded to push in a mounted attack on the left of our Infantry against the enemy, who appeared to be massing (behind the 3rd and 4th lines) for a counter-attack.

"The right subsection, M.G.'s 'R,' was able to cover the advance of the Cavalry, until it crossed the trench-line, and a little beyond it. The two left subsections covered the left flank. Ranges from 1000 to 2000 yards.

"Another subsection, 'T,' followed the Hussars, mounted and wheeling to the left, came into action under very hot fire about 1500 yards from enemy's line, and pushed forward, its guns dismounted, the horses being sent right back. This subsection was able to bring covering fire to bear on the flanks of the advancing Hussars.

"During the mêlée no fire action was feasible, owing to the large amount of dust and general compression which was inevitable.

"The whole of this action was carried out on more or less open ground, and under continuous gun and rifle fire from enemy's rear positions. The enemy machine-gun fire was fairly heavy, and fired at extreme range, as the angle of descent testified, causing casualties behind fairly good cover to our M.G. subsections.

"No. of rounds fired, 10,500.

"Men and horses were all day without water, and did not obtain same until midnight, at which time the Infantry had taken over from the Cavalry. Please find attached sketch of position."

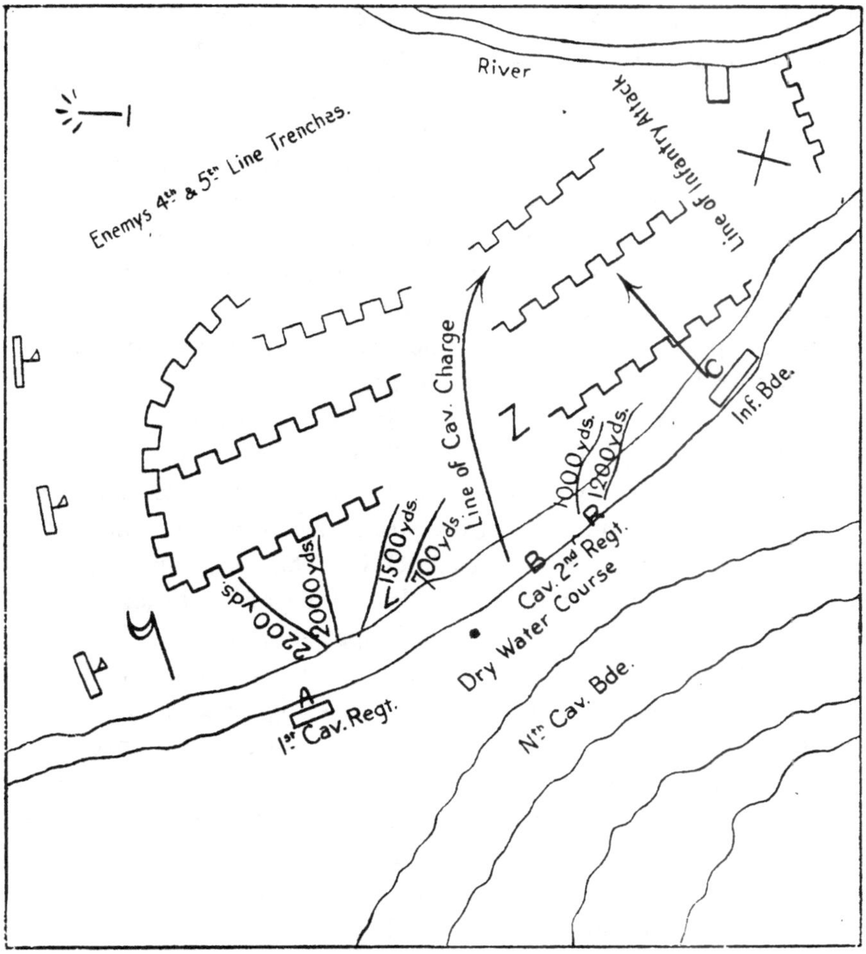

Extract from letter written by Private Hugh H. Mortimer, M.M., relating to the charge at Tekrit, 5th November 1917, for which he received the M.M.

"I must have the luck of the devil, as my horse got a flesh-wound—bullet—in the leg, but is practically all right again now.

"We've got six honours in the Regt., my squadron leader and a Lt. in B—M.C.'s, a sergt. in my troop D.C.M., 2 M.M.'s in other sqdns., and—hold your breath—they've given me one—Military Medal. So I suppose you'll have to put that in 'The Times' now—

eh! what! It seemed to be nothing to me when I was in it. We charged about 300 Infantry on the road and in the trenches, with machine-guns and artillery behind 'em. We caught 'em weak—just about 100 of us—2 weak sqdns.—in the mist of dusk—and, my God! it seemed as if we were riding into Hades. Funny thing tho', I didn't have wind up then, was feeling rather elated cos we'd waited, and moved about, practically all day, for a real rut at them, and been bombed from aeroplanes, shelled and sniped, till we were just itching to get at 'em. Well, they got pukka wind up when we got in amongst 'em, and started firing wildly, some trying to use bayonets, others holding up their mitts and shouting 'Kamerad.' Then we got order to rally as we weren't strong enough for pursuit. Starting to come back, I spotted a youngster badly wounded in the foot and horse shot, so I dismounted and put him up on my steed, which was too bobbery to carry the two of us. He got back O.K. Then my fun started. I had my revolver and about 14 rounds and began to think about getting back somehow myself, when I spotted a chum of mine lying about 60 yards from the trenches. Well, two of us tried to get him out of it, but he would not have it—thigh shattered, blood in spurts, mad with pain. J. Turk, seeing us retiring, heartened up again and potted at us like blazes, shrapnel, machine-guns, and every bally thing. At the finish I had to leave him, to my sorrow, altho' we got him away afterwards. Still he suffered a hell of a lot from exposure, as the devils came out and stripped him—he'd be unconscious by then—in the dark. I hear that he's very bad in hospital, but likely to recover.

"I haven't got the gong yet, owing, I suppose, to Gen. Maude's death, but they've made as much fuss over it as if we were—millions of V.C.'s. I've shaken hands with two Generals, and been in about half a dozen parades till I'm fairly fed up, and nearly said, 'Keep the old gong and let's have a bit of peace.'

"Well, I'll think about drying up now. Am in the pink and quite fit; hope you are all the same, and that you, dear mother, are carrying on."

BAGHDAD CEMETERY

CAPT. F. NORMAN PAYNE'S GRAVE IN BAGHDAD CEMETERY

Extract of letter from Colonel Richardson, dated 7th Cavalry Brigade, M.E.F., 15th November 1917.

". . . We have just got back into camp for a short rest to re-equip, reclothe, and rehorse after two months of continuous trekking — we estimate that we have covered at least 500 miles since September 22nd—which culminated in a big battle on the 5th of this month, in which the Regiment again charged Turkish trenches, and two of the squadrons did very brilliant work.

"I have to report with great regret the death yesterday in Baghdad from cholera of Captain Payne, who last winter, you may remember, was in Billy's squadron, and whose letters to his wife were of great interest."

CHAPTER XIX.

WINTER OF 1917–18.

AFTER the lamented death of General Maude the command of the British army in Mesopotamia devolved upon Lieut.-General Sir W. R. Marshall, K.C.B., K.C.S.I., who had been one of the two corps commanders during the advance on Baghdad.

In a despatch giving an account of the operations of the force during the winter which followed General Maude's death, General Marshall explains in a few words the situation with which he had to deal when he took over charge. After paying a warm tribute to the work and character of his dead Chief, he writes: "When, therefore, I had the honour of being appointed as his successor, the *moral* of the army was magnificent, while organisation and training had reached a high level of efficiency. The Turkish army, on the contrary, was low in *moral*, and desertions from it were numerous and frequent; on the Tigris and Euphrates they had retreated out of rapid striking distance, and only on our right flank was there a good opportunity of hitting them."

General Marshall does not take up at this point the consideration of any general questions outside the immediate purview of the Mesopotamian force, but it may be observed here that the state of affairs in the various theatres of war had of late greatly altered. The collapse of Russia had made an immense difference in Europe, and also in Asia, where Great Britain was now left practically alone in face of the Turkish military power; and on the left of the great Turkish battle-line the enemy had made a notable advance, threatening the Caucasus and Northern Persia. On the other hand, it had now become fairly clear that the long strain of the war was telling on the Turks, who, it must be remembered, had entered into it when

Lieut. M. R. Farrer

nd Lieut. A. C. Barrington

Lieut. F. G. Lawrence

Lieut. E. Goodman

Lieut. W. W. N. Davies

Lieut. H. G. Keswick

2nd Lieut. E. P. Barrett

still shaken by their recent defeat at the hands of Bulgaria. The defence of the Dardanelles, too, though successful, had tried them more than the world suspected, and they had not shown themselves capable of throwing forward from their base in Asia Minor, either against General Allenby in Palestine on their right, or against General Maude on their centre, such masses of troops as some had expected. Indeed, on both these fronts, the British now seemed to be numerically superior. But to revert to the operations on the latter front.

In the circumstances presented to General Marshall he decided to attack at once, while the weather was favourable, the Turkish 13th Corps, in the country north-east of Baghdad, which had been so roughly handled earlier in the year. The attack was successful; the Turks in this part of the country were driven northwards with loss through Kara Tepe and Kifri, and the British hold on the great main road into Persia was securely established, whilst further progress was made in the Jebel Hamrin range. The Turks gave considerable trouble by flooding some tracts of land, and they, of course, resisted the British columns; but the fighting was not very severe, and it was soon over. On the 8th of December the bulk of the troops on this side were withdrawn into winter camps. It will be understood that they had now reached high rough country, where winter campaigning was not easy, and the enemy in this quarter also had got out of reach.

The winter months were spent by General Marshall in developing supply districts, in thoroughly opening up and establishing military control over the Lower Euphrates, from the neighbourhood of Baghdad to the sea, in relieving the state of famine to which the Turks had reduced some of the western provinces of Persia, and in numerous minor operations at various points of his widely extended charge. These, he writes, consisted generally of reconnaissances by Cavalry and light-armoured motor-cars, and of bombing raids by the Flying Corps. It was a winter of consistent rain in the lowlands, with snow on the higher ground, which made military movements difficult and threw a heavy strain on the lines of communication.

On the other hand, as General Marshall reports, "by the middle of December 1917 . . . the military position had completely changed owing to the magnificent successes gained by General Allenby in

Palestine," and this greatly strengthened his hands. Much, therefore, was accomplished.

Finally, when the weather began to improve with the early spring, it was decided that another blow must be struck at the Turks on the left—that is, on the Upper Euphrates. In this quarter they had been reinforced, and during the months of December and January they began pushing their patrols down stream from Hit, where the bulk of their troops were concentrated. General Marshall gave orders that a column should march on Hit and drive away the Turks, and that it should be aided in doing so by a Brigade of Cavalry. Hit was occupied on the 9th of March, and the Turks fell back to a stronger position some miles up the river. Here again they were attacked, and this time with great effect. As at Ramadie, in September, the Cavalry succeeded in getting right round into the Turkish rear and cutting off their retreat. By the 30th of March the Turkish force had been completely beaten, and the remains of it had fled to a great distance, leaving over five thousand prisoners in British hands.

The Thirteenth Hussars had no share in what General Marshall calls "the magnificent work done by the Cavalry Brigade" on this occasion. After their exertions in the autumn and early winter they were given a rest, and remained from the middle of December 1917 until April of the following year at a winter camp on the Tigris forty miles above Baghdad. There they found plenty to do in training drafts and remounts and other work of the kind, for the Regiment had suffered considerable loss in men and horses and equipment; but for about four months they saw no more fighting or marching, and were able to enjoy a comparatively quiet time with some sport and amusement.

A few extracts relating to this period are given below.

Extract from letter from Lieutenant G. R. Pedder, dated 4th December 1917.

"We are very messed about again—right on the right flank, and are at [? Chai Khanna]. I am very fit. It's pretty cold at night now on these stunts, and I shan't be sorry when we get back to our permanent camp.

"There are lots of geese, duck, and grouse up here; but you get no time to go out shooting, but get shot at yourself most of the time.

"I expect Jeffrey is well on the way back by now. The daytime is perfect now, and it is scarcely necessary to wear a topee."

From Captain Chrystall.

"*5th December* 1917.

"We have been out on operations nearly a week, after a few days' rest in camp. As I write this we are on outpost against the Turks in the hills, and our R.H.A. is shelling their positions, and we are getting a few in R.S.V.P.! I am feeling very fit now in spite of the cold, which has been very severe the last two or three nights. We are being relieved some time this afternoon by another Brigade, and shall be back into camp and have a little more comfort. Whilst out here, miles away from Baghdad, a wireless message came for me, making an appointment at the dentist!—in Baghdad to-day, a thing I had asked for about a week ago. A great thing war is now; fancy receiving this about 120 miles out in the blue and whilst in action."

"*29th December* 1917.

"I spent a very happy Xmas, and was able to arrange quite a decent dinner for the men. We are now in winter quarters, refitting and training again: it is so nice to have a comfortable tent to sleep in and on a bed. The weather has been very cold, 12 degrees of frost on 2 days last week, and snow down to the bottom of the foot-hills. The army is very well done out here now. We had frozen beef and plum-pudding and even potatoes in rations for Xmas day. We ourselves managed to secure a real live turkey from a neighbouring village. All you people at home must be far worse off for food than we are, and your experiences of air-raids cannot be any too pleasant.

"I have been playing football this afternoon for the Indian ranks of the squadron, and got a severe kick on the ankle for my pains; it will lay me up for a day or two."

Extract from letter from Lieutenant G. R. Pedder, dated 18th December 1917.

"We have been back here about a fortnight now, and I don't think we shall move again till after Xmas, at least I hope so, as it is frightfully cold at night now, much colder than this time last year; also, I am riding in the Baghdad Plate on the 27th and the Grand Handicap, also 7 furlongs, on the 29th, and I am showing 'Matches' in the Horse Show on the 30th. There's absolutely no news whatever. Moreover, we haven't got an English mail for over a month now. It must have been sunk.

"'A' and 'D' have amalgamated messes, and it is much pleasanter than having a squadron mess.

"Fitzgibbon and Le Patourel have gone into Baghdad for a week, as only eight officers are allowed to go there for Xmas, and those eight are all competitors — Hartigan, Keswick, Self, Welstead, Stirling, Ormrod, Norton, Twist.

"I am riding 'The Witch' in two races, the horse my servant rode and won on during the first meeting. Box (Lieutenant Jeffrey) hasn't got back yet."

Extract from letter from Lieutenant G. R. Pedder, dated 19th December 1917.

"Here we are back again after what ended up in quite a mild show, although at first we all thought it was going to be a very murky business.

"It has turned frightfully cold. This morning I woke up to find a thick layer of ice on the water in my water-bucket inside the tent. Thank goodness we are back here in our standing camp, with tents galore and plenty of clothes and food to help keep out the cold. Personally I like it, although you have got to wear a thick greatcoat all day.

"The Colonel is down river, sick with an ulcer or something, so Twist is commanding.

"There is a tremendous show on in Baghdad for Xmas this year. Two days' racing, 26th and 27th — I am entering two horses for three races,—a duck shoot, which I shall also attend, or hope to, golf tournament, cricket, footer, hockey, &c., &c. Of course, if the rain comes it will all probably be off."

Extract from Lance-Corporal Bowie's Diary.

"Some 12 days later, on the 30th instant, we received sudden orders to intercept a force of the enemy who were advancing through the Sakaltutan Pass. After having covered some 60 miles in that direction we came into contact with the 13th Turkish Army Corps again at about 6 P.M. on the evening of the 2nd December. We immediately deployed and attacked them, and, after having driven in their outposts, we succeeded in advancing close up to the mouth of the Pass, where the Brigade held them until about 12 noon the following day, when we were relieved by the 6th Cavalry Brigade who had been resting in bivouac some 10 miles back. Simultaneously with this action our Infantry, under Lieut.-General Egerton, had advanced up the right bank of the Diala river and had cleared the whole area south of Qarah Tappah, and were advancing to attack the Turks at the other end of the Pass. The following morning we crossed the Shatt-el-Adhaim river and advanced up the right bank as far as Abu-Ghuraib, where we threatened the Turkish right flank, after which we returned in the afternoon and relieved the 6th Cavalry Brigade, who were still holding the Pass. This Brigade were bombed in their bivouac early the next morning by two enemy airmen; this prevented them from relieving us for some hours. Meanwhile, General Egerton's force, co-operating with us, had attacked the enemy in the rear, inflicting heavy casualties among them. The enemy thus being attacked at both ends, he succeeded in escaping over the hills (no small feat in itself), under cover of the night. The Russian force of Cossacks under Lieut.-Colonel Bicharakoff co-operated with us on the right flank throughout these operations. The total British captures were 6 Turkish officers, including a Brigadier Commander, 121 other ranks, and 2 guns.

"On the morning of the 6th we commenced our march back to Sadiyah, going into bivouac for the night after doing some 12 miles. The following morning we sent out patrols to attack and disperse some small enemy detachments who were reported to be in the vicinity by our airmen. An exceptionally bad sand-storm raged the whole of this day, during which we remained in bivouac.

On our patrols returning and reporting all clear, we continued our march for some 25 miles, bivouacking for the night at Sinajah. Crossing the Tigris early the following morning, we arrived at our former camp at about 2 P.M. on the 9th instant.

"A few days later we were issued with our winter clothing, and about this time, too, the enemy airmen became very active, and bombed in turn Sammara, Akab, Baghdad, and many of our camps in between these places. They inflicted so many casualties among our troops that elaborate precautions were taken against surprise attacks. A system of trenches was connected all round the Cavalry horse lines, which enabled us in the event of attack to untie our horses from the lines, turn them about, and take cover ourselves in the trenches whilst holding them. These precautions undoubtedly saved us from a stampede when the enemy airmen paid us New Year Day's visit and bombed us incessantly until their supplies became exhausted. They then flew merrily away in spite of all our batteries' efforts and the 2 Archies, which were dug in down by the river bank to protect us from these raids.

"Just previous to this raid, towards the end of December, we received the good news that our Infantry had occupied Khaniquin without opposition on the 7th December, and also that our troops in Palestine, under General Allenby, had captured Jerusalem on the same date. Whilst on January 2nd we received word that our troops had occupied Qasr-i-Shirin without opposition.

"On January 15th the Fourteenth Hussars left the 6th Cavalry Brigade which was with us at the time, under the command of General Dunsterville, to operate in the neighbourhood of Baku, which is situated on the borders of the Black Sea.

"On the 19th instant the Regiment held some sports, which were the commencement of a series of sports and games, organised to help break the monotony of camp life in this country, which so depresses one's spirit, especially at this time of the year. On this day some excellent sports were witnessed, everybody doing their utmost to be left in for the finals, which were to be held on the 27th instant, whilst in the evening our Concert Party gave a show which proved a great success. The following day our football team played the team of the monitor *Mantis*, which was anchored close to our camp at the time, the result of the match being 1 goal to

Capt. G. W. Rose, R.A.M.C.

Capt. E. Wordley, R.A.M.C.

Capt. D. M. Methven
(21st Lancers)

Lieut. A. C. J. Elkan

Lieut. J. Hampson

nil in our favour. We invited the whole of the crew to dinner at our camp in the evening, every one spending a jolly time in consequence.

"The following day we were again visited by enemy airmen, but our batteries and Hotchkiss guns opened such a rapid and effective fire that after vainly trying to bomb us for some time they finally flew off in disgust, amidst the frantic cheers of the boys who were interested spectators.

"Soon after this we commenced our sports, in which every event was most keenly contested, and at the conclusion of the two days the results were as follows: Individual Tent-Pegging was won by our R.S.M., Mr Seekins; Skill at Arms by S.Q.M.S. Higgs; the best troop-horse was C98, ridden by Pte. Shaw; Officers Jumping went to Mr Macdonald; the Section Tent-Pegging to 'B' Squadron. In the dismounted events, the 100 yards was won by L.-Cpl. Carter in very good time, while the 1½ mile relay race was won rather easily by the team of 'C' Squadron, which included Sgt. Burgess, Sgt. Moss, L.-Cpl. Carter, L.-Cpl. Smithers, L.-Cpl. Bush, and Pte. Bowie. The tug-of-war was won by the 'D' Squadron team after an exceptionally fine pull, which concluded the programme. At about 8 P.M. the same evening our party gave another concert, this time to the whole Division, being assisted in their efforts by some of the crew of the monitor. This show proved a great success, every one thoroughly enjoying a good evening's entertainment. On January 26th some Divisional races were held on a course we had made some 2 miles west of our camp, the ground here being splendidly suited for our purpose, being as flat as a billiard-table for many miles.

"On January 27th General Dunsterville's force left Baghdad, the following officers and N.C.O.'s from our Regiment accompanying the force: Captain Annett, Lt. Cochrane, Sgt. Vernon, Sgt. Taylor, and Sgt. Saunders.

"Commencing on February 22nd, a 3-days' race meeting was held at Akab by the 17th Infantry Division. The Regiment entered several horses, only one of which was successful—namely, Captain Blythe's Army Pattern; while our Concert Party gave a show there on the nights of the 22nd and 23rd, which were greatly admired by all who witnessed them, the G.O.C. of that Division sending a letter of congratulation to our Colonel in having such a talented party.

Shortly after this, on the 28th inst., the Cavalry Division held some athletic sports, in which several of the events were open to all comers. In this the Regiment secured several prizes—namely, 1st in the 100 yards for officers, Mr Osmond; the team of 'C' Squadron, L.-Cpl. Carter, L.-Cpl. Bush, Pte. Jones, and Pte. Bowie gaining 2nd prize in the Relay Race. L.-Cpl. Bush also 2nd in the Obstacle Race, while another 2nd prize was won by Pte. Bowie in the Open Mile, our tug-of-war team of 'C' Squadron being knocked out in the semi-final after a very game struggle. On March 4th we turned out a Rugby 15 to play the Australian Wireless Section's 15, who had played so far an unbeaten record, and after a most exciting game the match ending in a draw, neither side being able to score.

"The following day being the anniversary of our charge at Lajj, the Regiment held a singing competition in the evening. This proved to be highly amusing, judging by the vociferous cheering which greeted the competitors' efforts, the first prize for sentimental songs being won by Sgt. Burgess, and the first for comic by Pte. Bowie. At the conclusion of this, our Colonel made a speech in honour of our comrades who fell in the 'Charge.' On reading the names out the Colonel was almost overcome with emotion, and it was only with difficulty that he was able to complete.

"About this time we received the news that our forces operating on the Euphrates river had captured Hit on the 7th inst., and on the night of the 9th the enemy had evacuated Sahilijah, abandoning 2 mountain guns and much ammunition.

"On the 14th the 2nd Baghdad Race Meeting was held, also the semi-finals and finals of football, hockey, polo, golf, and boxing tournaments were contested during the week. Our football team was knocked out in the third round by the 25th Machine-gun Squadron, by one goal to *nil*, after a hard-fought game. The outstanding feature of this tournament was the match between the Worcesters and the Buffs, the first game they played being of 60 minutes' duration; but as neither side was able to score extra time was allowed, and they played on for three periods of 20 minutes each, and again neither side was able to gain the advantage. The match was then postponed until the following day. On the replay, which lasted with extra time, some 2 hours and 40 minutes with still no score being registered, both teams withdrew from the tournament. The Regiment ran several horses in the racing, Lt. Ormrod's 'Venus' again winning the Baghdad Cup. This time she was ridden by Frank Wootten, the famous jockey, who happened to be in Baghdad at the time.

"Some little time after this we received the news that our Column, commanded by Major-General Brooking, operating on the Euphrates river, had, on the 26th inst., attacked the 50th Turkish Division at Khan Baghdadi. The Cavalry cutting off the enemy's retreat along the Aleppo Road, they surrendered with their Commander, the total captures being 18 Germans, including 2 officers, 212 Turkish officers, and 5236 Turkish other ranks, 12 guns, 57 machine-guns, 2 German wireless stations. During the six days' fighting they pursued the fugitives of this force to a point some 73 miles north of Awah, destroying in this village some 300,000 rounds of gun ammunition and some 3¼ million rounds of rifle ammunition.

"On April 9th some Divisional Races were held on the race-course near our camp, and in these several of our horses proved successful, Lt. Ormrod's 'Venus' winning the ½-mile steeple-chase, and 'Whiskers' by the same owner was first past the post in the 6 furlongs, but on an objection being raised was disqualified later. Mr Welstead's 'Nightshade' with its owner up won the sixth race, while Mr Jones' 'B 3,' ridden by Sgt. Holloway, won the Novices' Flat Race.

"On the following day we received the welcome news that leave to India was granted, and a party consisting of 2 officers and 12 men left the same day *en route* for India to enjoy a well-earned 28 days' furlough.

"A few days later, on April 14th, we commenced an Inter-Football Tournament. The first game was between 'B' and 'D' Squadrons; this resulted in a win for 'B' by 2 goals to *nil*. The next match was between 'A' and 'C' Squadrons, and this was a win for 'A' by 2 goals to 1, while the following day 'B' played 'A.' This match was very keenly contested and aroused great excitement, the result, after extra time had been allowed, being a draw of 1 goal each. Replaying again the following day, the result this time was a draw of 2 goals, and this also after extra time."

CHAPTER XX.

THE SUMMER OF 1918—KULAWAND AND TUZ KERMATLI.

THE season of military operations in the country to the north-east of Baghdad, where the British and Turks now faced each other, may be said to begin with the month of April, and in that month General Marshall's troops were ready to move again.

Before any movements on a large scale began, however, General Marshall had to deal with two troublesome complications, one on his left and one on his right.

To the left, in the Mahomedan holy cities of Nejef and Kerbela, beyond the Euphrates, the British invasion of Mesopotamia had naturally caused some excitement, and this had been fostered by German intrigues and gold. The holy cities do not belong to the same branch of Islam as the Turks, being "Shia," not "Sunni," or, as one might say, Protestant, not Catholic; and having been very carefully treated by the British they had shown a friendly spirit enough. Still they were Mahomedan, and contained some fanatical elements. In the month of January British troops had been attacked near Nejef, and a few weeks later the British political officer was murdered. Some punishment had to be inflicted for these offences, and though General Marshall would not attack the sacred town, it was blockaded and called upon to deliver up the murderers. Happily the well-behaved inhabitants, under the guidance of their religious leaders, decided to accept the British terms, and the whole affair was satisfactorily settled.

At the same time as this trouble arose to the westward, some Persian tribes on the east began to respond to the incitement and gold of German emissaries, and it was found necessary, as General

S.Q.-M.S. S. B. HAINES, D.C.M.

R.S.-M. (late Lieutenant)
M. CHURCHHOUSE

Sgt. S. G. STRAWBRIDGE, D.C.M.

Lieut. W. J. L. NORWOOD

2nd Lieut. J. H. LUCAS

Marshall says, to "give them a lesson." But this trouble also proved to be short-lived. Friendly tribesmen took action against the malcontents, and with the aid of a small British column completely defeated them. They lost heavily, "and our aeroplanes turned their defeat into a rout."

In the meantime, while these complications were being put straight, General Marshall had worked out his scheme of action against the Turks, and had decided that they should again be attacked to the north-east of Baghdad, so as to drive them farther away and make the Persian line of communication more secure. His intention was to expel them from the area about Kara Tepe, Kifri, and Tuz Kermatli, which will be found marked on the sketch-map, p. 201, and to occupy those points, thus turning and subjugating the troublesome range of Jebel Hamrin.

It was a difficult operation if the Turks fought well, and General Marshall's scheme of action was carefully worked out. Several columns were to move from different points, and it is not easy to follow in detail the movements of each. It will suffice to say that during the latter part of April and the earlier part of May 1918, in spite of mountainous ground and torrents of rain, which caused heavy floods in the rivers and washed away temporary bridges—in spite too of some stubborn fighting on the part of the enemy—the whole area was completely cleared and the country was occupied as far north as Kirkuk, one hundred and thirty miles from the railhead and considerably more than that distance from Baghdad. The British losses were very small—only twenty-six killed and two hundred and ten wounded—while the Turks lost heavily, over three thousand in prisoners alone.

General Marshall had good reason to be satisfied with this result. "Owing," he says, "to the distance covered by these operations, and to the bad weather which prevailed throughout, the work of the troops was very arduous, and I have nothing but admiration for the good work put in by all arms and the excellent co-operation which was maintained between them." This co-operation included much continuous work done by the Cavalry, and on at least two occasions they had the chance of attacking the enemy as Cavalry, mounted and sword in hand.

On both these occasions the Thirteenth were well to the fore.

They were no longer with their own Brigade, but had temporarily taken the place of their old comrades of the Fourteenth Hussars in the 6th Brigade, then commanded by Brigadier-General Holland Pryor. General Marshall describes the first action in the following words: "The 6th Cavalry Brigade . . . after an arduous night march overtook the Turks who had . . . taken up a defensive position at Kulawand. . . . Feigning a frontal attack, whilst gradually working round the enemy's right flank, our Cavalry cut the enemy's lines of retreat and then charged right through his Infantry, killing some two hundred, including two battalion commanders, and capturing 565 prisoners, a mountain-gun, and considerable booty." This was a fine performance, and brought the Regiment much credit. It was followed up by some skilful manœuvring which induced the Turks to hold a position at Tuz Kermatli, where a day or two later there was another considerable action. Here again the Cavalry did well, and the Thirteenth got another chance. This time the actual charge was a smaller affair, confined to a couple of troops under Lieutenant Macdonald, but it was spirited and successful, the two troops galloping a nest of machine-guns, and taking a considerable number of prisoners with little loss to themselves.

On the 7th of May Kirkuk was taken, and the short campaign came to an end. "Difficulties of supply," says General Marshall, "due to distance and state of ground, now made it imperative to stop further pursuit," and a few days later most of the British troops were withdrawn to their summer camps. The Thirteenth returned to their old camp at Chaldari, near Baghdad, arriving on the 22nd of May, and there they remained throughout the summer months.

It proved to be an important and anxious time for the Commander of the Army, as a force under Major-General Dunsterville was sent into Persia, and eventually crossed the Caspian to Baku, where there was some severe fighting. There was much work to be done also in Mesopotamia itself, and so far as the climate permitted, military training went on steadily, a special feature being the development of schools of instruction. But owing to the extreme heat the summer was for the bulk of the troops a period of inactivity, and it was found possible to send a large number of officers and men on leave to India. They wanted all

AFTER THE CHARGE AT KULAWAND

PRISONERS TAKEN AT KULAWAND

the rest they could get, for they still had some hard marching and fighting before them, and the Thirteenth, in their pleasant camp at Chaldari, thoroughly enjoyed the respite.

The following extracts refer chiefly to their doings at Kulawand and Tuz.

Extracts from narratives by Captain Gowan, enclosed in a letter from Colonel Richardson, dated 27th July 1918.

"*April* 26-27, 1918.—The column moved at midnight and marched till 4.45 A.M., arriving at Tel Sharif at that time. The road was difficult to find in many places, and in the numerous nullahs which were crossed the going was deep. When the position had been fixed at dawn, the column moved forward for 2 miles, and then fed at 6 A.M. After ½ an hour halt the Regiment with advance-guard of C. and C. moved to M. 27 Central, and halted till 9 A.M. At that time the Regiment was ordered to move on the Kulawund position, and with L.A.M. cars and Lewis guns and Ford vans turn it from the W. end by enfilade fire. Some difficulty was found in locating the position, and it was not till 11.30 that it was reached and found unoccupied. The L.A.M. cars and Fords had been unable to get forward owing to water-logged country and nullahs. A number of Turks with some tents were seen to be in position on the Tuz-Kifri road, close to the hills, and their numbers were estimated by an aeroplane at 250. By 12 noon the G.O.C. Brigade had moved forward, and arrangements were made for a simultaneous attack by the 3 Regiments. At about 12.20 a move forward was made, the Brigade advancing in line of column of wings, in the order 22nd, 21st, 13th from the right. I understand that both Indian Cavalry Regiments advanced in line whilst we were in column of wings: I am not sure whether the 22nd had all four squadrons or only two. The Regiment had 'C' and 'D' Squadrons in front line under Captain Godfree, 'B' and half 'A' in the second, at about 200 yards' distance, the remaining troop of 'A' being left as guard to our left flank, dressing station, &c., and the H.G. guns being sent to the flanks. The leading wing on nearing the road shouldered to the right and were in a position to co-operate with the 21st Cav. when Captain

Godfree received orders from the Brigade to take no further part in the attack, but to join the Headquarters near Kulawund village. While he did this the 2nd Wing, leaving a troop of 'A' as escort to the Regimental Aid Post, moved to the Kifri-Tuz road and scattered along it to the left or Tuz direction, and succeeded in collecting about 50 prisoners. Two or three machine-guns were in action against us until we were within 300 yards of the road, and were replied to by the H.G. troop of 'D' Squadron from our left flank. A number of the enemy succeeded in getting away into the ridge of hills E. of the road where Cavalry could not pursue, and after despatching the prisoners to Brigade H.Q. the wing rallied and moved to a mound at the edge of the Kulawund position. From captured officers' statements it was clear that this Cavalry attack took the enemy entirely by surprise. The enemy troops engaged consisted of the 5th Regiment, which had received orders on the evening of the 26th to withdraw from Tuz. Expecting only an Infantry attack, on the following morning they evacuated their positions without undue haste, and after halting several hours . . . were on their way to Tuz on the 27th, when they were overtaken.

"*April* 29.—We drew back and circled to the W. behind the battery, and were ordered to get touch with the 21st and 22nd Cav. and then attempt to get across the Tuz-Tan road to stop the enemy's retreat. Moving W., it was found that the 22nd and 21st were held up by a nest of M. guns on a small knoll. At the time of our arrival these were not directed on us, and appeared to have left their position. Two troops of 'B' were accordingly pushed forward, and led off so as to leave this knoll on their right. After a short distance they came under M.G. fire, and 2nd Lt. MacDonald wheeled the two troops to the right, thus putting his own troop in front and that of Sgt. Brencher in rear. He galloped the hill and took it—the enemy, except two, escaping being sabred by putting their hands up when the men were on them. This action opened the way across the plain, and the three Regiments, 21st, 13th, 22nd from the right, moved across, meeting small bodies of the enemy, and killing or taking them prisoners."

Letter from Colonel Richardson to Sir R. S. Baden-Powell.

"7TH CAVALRY BRIGADE, M.E.F.,
IN THE FIELD, *1st May* 1918.

"MY DEAR GENERAL,—Your note with enclosures regarding Regimental Monuments and Gravestones reached me only a few days ago, when we were already on the trek of our present stunt. . . .

"I am sorry to say I quite forget the contents of my letter to Symons to which you refer, but I imagine it dealt with our part in the battle of Tekrit; since that time we have been almost continuously in one camp, on the Tigris, about 40 miles above Baghdad, and although for more than four months we did not have a shot fired in anger we were far from idle—*i.e.*, when the wet weather made it possible to be otherwise. Not only were we able to put in a lot of much-needed—on account of the numerous officers and other reinforcements during the past six months—instruction in equitation and elementary and secondary drill and training, but we were able to have games and sports—both mounted and dismounted—for the men, and polo for the officers; we also had the opportunity of getting fitter than at any time since we have been in this country, and I doubt very much whether men or horses have ever been fitter physically than when we started out, ten days ago, for this stunt. Only one Brigade of Cavalry is being utilised, and it is not the one to which we properly belong, but we are taking the place of the 14th, who are scattered about on an independent mission. Our march to the rendezvous was not propitious, as we were dogged by heavy rain and thunder, which made life on operation scale, without tents, a delusion and a snare for four days, but we got through with only one case of pneumonia. Once we got started on this stunt—the object of which has been to clear the whole of our north-easterly front of the enemy—we began to get into a country of magnificent crops and glorious grazing, in which the horses have been revelling, and growing visibly bigger. The operations themselves have partaken more of the nature of South African drives, and Johnny Turk was, I think, caught napping by the largeness of the scale of the operations, and he has been so badly fed lately that he could not travel sufficiently quickly to escape us; however, he put up, as he always does, a jolly good fight when cornered, and as we were unable to close the fourth side to the square—*i.e.*, on the north, where a steep ledge of rock, backed by hilly country, discounted our mobility—a good few got away. However, we have accounted for nearly 2000 prisoners, ten large and very many machine guns, and our casualties have been less than 250 (in the whole force), so I suppose we may reckon the operations as having been fairly successful. Of course, from the nature of things, the principal rôle fell to the Cavalry, and even the Pan-Infantry men were fain to admit that we have done it well. On the first day we came into touch with the enemy—mostly troops retiring from outside positions, from which they had been driven by the Infantry; but, covered by the garrison and guns of the aforesaid rocky ledge, the three Regiments of the Brigade galloped in two lines of wings over about 1½ miles of country towards the ledge, and swung right-handed along the foot of it, the net result being about 600 prisoners and several machine-guns. The next day there was a pause, as the main position, which was pretty well served by big guns, still remained, and we could not hope to take it without the help of the Infantry; but next morning the latter came through our outposts, and we concentrated on the left flank and began the advance, which I may fairly claim the Regiment initiated, which put the finishing touch to the operations. A good few of the enemy no doubt escaped through the hill country, and perhaps he got a gun or two away, but certainly a thousand prisoners and ten guns fell into our hands.

"When I claim that the Regiment initiated the final advance, I must explain the situation in more detail: over night 2 squadrons were on outpost duty on the left bank of the river, and two troops of a third squadron were away with the first batch of 500 prisoners; the balance of the Regiment with which I was, was on the right of the Cavalry Brigade, advancing up the right bank of the river; we were held up at dusk by the enemy in an entrenched position, and remained where we stood all night; soon after daybreak the Infantry took over our line, and the Brigadier sent me off to the left flank to try and get across the path of the retiring enemy. I found that the two Regiments of Indian Cavalry

Capt. D. W. Godfree, M.C.
(21st Lancers)

Lieut. H. Macdonald, M.C.
(*Murdered 14th July 1919 at Amadia by Kurdish gendarmes*)

Lieut. C. A. G. McLagan
(*Wounded at Tuz Kermatli, 21st April 1918*)

Capt. B. W. D. Cochrane

Sergt. W. Matthews, M.M.
(*Killed at Tuz Kermatli, 28th April 1918*)

Lieut. L. Osmond

had been trying to advance with the same object for a couple of hours, but were held up by small parties of enemy entrenched on knolls and well supplied with machine-guns. I determined to advance, and put one of my few remaining sabre troops into the scouting line under MacDonald, a lad who only joined us 6 or 8 months ago. When we got about 800 yards from the nearest knoll he met very heavy machine-gun fire, but as it was apparently directed against his left flank, he at once gathered his troops to the right, and following a circular path he galloped the hill, taking the gunners in flank and reverse, and in less than five minutes had the whole party of Turks killed or marching as prisoners to the rear, and 4 machine-guns captured. He had only one man badly wounded. It was, I think, without exception the quickest and prettiest—not to say the most successful—small piece of work it has been my fortune to witness. This little episode put great heart into the Indians, and thereafter it was almost a race for the foot-hills; all opposition went like a house of cards, and it was only a question of mopping up. I, on behalf of the Regiment and MacDonald, have received congratulations on all sides, and the Brigade and Divisional Commanders have promised to support my recommendation for immediate rewards, so I hope MacDonald and his troop may get the honours they have earned so well.

"With the exception of Twist and of my Adjutant Gowan, I have no other officers with pre-war training and experience, but we have been most exceptionally lucky in our temporary officers; for this, I think, I have to thank Ronnie Brook chiefly, and he would, I am sure, be interested to see the account of the operations, should you have it copied.—With all good wishes, yours sincerely,

(Signed) J. J. RICHARDSON.

"*P.S.*—The following coincidence of special interest has come to light. Among material captured by MacDonald on the knoll was a signaller's telescope, which he gave to his signaller, Sgt. Burgess. The latter at once recognised it as the one he himself lost at Lajj on March 5th last year, when his horse was shot and he had to abandon his gun."

Letter from Colonel Richardson, dated 7th Cavalry Brigade, M.E.F., 3rd May 1918.

"For the last fortnight we have been operating with the other Brigade in place of Col. Bridge's Regiment, which is engaged on an independent mission. . . .

"The Regiment has seen some smart fighting, including two charges, in the past week, and the outstanding feature of the operations, which has been very successful, and of which I am sending an account to B.P. [Sir Robert Baden-Powell], has been a gallant and brilliant charge by a troop of the 13th under MacDonald, a Scotch lad who joined us about 6 months ago.

"It is a matter of greatest pride, mingled with pleasant surprise to me, how they continue to roll up with these first-rate performances, once the opportunity presents itself, and I only hope I may never let them down."

Extract from letter from Colonel Richardson, dated 7th Cavalry Brigade, M.E.F., 25th May 1918.

"After writing to you on the third of this month we continued our drive northwards for a matter of about 60 miles, the Turks pursuing their usual tactics—*i.e.*, holding us up at strong positions with their big guns until dark enabled them to get away their valuable papers and other important stuff. In this way we made them evacuate the country as far north as Altum Kupri.

"I do not hesitate to give names now, as I see Reuter has published the message from Candler, the correspondent, in almost the identical words that I gave him—he was messing with us during that period—which is within striking distance, not more than 70 miles, of Mosul, the big Turkish base.

"Owing to the difficulties of transport we had been on half rations, but we got most glorious grazing for the horses in most places, so they did not suffer much until we struck the Jebel Hamrin with its brackish water and no grazing. Right up there the climate was really good, and we could work all day, though, of course, it was hottish, and on one or two nights we got wet through, for, of course, we had no tents. But in this case also, as soon as we struck the

Jebel Hamrin on our return, the sun became very powerful and the sand-flies very bad.

"However, the operations were very interesting, although the Regiment had no opportunity of doing more than ordinary work, because Johnny Turk declined to come to grips; but as the Brigade put it: 'You did just what I wanted you to, I could not ask you to do more,' I don't think we had any cause to grumble. . . .

"When I wrote you from Mendali, in October, that we were short of food, I referred only to mess stores and extras; rations have always been good, especially when we were within measurable distance of the railway, when we got most excellent frozen beef. Indeed, I expect we are being much better fed than you poor people."

Extract from Lance-Corporal Bowie's Diary.

"Fate decreed that we were not to finish this tournament, as on the following morning we were ordered out to take part in operations again. This was the 18th of April, and we had to leave camp some few hours later than ordered; owing to the extremely bad weather and the high floods having broken the pontoon bridge over the river, it was impossible to cross the Tigris till midday. Heavy rain continued to fall throughout the day, which made the going very difficult, and we finally arrived at Baqubah at about 8 P.M., where we picketed down for the night. Our transport arrived in at 9.15 P.M. Every one being so thoroughly wet through, and the weather being so bitterly cold, a rum issue was ordered, after which we were fortunately able to obtain shelter for the night in a rest-camp near by. Owing to the heavy rains which had fallen the column did not march for Abu-Jisrah until 10.30 A.M. the following morning, and after some very heavy going we made that place at about 3.30 P.M. Our transport experienced great difficulty in getting along, and did not arrive until some two hours later. The heavy rains continued throughout the night. The following morning, April 20th, it was found impossible to move in consequence of the exceedingly heavy rains, which continued till 2 P.M., and made the ground a perfect quagmire, and absolutely impossible for wheels, the men during these two days having a most uncomfortable time owing to the impossibility of erecting any sort of shelter. An incident worthy of note here was an impromptu football match, played amid a perfect sea of mud, in which the ball not infrequently became imbedded, whilst the rain poured down in torrents. All this, however, did not deter the players, who, after a most strenuous game, did obtain some little warmth from this exercise. At 11.30 A.M. on the morning of the 21st instant we continued our march to Sharaban, where we arrived about 4 P.M. Here we joined up with the 6th Cavalry Brigade, who were in bivouac there. During the next two days we remained in bivouac, whilst we endeavoured to free ourselves of some of the mud with which every one was plastered, grazing our horses meanwhile. Here also our squadrons were arranged to conform with the formation of the 6th Brigade; each squadron was to consist of three sabre troops and one Hotchkiss-gun troop. On the morning of April 24th the Regiment paraded at 9.30 A.M., but owing to heavy rain falling again we did not move off until 11 A.M., when we marched at the head of the Brigade crossing the Cutthroat and Auction Bridges; we halted at Ruz for a short time, and marching on again we crossed the Table Mountain Bridge over the Diala river at 1.30 P.M. Here we watered and fed our horses, afterwards moving off again with the 22nd Native Cavalry, commanded by Lieut.-Colonel Young. We marched on to Abu-Haiar, arriving there about 5.30 P.M., and bivouacking for the night. The following morning at 11 A.M. we marched to Ain-Lailah, which we reached at 3.30 P.M. We bivouacked that night about 1½ miles south of the Pass. The Column commenced marching through the Pass at 3 A.M. the following morning, but the Regiment being Rear Guard did not march through till 7.30 A.M., and then proceeding on to the Nahrin river, where we watered and fed our horses. Resuming our march till 11 A.M., we joined the remainder of the Column, who were in bivouac at 12.15 P.M. Here one troop proceeded to reconnoitre the road over which we were to make a night march the same night, after which they returned to bivouac at dusk. Marching again at midnight, the Column proceeded to Tel-Sharia, arriving at about 4.45 A.M. The road was very difficult to find in many places, and the nullahs which we had to cross were very deep, many being almost full of water, owing to the recent heavy rains. After our patrols had located the enemy at dawn the Column moved forward some 2 miles and halted till 6.30 A.M., when the Regiment

LT.-COL. J. J. RICHARDSON, D.S.O., AFTER A HARD DAY AT TUZ

MACHINE GUNS CAPTURED BY THE REGIMENT AT TUZ

was ordered to advance another mile and halt. At 9 A.M. we received the order to advance on to the Kulawand position: accompanied by our armoured cars and some Lewis guns mounted on Ford cars, we proceeded to attack it at the west end. We experienced great difficulty in gaining the position, not through an enemy resistance, which strangely enough was not offered, but through the water-logged condition of the country and the very deep nullahs, and owing to this our armoured cars and Ford cars were unable to continue the advance. On us eventually gaining the position we found it to be unoccupied. The enemy, having retired during the night, were located by our airmen a little later, who reported them to be holding a position on the Tuz-Kifri road, close up to the foot of the hills. Although we had no support within 20 miles of us, it was decided to attack them at once. Having worked our way up to some large Moards [? mounds] some 1¾ miles from their main position, we took cover here to give [? save] our horses. Some few minutes later we drew our swords and trotted into the plain where the 10 squadrons were formed into an extended line in the formation of a scythe. As we rode on at this pace parties of the Turks started running down the road, while other groups took up positions in the rough ground and water-cuts, or kneeling in the high corn opened fire on us. Here the Turks showed some of their old spirit in the stand they made, but the sudden flash of our swords in the sun which made a wide arc of light seemed to take the heart out of them, and their fire was wild and high, and our casualties few. Our artillery kept up a creeping barrage on the village as we attacked, and our line maintaining this pace until within some 500 yards of the enemy, when we lowered our swords and charged into them. Our aeroplanes which took part in the action flew very low and increased the enemy's confusion with bombs and machine-gun fire. Over 150 Turks were killed and 538 prisoners were taken, also 2 mountain-guns, very few of them escaping in the hills. Meanwhile Kifri-ain-Faris and Chanan-Keurri had been occupied by our other columns without opposition, and our movements were so rapid that 2 battalions of the enemy had to take to the hills in order to escape the net we were spreading for them. We afterwards heard that this force had been robbed and disarmed by the Kurds, who never lose an opportunity of revenging themselves on the Turk, and even those who escaped us in the Kulawand charge returned and surrendered to us rather than face these tribesmen. After sending back our prisoners we moved forward to reconnoitre Tuz, and we soon came under some very heavy accurate gun-fire until about 4 P.M., when orders were received to withdraw. We then moved back some ten miles and bivouacked behind our line of outposts for the night. At 6 A.M. the following morning 'A' and 'D' Squadrons were sent to reconnoitre Tuz, the Brigade following some half an hour later, while the 21st Cavalry were despatched to find a ford over the river Aqsu at Khasradalah. 'A' and 'D' Squadrons soon came under fire from Tuz and suffered some casualties. Meanwhile the remainder moved on to Khasradalah in an endeavour to surround Tuz, but were held up by enemy machine-guns in position on the banks of the Aqsu river; and the armoured cars being unable to advance along the Kifri-Tuz road, no farther advance was possible. We halted here for the night after having crossed the river, while 'C' Squadron and a section of the Machine-Gun Squadron were thrown forward on each side of the track; 'B' Squadron, having all the horses, were some 1500 yards in rear down by the river. Intermittent firing was carried on throughout the night. At 4.30 A.M. we stood to arms and saddled up, receiving word that our 2 Infantry columns would advance at dawn, one column advancing up the right bank of the river, and the other up the left bank, the latter with the object of taking the Heights of the Naffdach, east of Tuz. The enemy had been reinforced during the night by a battalion from Tauq, and was holding a front of 7 miles, stretching from Yanhah-Buyuk to Tuz-Khurmatli, and at Tuz they held a position covering the road and extending over the hills. At daybreak the Hotchkiss-gun troop of 'C' Squadron opened fire from the trenches on the right bank of the river, and soon came under some pretty accurate shell-fire. Shortly after this B Battery, of the 66th Brigade R.F.A., galloped up beside them and opened up a heavy fire on the enemy's position at about 6 A.M., the Infantry having deployed for the 2 main attacks on Tanijah from the south-west, and on Tuz from the south. We were ordered to withdraw and make a detour to the west. We circled behind the battery to join up with the 21st and 22nd Cavalry and get across the Tuz-Tauq road to cut off the Turks, who were already falling back before our Infantry, who were advancing with steadiness under machine-gun and artillery fire, right up to the guns which were in action until they were within two hundred yards of them. Meanwhile we were held up for a time by a nest of machine-guns some 1200 yards away on a small knoll. Two troops of the 'B' Squadron, under Lt. MacDonald, were ordered out as a scout line. They led off to the left of the knoll, and having got in line with it, wheeled sharply to the right and charged the position in the rear and the flank, all the enemy except two throwing up their hands to escape being cut down. The taking of this knoll opened the way across the plain, and we continued to advance, charging bodies of Turks here and there, either killing them or taking them prisoners. The largest body we found was between the Tauq road and the hills, and these put up some resistance for a time until charged from two sides by us and the 22nd Cavalry, the 21st Cavalry having wheeled towards Tuz to deal with other bodies, who at that time were shelling 'A' and 'D' Squadrons, commanded by Captain Godfree. A number of the enemy made their way over the hills and were harassed by our fire. Soon after this the fire of the enemy died away, and by 9.30 A.M. all opposition had collapsed, the whole position having fallen to us inside 4 hours with a very slight loss. 1200 prisoners were taken, 25 machine-guns, and 12 field-guns that were in the position. The action having passed, we rested for some time where we were, and on 'A' and 'D' Squadrons rejoining us in the afternoon we went into bivouac opposite Tuz at about 3 P.M. We remained in bivouac here for five days, during which we rested and grazed our horses. Whilst here, our Brigade Commander received the following wire from General Egerton: 'Many congratulations to troops. Please convey my thanks to General Caley and General Hollen-Prior for their good work, and that of their officers and men. Cavalry charge must have been magnificent. Please accept my repeated appreciation of the performances of all ranks under your command.' On the 5th May we moved off at the head of the Brigade at 8 A.M., and after crossing the bridge of the Tauq-Chai at about 5 P.M., we bivouacked for the night some 2 miles north-west of it. We moved off again at 8.30 A.M. the following morning in the direction of Taza, when shortly after our aeroplanes reported that the enemy had evacuated their position there. On receiving this information we were ordered to advance to catch the retreating enemy at the village of Taza-Kharatli. On arriving here at 11.30 A.M., after crossing the Oazah Chia on the Kirkuk road, we found the village unoccupied. Making a short halt here we moved off again at 1.45 P.M., marching on to Kirkuk. The armoured cars, after having been towed through the river, had preceded us along the Kirkuk road. At about 3.39 P.M., nearing Balban, we came under some shell-fire at long range, this we avoided by making a detour to the east. After sending out patrols, who were met by a very heavy fire, we galloped forward and dismounted for action against the enemy holding the trenches just outside Kirkuk. Finding the enemy so strongly entrenched, we were unable to advance, but hotly engaged them until dusk, when we were ordered to withdraw and form an outpost line about 3 miles from the town. This we did about 7 P.M., when a heavy storm broke out and very heavy rains continued to fall throughout the night, which, combined with the intense cold, made rest for us that night an impossibility. (The following names appeared on the Order for the day for gallantry in action: Capt. J. A. Jeffrey, Capt. A. M. Sassoon, Lt. J. J. Crystall, and R.S.M. Seekins, all being awarded the M.C.) Our outpost line was formed near Balban, on a track running east and west. The night was fairly quiet, except for several loud explosions which occurred in and near Kirkuk. The morning found the ground more or less a quagmire, and at 6.30 A.M. we received orders to advance on the town. 'A' Squadron moved off at 8 A.M., followed by the Regiment and the 21st Cavalry some half an hour later. No resistance was encountered, and at 10.30 A.M. the Regiment entered the town, 'C' and 'D' Squadrons, under Capt. Godfree, moving on to the high ground covering the north of the town. Several large hospitals were found here, one containing 358 patients, the population being in a starved condition. There were also several corpses of Turkish soldiers horribly mutilated. This was said to have been done by Arabs to avenge the raping of their women-folk. The whole place was in a filthy condition. At 3 P.M. the Column moved out of Kirkuk and bivouacked near Taziyan, some 3 miles down the Taza road, a heavy storm breaking over us just as we were pegging down our horses. Meanwhile 'C' and 'D' Squadrons were sent forward to some hills some 4 miles north of the town

GUNS CAPTURED AT TUZ

THE BRIDGE AT MOSUL AND "THE BAGHDAD"

to harass the retreating enemy, and to protect our own main body from attack. These two squadrons, after successfully hastening the enemy's retreat, reconnoitred the road to Yarvali, rejoining the Column in bivouac about 5 P.M., bringing with them 2 prisoners. The Column remained in bivouac the following day, and at 12.30 A.M. we received orders to send 2 squadrons to reconnoitre the northern road to Alton-Kupri as far as Sakizli, 'B' and 'D' Squadrons doing this commanded by Capt. Godfree. They were also accompanied by a Company of the R.E.'s for the purpose of destroying enemy shell-dumps which were known to be in the vicinity. Two such dumps were located and destroyed near the village of Daraman, and at Sakizli 7 large motor-lorries with engines destroyed were found. These contained 2 mountain-guns, 2 machine-guns, and a large quantity of shell ammunition of various sizes. At 5 P.M. orders were received by heliograph to proceed to Yarvali and search it for any documents the enemy may have left. Arriving there about 8 P.M., we discovered a number of documents, and also several aeroplane engines and drums of petrol, and a quantity of shell ammunition. It being necessary to destroy this material, the wing returned and bivouacked at 11 P.M. The Column rested, and we grazed our horses all the following day, and at 6 A.M. on the 10th May we marched on Alton-Kupri with the intention of driving the enemy over the Lesser Zab river and destroying the bridge there. We were preceded by 2 of our armoured cars, who at about 11.15 A.M. were held up by the enemy, who were in position on Guk-Tappah Hill, while the road was shelled by the enemy's batteries in Alton-Kupri. The armoured cars succeeded in driving the enemy from Guk-Tappah, but were quite unable to advance farther owing to a deep canal which cut the road at this point. After observations had been made from Guk-Tappah, the Regiment received orders to make an encircling movement to the right, whilst the 21st Cavalry did the same to the left. The 22nd Cavalry advancing in the centre succeeded, by covered approaches, in getting within 2 miles of the village which was on the farther side of the river. The Regiment experienced considerable difficulty in advancing, owing to a considerable number of very deep nullahs, which were full of water at the time. The village was hidden from view by the high ground on this side of the river. We advanced by wings 'A' and 'B' leading in extended order. After advancing about 1 mile we were heavily shelled by mountain-guns, 14 pounders and 4.53, and considerable rifle and machine-gun fire was opened on us by enemy Infantry who were holding a series of trenches and rifle-pits on this side of the river. The Regiment immediately dismounted for action. The Hotchkiss-gun troop of 'A' Squadron moving forward opened a very effective fire which was well supported by our batteries. This proved too much for the enemy, who retired to their second line of defence. On this side the Hotchkiss-gun troop of 'D' Squadron advanced about 1000 yards with 'C' Hotchkiss-gun troop in support. Meanwhile 'B' Squadron were endeavouring to get down to the river bank, but were unsuccessful owing to the heavy rains and accurate firing of the enemy's batteries who were dug in on the opposite bank. We continued to engage the enemy until about 6 P.M., when orders were received to withdraw to Daraman. During our retirement the enemy shelled us with increased violence, and we sustained several casualties, but luckily the number was small in comparison with the terrific fire we were subjected to. One troop of 'A' Squadron was left near Guk-Tappah as escort to our casualties, whilst we continued our march back till 10.45 P.M., when we bivouacked for the night. Orders were received the next morning to return to Kirkuk. We did not move off until 2 P.M., and arrived at our previous bivouac near Kirkuk at about 6 P.M. The following morning a strong patrol left bivouac at 7 A.M. to reconnoitre a vast stretch of unexplored country between Taza and Telabw-Shahanah, whilst the remainder of the Regiment, moving off some half an hour later marched to Taza, some 9 miles distant, where they went into bivouac. It was whilst here that a small, but nevertheless fierce, caste riot broke out among the Indians who had the handling of our transport. This necessitated the whole of the Regiment having to turn out fully armed before it could be quelled. At 5.30 A.M. the following morning, 13th May, the Regiment commenced to march to Cemetery, where we were to join the 7th Cavalry Brigade again. To reach the above place we now had to cross some hitherto unexplored country, and so far not having received any reports from our patrol which had left us the previous day to reconnoitre this part of the country, we had to proceed with caution. Halting at about 10 A.M. to water and feed our horses, we moved off again at 11.30 A.M. Meanwhile our patrols had completed their reconnaissance, and got into communication with us by helio, sending us a message to the effect that an enemy column were rapidly disappearing into the hills. The Regiment not being strong enough to intercept them, the patrol was recalled and we continued our march, arriving at Cemetery, and joined the 7th Cavalry Brigade at 8 P.M., after marching some 50 miles. Resting till 4 P.M. the following afternoon, we moved through the Pass to Ain Nakhailah, some 4 miles distant, bivouacking about 1 mile south of the salt springs there. The following morning we continued our march on to Daur, which was some 20 miles over a waterless desert. This proved a very trying time for us, the little water which we were able to obtain before leaving the springs being very brackish and salty, which, combined with the intense [heat ?], produced a thirst which was well-nigh unendurable. These arduous marches, coming on the top of our very recent hard fighting, tested our endurance to the utmost, several of our men falling out of their saddles from sheer weakness from want of food and rest. The sight of the river Tigris had a noticeable and stimulating effect on both man and horse, and on reaching the river both were jostling each other in their frantic efforts to quench their burning thirst. On bivouacking here many of us took the opportunity of having a much-needed dip in the river. Our way now lay alongside the river, and owing to the condition of both men and horses, it was decided to finish our march by easy stages. Accordingly, at 4.30 A.M. the following morning we trekked to Samarah, a distance of 12 miles, and bivouacked some 2 miles north of the town. Moving off at the same time the following morning to avoid the intense noonday heat, we marched a farther 13 miles and bivouacked at Khan Mifraji at 10.30 A.M. We trekked on to Akab, a distance of 15 miles, the following morning, crossing the pontoon bridge over the river Adhaim, and bivouacked about a mile distant on the other side. The morning after this, commencing at 5 A.M., we did another 17 miles on to Ktuwair, and went into bivouac there at about 12 noon. From here we moved off at 4.30 A.M. and marched to Baqubah, another 16 miles distant, where we arrived at about 11.30 A.M. Trekking a farther 12 miles the following day, we bivouacked at Conningham's Post at 10.30 A.M.; the following day, the 22nd May, we again moved off at 4.30 A.M., and covered the remaining 16 miles to our summer camp at Chaldari, where we arrived at about 11 A.M., our strength on marching in being 19 officers, 315 other ranks, and 445 horses. On the morrow the following was published in Regimental Orders: 'The Commanding Officer wishes to take the opportunity of expressing his keen appreciation of the conduct of all ranks during the recent operations on the north-eastern front which have so successfully attained their object. During the course of these operations the Regiment has encountered, apart from the enemy, many hardships and difficulties, and it is the cheery spirit in which these hardships have been met, and the constant courage with which these difficulties have been overcome, that have elicited once more the Commanding Officer's admiration and pride.'

"The same day the General ordered that our horses were to have three weeks' rest. On the 25th May the following extract from a letter to our Commanding Officer, Lt.-Col. J. J. Richardson, from our Colonel-in-Chief, Lt.-General Sir R. S. S. Baden-Powell: 'I was delighted to see so many of the Regiment mentioned in despatches the other day, and I want to offer my best congratulations and good wishes to Twist, Steele, Barrett, Gowan, and Williams, and also to the N.C.O.'s and men who gained distinction. It is a great delight to us all at home to know that the Regiment has done so well.'

"The following extract also appeared in the same orders. Extract from Order of the Day, No. 105, dated 18/5/18: 'Awarded the M.C. Lt. H. MacDonald. Awarded the M.M. Sgt. J. Cunningham, L.-Cpl. Kitcher, and Pte. Owen.' Two days later after this we received the news that Pte. Owen had died from his wounds."

CHAPTER XXI.

THE AUTUMN OF 1918—LAST BATTLE WITH THE TURKS—CLOSE OF THE WAR IN MESOPOTAMIA.

THE summer of 1918 having come to an end, the British army in Mesopotamia prepared for another advance.

The general aspect of the Great War had by that time undergone a sudden and promising change, for the Germans had failed in their fierce offensive against the Allies on the Western Front, Austria had shattered herself against their lines in Northern Italy, and Bulgaria was firmly held by the Salonika force, while in Asia the exhaustion of the Turks was now very evident. They had completely failed to stop General Allenby's victorious advance in Palestine and Syria, and were hastily withdrawing troops from the Persian frontier and the Caucasus. In Mesopotamia they were still trying to show a bold front, but it seemed clear that they were no longer in strength sufficient to withstand a determined attack by the large and eager British army opposed to them.

"This, then," says General Marshall, "seemed the moment to take the offensive against the Turkish 6th Army, covering the approaches to Mosul, and it was with great pleasure that on the 7th October I received orders to carry out this movement."

The main body of the enemy's troops, now reduced from long fighting and other causes to about 9000 men with 59 guns, held a formidable position astride the Tigris, at the Fatha Gorge — some thirty-five miles above the British railhead at Tekrit. The position had been thoroughly prepared for defence during the past eighteen months, and was of great natural strength. The Turkish right flank rested securely on two ranges of rugged hills which projected into the waterless desert, and therefore could not be turned by slow-moving Infantry. Their left extended for about five miles along the crest of the Jebel Hamrin, and was strongly entrenched.

To attack such a position by direct advance up the Tigris was a difficult matter, and General Marshall would have preferred to move his main force by way of Kirkuk in the north-east, but this was impossible. A large part of his transport had been taken

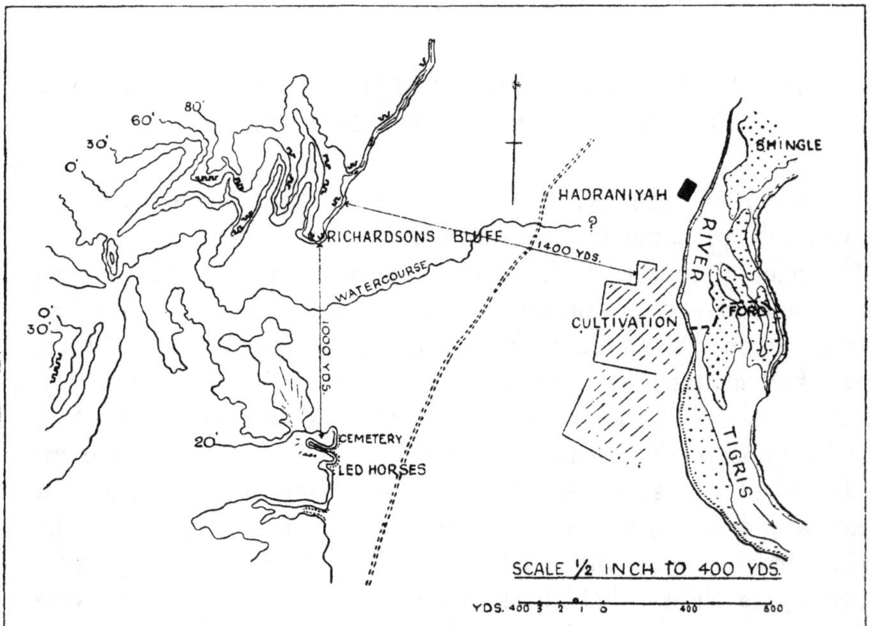

APPROXIMATE SKETCH OF RICHARDSON'S BLUFF POSITION

away by the operations in Northern Persia and the necessity for collecting the harvest, so that there was not enough to subsist a large body of troops at such a distance from the railway. The only available route for the main force, therefore, was the line of the Tigris, by which supplies could be pushed up, and this was perforce the one selected. Even so there was only transport enough to make mobile his Cavalry and half his Infantry.

The plan adopted was to work up the eastern bank of the river and turn or force the enemy's left. If that could be done, the

LT.-COL. J. J. RICHARDSON, D.S.O. SEPTEMBER 1918

intention was to cut his line of retreat on Mosul by means of Cavalry working round his left, and light-armoured motor-cars moving round his right through the desert. A small column was sent up towards Kirkuk under Brigadier-General Lewin to keep off any attack upon the British right.

But the reduction of the Turkish position proved to be an easier task than any one had expected. On the night of the 23rd October, when all was ready for the general attack on the Fatha Gorge, a British column pushed along the crest of the Jebel Hamrin against the enemy's left, while the 7th Cavalry Brigade, to which the Thirteenth had now returned, moved round the north of the range. "The vigorous action of these two columns," says General Marshall, "caused the Turks to abandon the very strong Fatha position before daylight," and to retreat up-stream.

Then followed some days of manœuvring and fighting, the Infantry pressing steadily upon the retreating Turks in spite of great difficulties, for the enemy had destroyed the road through the hills on both banks of the river, and a Brigade of Cavalry under General Cassels trying to work round into the enemy's rear. In this attempt the Cavalry were successful. Moving wide of the Tigris, they rode right round the enemy's left, and on the 26th October, coming back to the river, they struck a ford opposite Hadraniya, fifteen miles above Shergat, the point where the Turks intended to make their real stand. This movement was masked by the other Brigade, the 7th, under Brigadier-General Norton, who demonstrated on the line of the Tigris.

The Hadraniya ford was a bad one, for it crossed three channels of the river, one of which was nearly five feet deep, with a strong current; but Cassels got his regiments over, and, directly he had done so, sent one of them at a gallop to seize an important gorge five miles down stream, between him and the Turks, who were retreating upon him. By the morning of the 27th October the bulk of the Brigade had taken up a strong position blocking the road to Mosul, and it had been joined by the light motor-cars, which had got round through the desert.

It was a daring move, for Cassels had only three weak regiments of Cavalry and a battery of Horse Artillery, while below him was the retreating Turkish force, and above him to the north an unknown number of Turkish reinforcements from Mosul coming down to join it. But this was the one chance of capturing the main body of the enemy, and the Brigade meant to hold its position at all costs until supported.

For the next two days, the 27th and 28th October, Cassels was in serious danger, for though to conceal his weakness he boldly attacked, he was soon attacked himself from the south by a very superior force of Infantry and guns, which made repeated attempts to break through, while the Turkish reinforcements from Mosul moved against his rear. But though hard pressed he managed to hold his ground, and on the afternoon and evening of the second day support began to arrive. First came a battalion of Infantry which had made a forced march of thirty-three miles from the eastward, and then after dark the 7th Cavalry Brigade, which after a march of forty-three miles crossed the Tigris by the difficult Hadraniya Ford. In doing so they lost several men and horses drowned, but their arrival made all the difference. Cassels was now too strong to be easily overwhelmed.

Still the position was critical, for the main body of the Turks, though retreating, had not been beaten, and the British troops were nearly worn out. The Infantry "had been marching and fighting for four days under most arduous conditions," while Cassels and his own Cavalry Brigade had been continuously in action for seventy-two hours, and all the force now at his disposal had made long marches. But if the Turkish force was to be held, there must be no rest for the troops, in spite of exhaustion and darkness, for the enemy saw that he was trapped, and would be sure to fight desperately for freedom.

He certainly did so. During the night of the 28th he tried agai and again to break out northwards, and though he did not succeed, he fought on all through the 29th, and all through the night that followed; and some of his counter-attacks against the British Infantry who were pressing him from the south were by no means easily stopped. It was a splendid defence, and rightly called forth the admiration of his assailants.

The end of the battle may best be told in General Marshall's words:—

"Meanwhile a serious threat from Turkish reinforcements (5th

Division troops) moving down from the Mosul direction developed against Cassels' right flank. These troops established themselves with guns and machine-guns on the high bluffs near Hadraniya, but were promptly dealt with by the 7th Cavalry Brigade. The Thirteenth Hussars galloped across the open, dismounted under the bluffs, and, led by their Colonel, carried the position by assault, many Turks being accounted for with the bayonet.

"Mounted pursuit by the remainder of the 7th Cavalry Brigade cleared away further menace from the north, and resulted in the capture of 1000 prisoners, with 2 guns and 12 machine-guns.

"The enemy's avenues of escape northwards were now effectively blocked, thanks to the skilful and resolute handling of his detachment by General Cassels, and the firm determination of his troops, who repulsed every attempt by the enemy to break through.

"Gripped as in a vice, with his men packed in ravines which were raked by our guns from across the Tigris, Ismail Hakki, the Turkish Commander, found himself in a hopeless position. All attempts to break through had failed, communication with the Turkish Commander-in-Chief at Mosul had been rendered impossible, and no relief was in sight. At dawn on October 30th, just as our troops were about to renew the attack, white flags appeared all along the Turkish lines, and later on Ismail Hakki surrendered in person."

So ended the last battle fought in the Great War by a Turkish army. It had been gallantly contested, but the victory was as complete as a victory could be. The Turkish 6th Army had been not only defeated but captured.

Immediately after the surrender, the 7th Cavalry Brigade and armoured cars pushed rapidly along the Mosul road and made a considerable addition to the number of prisoners. The 14th (Indian) Lancers, charging with the lance, captured a convoy and 500 men, while the 13th Lancers captured a Turkish battalion. Altogether the Turks lost on the 30th October over 11,000 prisoners and 51 guns, besides great quantities of machine-guns, animals, and military material of all kinds.

By the first of November both Cavalry Brigades had got within twelve miles of Mosul, when they were met by a flag of truce informing them that an armistice had been concluded between Great

Britain and Turkey. Hostilities ceased, and the long campaign in Mesopotamia had closed with a dramatic success.

At Shergat, as elsewhere, the bulk of the close fighting had of course been done by the Infantry, but General Marshall reported that "the daring and brilliant leadership of the Cavalry Brigades under Brigadier-Generals Cassels and Norton contributed largely to the final result," and it may safely be affirmed that no Regiment in those Brigades had more right to be proud of its share in the work than the Thirteenth Hussars.

The extracts appended give details of its movements during the last period of the campaign, and especially of the affair at "Richardson's Bluff," as the scene of the fight at Hadraniya came to be called.

Account by General Norton, commanding 7th Brigade.

"The Brigade left Fathah at one o'clock on the night of the 27th-28th October, crossed the Lesser Zab at about 8 A.M., and reached the Hadrani-Yah Ford over the Tigris at 4.30 P.M. on the 28th, having marched about 50 miles since 1 A.M., the last 30 under a very hot sun, and about 70 miles in the last 36 hours.

"The leading Regiment (13th Lancers) immediately commenced to cross to the right bank, and got over without accident, but as the ford was a difficult one, there was considerable delay, and, by the time it came to the turn of the 13th Hussars, it was practically dark. The ford, which had been difficult in daylight, became positively dangerous after dusk, and a number of men, particularly those leading pack-horses, got into difficulties, and were carried down-stream. Hampered as they were by their ammunition and other equipment, their situation was a very precarious one, but most of them fortunately managed to scramble out on one bank or the other.

"When the roll was eventually called, it was found that three men, viz.:—

No. 8267, Lance-Corpl. Bartlett,
9196, Private Donoghue,
31571, Private Newman,

and two horses, were missing. With the exception of Private

Newman, whose dead body was found by the Infantry several days later some distance down-stream, no trace of these men has ever been found.

"In the whole Brigade, 21 men were carried away by the strong current, and drowned while crossing this ford.

"Only three squadrons of the Regiment crossed on this night, the remaining squadron ('A' Squadron, under Lieutenant H. C. D. FitzGibbon, M.C.) being ordered to remain on the left bank as escort to the Battery, which could not be got across in the dark.

"The Brigade, less the battery, this squadron, and the 13th D.C. Lancers, bivouacked on the river bank; the 13th Lancers, as soon as they were across, had been sent forward to support the right of the 11th Brigade, which was occupying a ridge of hills S.W. of the ford, parallel to the river, and about 3000 yards away from it.

"The night passed without incident, although there was a considerable amount of firing from the position occupied by the 11th Brigade and the 13th D.C. Lancers.

"Before daybreak on the 29th, the Brigade was moved into a covered position under the river bank, and orders were issued for the battery and the remaining squadron of the 13th Hussars to cross to the right bank as soon as there was sufficient daylight.

"The moment the squadron commenced to cross, the Turks shelled the ford heavily and accurately from 2 guns on high ground about 3000 yards to the N.W.

"The squadron managed to get across, fortunately without casualties, but it was obviously impossible for the battery to do so until the hostile fire had been subdued.

"About 7 A.M., V Battery, R.H.A., came into action on the left bank, and the Turks shortly afterwards withdrew their guns.

"About this time a squadron of the 13th D.C. Lancers, which had been ordered to push up the Mosul road at daybreak, encountered heavy shell and M.G. fire, and was compelled to fall back to the Cemetery, where it held its own with difficulty.

"It was clear now that the enemy were holding the high ground to the N.W. in considerable force, and did not consist, as reported by aero reconnaissances, of merely a few scattered groups of Infantry.

"Their position was on a range of hills approximately parallel to the river, and at about 3000 yards from it. At the southern extremity of this range was a high bluff, which will be referred to as Hadraniyah Bluff. Farther to the south again was the range held by the 11th Brigade and 13th D.C. Lancers, and between Hadraniyah Bluff and the river, and some 1000 yards to the S.E. of the former, was a fair-sized isolated hill on which was a Mahomedan cemetery.

"After a personal reconnaissance of the position, so far as was possible, the Brigadier at 8.15 A.M. issued orders for the 13th Hussars, supported by the 24th Jat Lancers, and covered by the fire of V Battery, R.H.A., from across the river, and the Vickers guns, a subsection of which was now in action on the Cemetery Hill, to gallop across the open from their covered position near the river, dismount under the Bluff, and carry the position by a dismounted attack.

"It should be explained that the contour of the hill made it obvious, that unless the Turks came out of their trenches, which were none too well sited, and advanced a short distance down the hill, there must be a considerable amount of dead ground at the foot of the slope, which would give cover to the horses, and enable the troops to form for the dismounted attack. It was hoped that the gun-fire would make it impossible for the Turks to leave their trenches for the purpose of covering this dead ground.

"On debouching from their covered position on the river bank, in accordance with the above orders, the leading squadron at once came under a hot fire from the Turkish machine-guns on the hill, and the squadron leader (Captain Evans, 21st Lancers) fell severely wounded.

"This caused the squadron to lose its direction: it inclined slightly to its left, and halted under the cover of Cemetery Hill.

"The remaining squadrons followed it, and eventually the whole Regiment was concentrated behind this hill.

"By this time, it was clear that the guns had not yet succeeded in subduing the Turkish fire, and the further prosecution of the attack was held up to allow a further artillery preparation.

"Brigade Headquarters was now on Cemetery Hill, and the F.O.O. of the Battery was also established there with his signallers.

"About this time—viz., 10 A.M.—a Field Battery, No. A/33, R.F.A., having crossed the ford, was placed at the disposal of the Brigadier, and came into action some 2000 yards to the S.E. of Cemetery Hill.

"At 12.30 P.M. the Brigadier personally gave orders to Lieut.-Colonel Richardson to renew the attack on the Bluff at 1.20 P.M., and detailed arrangements were made for the attack to be closely supported by both Batteries and all available Vickers guns.

"In accordance with these orders, the Regiment, in column of wings with extended files, debouched from under cover of Cemetery Hill, and galloped straight for the Bluff. The ground was flat and open, the only obstacle being a rather deep dry nullah about half-way, which caused a little crowding and delay. The moment the leading squadron debouched, the Turks opened a heavy fire with all their machine-guns, but the accurate fire of our two batteries and the Vickers guns made it impossible for them to leave their trenches, and the squadrons soon got into the dead ground under the Bluff.

"During this phase of the attack, although these galloping squadrons were absolutely exposed for at least 500 yards to the fire of 12 machine-guns and 1000 Infantry, at ranges from 1500 to 500 yards, not a single man was hit, and only one or two horses.

"Had the Turks put down a machine-gun barrage on some point, such as the afore-mentioned nullah, and kept it there, our casualties must have been considerable; but they were evidently disconcerted by the heavy shelling, and there appeared to be no central control, each machine-gunner firing as he pleased.

"Having reached the foot of the Bluff, the Hussars lost no time in dismounting and forming for attack. Having fixed bayonets, they swarmed up the almost precipitous hill, gallantly led by their Commanding Officer.

"All this time, both batteries, using 'gun-fire,' were bursting their shells with beautiful accuracy along the crest of the Hill, and the Vickers guns were also keeping up a heavy fire on the trenches.

"This barrage did not lift until the Hussars were within a few yards of the crest, and there is no doubt that the splendid accuracy and rapidity of fire of these guns was a most important factor in making the attack the success it was.

"From Brigade Headquarters on Cemetery Hill, it was clear that the Turks in the trenches, having seen the squadrons moving towards them at a gallop, and being unable to see what was happening at the foot of the hill, were becoming anxious. Individual men and small parties could be seen getting out of the advanced trenches and moving to the rear, many being caught by the barrage as they did so.

"As the barrage lifted, the Hussars, still led by their Colonel, charged over the crest, shot or bayoneted such Turks as still confronted them in the trenches, and then rapidly forming line to the right, swept along the crest to deal with the Turks who were still holding their position at the north end of the hill.

"As soon as the Hussars' attack had gained the crest, the 13th D.C. and 14th Jat Lancers quickly mounted, and galloped round to the west and north, captured the enemy's two guns, and cut off the retreat of the Infantry retiring before the attack of the 13th Hussars.

"Seeing their retreat entirely cut off, the whole Turkish force surrendered.

"In this successful operation, 2 mountain-guns, 12 machine-guns, and 982 unwounded prisoners were taken, and there was a considerable number of Turkish dead and wounded on the hill.

"This enemy force consisted of the whole of the 13th Regiment which had been sent from the 5th Division in Persia to reinforce the army of Haqi Bey."

Extract from Colonel Richardson's letter to Sir R. S. Baden-Powell.

"7TH CAVALRY BRIGADE,
IN CAMP OUTSIDE MOSUL,
4th Nov. 1918.

"MY DEAR GENERAL,—I must send you a line to tell you of another fine performance by the Regiment—a performance which our Brigadier (Gen. Charles Norton), the Commanding Officers of the other Regiments of the Brigade, and other eye-witnesses describe to me as the finest thing they have ever seen. I myself agree that not only as a spectacle but in its complete and immediate success it is the finest performance put up by the Regiment during the war,

Sergt. R. Holloway
(*Killed at Hadraniyeh, 29th October 1918*)

L.-Sergt. Wm. Leeman
(*Killed at Hadraniyeh, 29th October 1918*)

S.-Sergt. J. F. Couch, D.C.M.
(*Died at Chaldari, 4th July 1918*)

2nd Lieut. G. R. Russell
(*Wounded at Hadraniyeh, 29th October 1918*)

Cpl. A. Vinall, D.C.M.
(*Died of wounds, 31st October 1918*)

but as a test of discipline and tenacity I would place Lajj first: there we were on the verge of annihilation, and yet managed to put up a fighting front to the enemy; here it is true we were foiled in our first attempt, but the sqdns. reached cover in perfect order, and the failure of the early morning only made them the keener to score in the afternoon. I enclose a copy of the dry sober account of the day, which I made in my note-book two days after the occurrence, so I will only give you here a brief outline of the events which led up to our being where we were, and of the few days after the 29th which led us to the occupation of Mosul.

"Since an armistice, in the nature of a complete military surrender by Turkey, has been arranged and signed, I hardly suppose that the censor will seriously object to numbers of Brigades and Divisions being given, but I will not risk more deletion than is necessary for the tale.

"Well, the original objective was the reduction of the Fathah position—a very strong natural position formed by the passage of the Tigris through the range of hills known on the left bank as the Jebel Hamrin and on the right bank as Jebel Makhul—about thirty miles N. of Tekrit, held on both banks by some 3000 rifles and 40 guns. Two Cavalry Bgds., ours and Cassels', had made a long turning movement over the Hamrin, and were in position to the N. and N.E. of the Turkish position by dawn of day, and our feelings were rather mixed when we learnt about 8 A.M. that Johnny Turk had evacuated the position a day or two earlier, and retired mostly along the main road to Mosul on the right bank, which ran through a series of strong natural defensive positions against which the Infantry could make only slow progress against the Turk, who is a past-master in the art of rearguard fighting: so obstinate was he in holding on to a position opposite the confluence of the Lesser Zab river with the Tigris that we were hurried one night to the Fathah Gorge, which we bridged with a pontoon, to go up the right bank and create a diversion in favour of our Infantry; on arriving at Fathah, however, the position had changed—the Turk had retired, and we were no longer required on the right bank; but in the meantime Cassels' Bgde. had crossed the Tigris some twenty-five miles north of the Lesser Zab, and had taken up a position astride the line of retreat of the Turks, and reinforcements to the latter were said to be closing on Cassels' rear, so after a few hours' halt we were hurried back to the Zab and beyond it to the ford, which we reached at dusk on the 28th, having covered 40 miles since 1 A.M. that day. The ford was deep and unmarked; I was riding my biggest horse, and the water was over my knees; on many men it was up to the waist, and we were unfortunate in losing three men, of whom no trace could be found in the dark.

"The events of the 29th I have described in the attached account. The Bgde. that night (29th) bivouacked on its fighting ground, and next morning moved S.W. to a position on Cassels' right, where about 10 A.M. we learnt that the Turks caught between the Infantry and Cassels had surrendered: we therefore turned our heads northward, and about 2.30 P.M. came up with the Turkish rearguard, who surrendered after a few rounds from the battery, and the 13th Lancers pursuing brought back a lot more, making the total captures of the Bgde. up to about 2500 for the two days.

"Next day we continued our advance northward toward Mosul, but about 11 A.M. met Turkish envoys in a car, who came to tell us that an armistice had been signed.

"We have been on half rations or less for many days, and hardly get a scrap of outside news, but these days must be fraught with intense interest and excitement at home, and these details must in comparison fade into insignificance; but, hereafter, they will be of interest to all concerned with the Regiment, so I will not labour at further excuses for writing."

Enclosure in Colonel Richardson's letter to Sir R. S. Baden-Powell, of the 4th November 1918.

"THE STORMING OF THE CLIFF AND PLATEAU NEAR HADRANIYAH, AFTERWARDS CALLED (IN BRIGADE ORDERS) RICHARDSON'S BLUFF, BY THE THIRTEENTH HUSSARS, ON THE 29TH OCTOBER 1918.

"The night of the 28th-29th passed rather cheerlessly, most of the men having got wet to the waist, and in many cases their blankets and warm coats also having been soaked in fording the river the previous evening, and before dawn we moved down under cover of the river bank, where we were practically immune from the shell-fire which was opened on us soon after daylight and continued until about 7 A.M. About this time I received orders to gallop the Regiment and get a footing on a plateau rising almost sheer from the plain, on which several emplacements could be seen. I asked for and got the co-operation of V Battery, which put a strong barrage on the plateau from 8 A.M. to 8.15 A.M., at which hour we emerged from the river bank, supported on either flank by a subsection of 18 M.G.'s, commanded respectively by Lt. Noel and Lt. Foley. The following is a rough sketch showing disposition of Regiment and ground to be covered.

"As the Regiment reached the open, the four squadrons took up a diamond formation—'B' Squadron under Capt. Evans, 21st Lancers, leading, 'C' on left flank, 'D' on right flank, and 'A' Squadron (Lt. Fitzgibbon) bringing up the rear. I rode in the centre, and on either flank were the Hotchkiss guns of two squadrons co-operating with the subsection of machine-guns—*i.e.*, 'B' and 'D' on right, under Lt. Parkes, and 'A' and 'C' on left, under Lt. Hartigan, M.C.

"As soon as the Regiment came into sight in the open a very hot fire from M.G.'s, rifles, and camel guns (9-pounders) was opened on it, and when about 1000 yds. from its objective, Capt. Evans was very seriously wounded, and the troop leaders of the Leading Squadron led their troops to the cover of the Cemetery Hill, where the remainder of the Regiment quickly concentrated. I quite agreed with their action at the time, and looking back two days afterwards with the knowledge gained since, I am convinced that it was the right course to take. The crest of the plateau was very much more strongly held than had ever been anticipated, and sufficient time had not been allowed for the artillery barrage.

"We occupied the Cemetery Hill and tried to gain ground to the west, but the glacis was quite bare and exposed to M.G. fire from the plateau and from trenches to the west. The Brigadier visited us about 11.30 A.M., and I suggested to him that we should try again to get the hill if the artillery would plaster the plateau and the known trenches until we had gained a footing on the crest. This they did with great effect at intervals until 1.30 P.M., at which time I led the Regiment from the cover of Cemetery Hill in two waves, 'A' and 'D' Squadrons (Mr Fitzgibbon and Capt. Jeffrey, M.C.)

in the first, and 'C' (Mr Norton) and 'B' (Capt. Howey) in the second under Capt. Godfree.

"The first wave reached the foot of the bluff with only one casualty, a man whose horse fell in jumping a deepish nullah which lay across part of our course—dismounted to climb the almost precipitous cliff-side, the footing on which was made the worse by the loose rolling gravel surface; how the men scaled that cliff carrying their Hotchkiss guns and ammunition-bags will always remain a marvel to me.

"Meantime the second wave had reached the hill, also without casualties, and taken position on the right of the first. As we appeared on the crest fire broke on us from all sides, but the *moral* of the Turks had been so severely shaken by the Artillery that our casualties were very slight, and by bringing our right forward we were soon able to bring enfilade fire on the trenches to our front, and about 1.45 the first white flag was hoisted. The other trenches did not take long to follow their example, and in less than half an hour after leaving Cemetery Hill we were in complete possession of the plateau, and prisoners, whose numbers eventually mounted to 722, were being marched in batches from every side to a central spot. A few undoubtedly escaped, but the Indian Cavalry Regiments took up the pursuit and brought back 200 or 300 and a camel gun. The booty in arms was very large, at least 20 machine-guns and several thousands of rounds being taken.

"Our casualties were one, Sergt. Holloway, killed, and the following N.C.O.'s and men wounded—viz., Sergt. Barnard, Cpls. Vinall and Hill, Ptes. Little, Patrick, and Tranter (Cpl. Vinall has since died of wounds).

"Sergt. Leeman had been killed with the following wounded whilst we were on Cemetery Hill: 2nd Lt. Russell, Ptes. Powell, Gerrard, and Winter.

"Pte. Harvey was wounded by shell-fire at the ford early in the morning, and the following were drowned whilst crossing the river on the previous evening—viz., L.-Cpl. Bartlett, Ptes. Newman and Donoghue.

"It transpired later that the Turkish troops opposed to us on the Bluff were the three Battns. of the 13th Regiment who had recently come down from Lake Urmia to reinforce the troops on the Fathah position, and that they were preparing to attack the ford at 4 P.M. that afternoon, when our assault destroyed their plans.

"The fight was short and sharp and the victory complete; we owed our almost complete immunity from loss to the excellent shooting of V Battery and A 337, and it was, I must admit, some consolation to know that the troops which brought off the coup were all British."

Letter from Colonel Richardson to Sir R. S. Baden-Powell, dated Mosul, 12th November 1918.

"MY DEAR GENERAL,—Herewith the sketch (from memory) of the position assaulted by the Regiment on October 29th and the numbers of the Regiment taking part in the assault; these were omitted from my letter of last week; I hope before we leave these parts to get a proper surveyed sketch of the position,—for the purposes of history.

"We heard to-day that the Armstice with the Huns had been signed and commenced yesterday, as we take it that this is tantamount to a surrender which is practically unconditional on their part. I hope it may not be long before we say good-bye to this God-forgotten country.—Yours sincerely,

(Signed) J. J. RICHARDSON.

"The numbers of the squadrons:—

	Officers.	N.C.O.'s and Men.	Hotchkiss guns.
A . . .	4	64	4
B . . .	3	52	4
C . . .	6[1]	63	4
D . . .	4	60	4
	17	239	16"

Letter from Colonel Richardson to Sir R. S. Baden-Powell, dated Zacho, 75 miles N. of Mosul, 30th November 1918.

"MY DEAR GENERAL,—Just a line in continuation of my letter of the 12th to tell you that we have been given the following Im-

[1] "Includes myself and Captain Godfree (2nd in command)."

mediate Rewards in recognition of our work on the 29th October, viz.:—

Myself	a bar to the D.S.O.
Capt. D. W. Godfree	M.C.
Lieut. G. H. Parkes	M.C.
Corporal Vinall (since died of wounds)	D.C.M.
Corporal Raynor	D.C.M.
Lance-Corporal Viner	M.M.
Corporal Aukett	M.M.
Corporal Whitington	M.M.
Private Humphrey	M.M.

"A batch of honours of which any Regiment may well be proud, as I think you will agree; and indeed we are all very proud of them, and myself not the least proud!

"We were pushed off here with 2 guns of V Battery and a subsection of M.G. at short notice on Monday last to stop reported rioting by retiring Turks and to see them over the border, which is only some 7 or 8 miles from here, and we did the 75 miles in 31 hours. The Turks had heard of our coming and quitted—very wisely; we are well in the hills here in a pretty valley, and snow-covered hills are less than 50 miles away. Rather a change from the Baghdad temperature of July!

"On the 26th we heard that the M.M. had been awarded to Sergt. W. H. Burgess and L.-Cpl. Lomas: these I gather are the outcome of the last despatch.—With all good wishes, yours sincerely, (Signed) J. J. RICHARDSON."

Letter from Lance-Corporal George Costello (56799) *of* 15*th Squadron M.G.C.,* 7*th Cavalry Brigade, to Mr and Mrs Wallace—parents of James Wallace.*

M.E.F.,
14*th November* 1918.

"DEAR MR AND MRS WALLACE,—It is with a sad heart, and with the deepest sympathy, that I write to you of Jim's presumed death. Jim and I were staunch friends together ever since we were in France, and shared one another's hardships and joys all the way through. Some time back 'A' Section was split up; Jim went to 'C' and I to 'B,' so at the actual time of crossing the river (Tigris) we were apart, as my section was about one hundred yards in front. From 12 midnight the night previous to crossing, we had been marching, and crossed at 6.30 P.M. the following evening, and during that period we had covered a distance of 57 miles, which was a pretty stiff march, and men and horses were feeling a bit done up. When we got in the water some of the horses were restless and started plunging and rearing. This is what happened to Jim: I learnt it from men who were near him at the time. His horse reared and Jim came off. The current of the river was very swift at the time, and it was also dark. We have lost four men in the squadron in the same way, and nothing has been heard of any of them since. The next morning we got into action! The 13th Hussars formed line and charged up a hill at the Turks. The hill was too steep to get up, so when they got half-way up they did a dismounted action, and within a quarter of an hour the Turks on the top of the hill showed the white flag, and over one thousand surrendered; by late in the afternoon the 7th Cavalry Brigade had beat the whole of that force, capturing nearly three thousand prisoners. The finest piece of work the Brigade has done during the war. I am sure poor Jim would have delighted in it. We then proceeded on towards Mosul, and came in contact with, and captured, two hundred and fifty reinforcements with 7 machine-guns and some big guns. Towards the evening, with one Regiment of Indians (14th Lancers) and one section of 16th Squadron M.G.C., we overtook and captured between two and three miles of convoy and thirteen hundred prisoners. We are now in Mosul. When about 20 miles from Mosul an aeroplane dropped a message which said that the Turks had surrendered Mosul, and that hostilities with Turkey had ceased. Jim has left behind him a memory which one can never forget. Loved and respected by all who knew him, as he was always ready to offer his help and sympathy to any one in trouble. And you have cause to be proud of having had a son who kept up the traditions of a Britisher to its highest level, and was a true and noble soldier to the core. You might please write to Mr Justice Eve and let him know, as I am sure that Jim would wish it, as he held his late squadron leader, Capt. Eve, 13th Hussars, in the highest esteem, and I know that a short time ago Jim wrote to Mr Justice Eve thanking him for the photo. I hope that Jim's brother who was wounded has quite recovered by now. Well, bear up, and try and look on it for the best, as I know that all your trouble has come in a heap lately. I unfortunately lost my mother in an air-raid since I have been out here.—From Jim's old Chum, 56799, GEORGE COSTELLO.

"*P.S.*—If I should gather any information I will write and let you know at once.

"56799 L.-Cpl. COSTELLO,
15th Squadron M.G.C.,
7th Cav. Brigade,
Mesopotamia."

Extracts from letter from Private E. Cleary (6759 "*B*" *Squadron*) *to Miss E. FitzPatrick of Accrington, Lancashire.*

"MESOPOTAMIA,
25*th November* 1918.

"DEAR LIZZIE,—I don't know how to start answering your letters—I have received that many from you for the last 4 days; they all came of a rush, the reason of it was they were in the country, and could not reach us on account of the shortage of transport. . . .

"Well, we had some hard fighting for about a fortnight, but we managed to get round them and capture the lot; we were very lucky, for they nearly captured us instead, but we managed to get a good mounted charge in on them, and that was the turning-point of our last battle out here. Our Regt. alone captured 1300 prisoners, but it was a good job they did surrender to us, for they could have wiped out every man of us if we fought on, because our Regt. was not near up to strength: we lost a good many. It was hard lines on them, for it was our last fight; they surrendered Mosul to us the next day, and a good job too, for we were about done; we went 2 and 3 days at a time with no rations except a pound of dry flour per man, and you may know what a big feed one can get out of a pound of flour (I don't think), and that flour was ground barley. You would laugh to see us cooking it: we used to get a bit of tin, make a fire under it, then mix the flour up with some water into a dough, then roll it out into little flat cakes and put it on a tin and bake it, so you may know what nice bread it was (I don't think), but still I was glad of it; they kept us going all right, and even now, although we are in Mosul, all we are living on is what we buy off the inhabitants, but of course everything will come right again as soon as we get settled down a bit. We were within 15 miles of Mosul when word came through that they had surrendered the place, so we stopped where we were until next day, and then news came through that Turkey had packed up altogether; it was taken very quietly, every one was that fed up, not even a cheer was given. Then we marched into the City and took over all that was there: they had plenty of stuff—food, guns, and men: it seemed so strange we were fighting them a few hours before, and then all that was to be seen was our officers and their officers shaking hands and making much of one another; but it wasn't so in the case of the men—if our fellows had their way they would have been scrapping them, for we haven't forgotten some of the tricks they done to our fellows; but still, thank God! it is all over. Well, we had been in Mosul for about a week when the news came through of Germany packing up: even that was taken very quietly, for it was raining as hard as it could, and we didn't have a bit of shelter, so the only remarks that were passed was, it's a bloody good job it's over; so that is how we enjoyed the good news: of course every one was delighted but didn't show it. Well, Lizzie, you will think I have gone mad on the war, so I will change the subject. I don't know if you have heard much about Mosul or not; well, it is supposed to be one of the great Sacred Cities out here—they say it's mentioned several times in the Bible (but as I don't know much about the Bible I cannot say); it is much the same as Baghdad, but a little more up to date: there is a couple of Catholic Chapels here; of course all Armenians are Catholics, and besides, there is some French here too, but give me one corner of Accrington and they can keep all this country. Just outside Mosul is the ruins of Ninevah, the great city in olden days that was swallowed up by the great earthquake: it is a big mound now, but there is big holes made in it that a man can crawl through, and you can see the ruins of the City beneath: I have been in them many a time, for we find Guards on it every day; so that is how we pass the time away going through it. If you ask one of the Priests they may be able to tell you about it (but I shall be glad to see the back of it). The climate up here is just like England—in fact it feels colder to me: it has rained ever since we've been here, and I can tell you it's none too comfortable, for all we've got is one blanket and no other shelter of any kind. Well, Lizzie, it's time now I came to thank you for all the trouble I have put you to; first of all, I don't know how to thank you for getting me that Roll of Honour, as you say it is a nice thing, but the reason I wanted it mostly was because I had a good few mates who were on there. God rest them! Mr Rolfe was my Troop Officer, he led us in the charge, and the Turk that killed him was also killed by one of our fellows, so that was one consolation; if you look on there you will see Private Stones' name; he came from Whally. Pte. M'Nulty came from Birkenhead, Pte. Cole from Halifax, and Tpr. Spicer from Dublin: he was only about 18 years old. I shall tell you all about them when I go to see you at No. 41. It was very kind of Mr Eve for writing such a nice letter; his son was a nice fellow and a fine leader. I seen him get killed, he was shot right through the head. It was good of you getting them framed. Again I thank you from the bottom of my heart. Now I must thank you for sending Xmas card, and above all, I was most delighted at getting poor Dan's Memory Card. God rest his soul. It's the only thing that grieves me, is he did not live so as to have the pleasure of meeting him again. I was always looking forward to having a good time with him if both of us had the luck, but as you say God knows best."

Extracts from letter from Captain H. C. D. FitzGibbon, Thirteenth Hussars.

"FIELD.

"MY DEAR OLIVE,—Well, my goodness, a lot of things have happened since my last letter to you. Peace! what an extraordinary word: it is hard to believe the whole thing is over. I will try and tell you all since my last letter.

"The Brigade was in its summer camp, just a few miles north of Baghdad, when I got back from leave. You remember I sent you a word saying we were going out once more. The idea was to take the Fahta position. The Turkish forces were then holding the gorge where the Tigris runs through the Jebel Hamrin. It was supposed, and is, a very strong position, with a desert on one side and a range of hills on the other. He, however, did not hold all the passes, enabling us (the Cavalry) to get over and behind his position. We night-marched round, and at dawn attacked, but to find he did not like the look of the thing and had gone in the night. We next met him on a rather strong position he had taken up on the Lesser Zab. He held us up there for about three days, and started to retreat on Mosul. We, or rather Cassels and his Brigade (with the 7th Hussars) were rather too quick for him, and had got between him and Mosul, dug themselves in, and held them up. A most gallant performance. We were then hurried up to reinforce him as he was not strong enough for the job, and would undoubtedly have failed had we not come up. Two days after we had the lot prisoners, about 10,000. My Regiment had rather a fine chance. About 900 Turks were holding a fairly strong position on some hills with a very large field of fire in front of them. We got Archies to clear up that position. The only way to do it was mounted at the gallop over the open and dismount under the crest of the hills. We formed up in two lines, Jeffrey's and my squadrons in the first line and the other two in the rear, and off we went. Those who saw it said it was as fine a show as they had ever seen. The line was as good as on parade. We crossed the open at full gallop, dismounted under the crest, fixed bayonets, and before you could say knife had a good footing on the crest. The whole thing was so quick that the Turk was rather taken unawares, and we had cleared up the whole position and taken the lot in no time. My squadron at the time, the same as the others, was only sixty strong—that is to say, about 40 dismounted men. The Colonel led the first line himself, and got a bar to his D.S.O. for it—not bad!!!

"After that we were hurried up the Mosul road. We took another 1800 prisoners, reached a place called Hammam Ali about 15 miles south of Mosul, and for the first time heard of the Turkish Armistice. I then had a rather interesting time, as I went into Mosul with General Cassels to interview the Turkish Commander. It was rather a unique experience, the whole thing in French. To cut a long story short, they were obliged by the terms to evacuate the Mosul Villayat. We marched into the town, a very dirty but rather interesting place. A lot of people speak French. We were camped outside with very few tents and a very wet and cold climate, and no warm clothes to put on. The Regiment was then detailed to send a squadron up to a place on the border called Zakho, about 80 miles north of Mosul, to keep order, and here I am Lord and Master of my little Kingdom.

"It is rather funny being so far away from everybody, but rather nice in a way. I have got two officers with me, Stirling and Hartigan. The first you know of, and the other, also, is one of the best. We are living for the first time for 2½ years in a house, not a bad one at all, and even my men have got them into a large building, so on the whole we are rather comfy. It is very pretty; the village is in a beautifully fertile valley, with high snow-capped mountains in front of me. Plenty to eat and lovely *water* to drink, and so good for one!!! and I am the Lord and Master of the district. We are well out of Mesopot, and well into Kurdistan. Kurdish is an extraordinary language, but quite a lot of people speak French. Their dress is quite pretty, in a way rather like the Dutch with big trousers. The day before I left Mosul, news came in that Germany had thrown in. It was early in the morning, we were in a very small tent; I was shaving at the time: it was received with great calm and indifference. News came in later the same day that owing to the rain carts could not get up to us, and we were put on half rations. Less indifference was shown to this bit of news.

"I can't send this letter for a day or two, so will go on with it to-morrow. Me voila encore une fois! . . .

"It is extraordinary to think war is over! though I am afraid it will be a long time before they let us all go. What a lot of mothers and wives must be thankful. I suppose the sights at home must have been wonderful.

"I had a very interesting time yesterday. The mine only just started is full of coal, and if worked with method I should say worth any amount of money. The tribes in the hills are the most extraordinary thing in the world: they know no law, and obey no master but the chief of their tribe. Every man is armed with a rifle, and a modern one at that. Their villages are in such places that only a goat can get at them, so they can laugh at the whole world in general. The chief I had lunch with is a well-known brigand who robs everybody who passes anywhere near him.

"I had rather a funny complaint last night: A man came round most beautifully dressed, started off with a long-winded speech about the British and how just they were, &c., ending up by saying that the headman's servants had found his wife in the bazaar and soundly beaten her, and did the British allow that sort of thing?"

The Thirteenth Hussars' Charge at Hadraniyah, by an Eye-Witness.

"It can be given to few to watch from a vantage-point, in near vicinity, a spirited cavalry charge. All cavalrymen live for the day when they may take part in this most exhilarating of all the movements of war, and next to being a participator is the privilege of seeing a charge take place.

"At midday on October 29th, 1918, the situation near Hadraniyah was critical—to the south, Cassels' Brigade was still engaged delaying the retreat of Haqui Bey's Army; linked with him 2½ miles to the north, Norton's Brigade had, since dawn, been opposing the advance of the reinforcements that were hurrying to the Turkish Commander's assistance.

"Almost due west of Hadraniyah Ford, in crossing which at nightfall on the 28th the Brigade had lost twenty men drowned, stands 'Richardson's Bluff.' This Bluff is the culminating point of the ridge which forms the old bank of the Tigris; it lies some 2000 yards from the present river bank, and rises almost precipitously from the plain to the height of some 150 feet. Just to the south of the Bluff there is a re-entrant in the ridge, sweeping round which the latter continues to the south, at first at a lower elevation, but gradually rising till it reaches Huwaish Gorge, along the edge of which was Cassels' position. The plain between the Bluff and the river is as flat as the palm of the hand, save for two mounds to the south and south-east, on the nearer of which, some 900 yards distant, is a Mahommedan cemetery. At dawn the reconnoitring squadron of the 13th Lancers had been driven back by gun and machine-gun fire from the Bluff, and had taken cover on Cemetery Hill.

"At 8.15 A.M. the Thirteenth Hussars had endeavoured to gallop the Bluff, but owing to the leading Squadron Commander, Captain Evans, having fallen severely wounded, his squadron had lost direction and, inclining to the left, had arrived at Cemetery Hill and relieved the squadron of the 13th Lancers, and there it was joined by the remainder of the Regiment.

"Shortly after this the left was prolonged by the 14th Lancers, who were disposed so as to intercept a junction between the enemy that were opposing the 7th Brigade and those that were engaged with Cassels.

"In support, in rear of the centre of our position, the 13th Lancers were placed, while on the farther bank of the river, 3000 yards to the east, was V Battery, R.H.A., in action against the enemy guns. At 10 A.M., having forded the river, 337 A Battery, R.F.A., came into action in our support, from about 2000 yards to the south of Cemetery Hill. The combined fire of the two batteries soon silenced the enemy guns, and General Norton, realising that the Bluff was the vital point, decided to make a further attempt to wrest it from the enemy.

"It was evident by this time that the hostile force on the Bluff was considerably stronger than had been previously reported. Our position on the lower ground was very precarious, and a successful decisive attack by the Turks would have driven us into the open, rolled up Cassels' flank, and taken him in the rear. The terrain prevented an accurate reconnaissance of the enemy's position being made. It remained for General Norton to take the responsibility of ordering a direct attack to be made on the Bluff, supported by the guns.

"Orders were issued, the Thirteenth Hussars concentrated on their right, while both batteries heavily shelled the enemy's position.

"The concentration of the Thirteenth Hussars was evidently seen by the enemy, and occasioned a heavy outburst of fire, which revealed the fact that they were in possession of more machine-guns than up to then had been calculated. In consequence of this General Norton ordered the attack to be postponed while further artillery preparation took place.

"At 12.45 P.M. the General decided that further delay would be dangerous, and that the effort must be made. Our helio winked a message to the batteries that the Thirteenth Hussars were to attack at 1.20 P.M., and that they were to increase their rates of fire. Both batteries redoubled their efforts — admirably directed by Lt. Sutton (V Battery, R.H.A.), both ranging and fuses being beautifully accurate; they plastered the Bluff with H.E. and shrapnel. A slight delay in the arrangements made it impossible for the attack to take place till 1.30 P.M. In the interval the Thirteenth Hussars quietly formed up in column of wings under cover of Cemetery Hill.

"As the zero time approached urgent messages came from Major Greenshields, commanding V Battery, that his ammunition was running low. It was an anxious moment: on the crest of Cemetery Hill the Brigade Headquarters Staff lay, taking what cover they could behind the gravestones—down below the Hussars were quietly falling in, resolute but serious, as all knew that much depended on the coming assault, while its result was in the hands of the gods—on our left the 13th and 14th Lancers were ready to 'Mount' and 'Pursue,' if the attack was successful, or extricate their comrades if it was a reverse.

"On the Bluff was a veritable inferno—the H.E. bursting in black clouds, the shrapnel in fleece-like puffs, both hurling death on the doomed Turks—one continuous and awful roar swept the threatened area.

"Deafened by the noise and rendered dizzy by the concussion, Turks in the advanced trenches on the crest of the hill could be seen getting up and running in a drunken manner to the rear; some caught by bullets while *en route*, fell; others disappeared in the burst of an H.E. shell, and nothing remained when the black cloud cleared away.

"And then the Hussars advanced; it was a glorious sight—the first line led by their gallant Colonel, in perfect order, as if one on a drill parade, swept round the edge of Cemetery Hill—the pace increased—on they went with a loud inspiring cheer. Then equally steady followed the second line.

"The enemy's machine-guns rattled furiously — ours replied viciously—on, on went the Hussars—then to our horror we saw a check, and we realised that, straight across the front of the Hussars, between them and the Bluff, ran a nullah.

"What is going to happen? Is it too wide? Can they cross it? Another Omdurman or Shabkadr? were among the agonising thoughts that flashed through our minds.

"A little crowding—a horse down—then a cheer rose to our lips as we saw the Colonel thread his way across, and then, on his heels, scrambling, hustling, thrusting, scarcely checking, the first line crossed and went on.

"By this time the onlookers on the Cemetery Hill were mad with excitement, and quite oblivious of the bullets still whistling over them, they were standing up wildly cheering on the gallant Hussars. Now they had reached the foot of the Bluff—so close was it to us that we could distinguish individuals.

"Off they get—fix bayonets—up the hill they scrambled, with their Colonel still at their head; the barrage lifts beautifully in time—and the Hussars clear the crest and, spreading like a fan, on they go. A machine-gun is rushed—then a Hussar was seen to fall—bayonet work begins. . . .

"'By God, they are counter-attacking!' bursts from the Brigade Major's lips, as a party of Turks drive back a troop of the Hussars; but, steady as rocks, the Hussars meet the new attack — a few shots, and then the bayonet. On our left there is a thunder of hoofs —a host of lance-pennons dotted o'er the countryside shows the 13th and 14th Lancers are hurrying forward in support, lance in rest ready for the pursuit. This is the final act in the drama, and then all sobriety is forgotten; our helmets fly in the air as we cheer and cheer again, as on all sides white flags of surrender are raised by the defeated Turks.

"From every direction, in a few minutes, small bodies of the enemy, marshalled by the gallant Hussars, can be seen marching in —and the affair was over.

"It was with a choke in the throat that, a few minutes later, we shook hands with the gallant 'Jimmy,' and congratulated him on his splendid success, and ourselves on having been privileged to watch and in some way partake of it."

Transport fording the Tigris, 1918. *Major A. Cooke, Q.M.*

"On October 31st, the transport, which had been left at Huwaish on the left bank of the river Tigris, received orders to march at 6 A.M. and join the Brigade at Quayirah.

"After marching about 4 miles the ford was reached—apparently the one described in Lieutenant-General Sir W. R. Marshall's despatch of 11th April 1919, by which the 11th Cavalry Brigade, under Br.-General R. A. Cassels, C.B., D.S.O., had crossed.

"Two channels of the river had to be crossed before reaching the main stream. Entering the latter, the ford followed a narrow ledge in mid-river (the water here was about 4½ feet deep) which ran for a distance of about 500 yards up-stream; it then turned to the left through fairly shallow water to the right bank. To the right of the ledge the water was at least 5½ feet deep, to the left flowed the rapid and deep main stream of the Tigris.

"If a vehicle got off the ledge into the deeper water on the right, it was extremely difficult to get the animals to start again; but if they got off to the left there was little or no hope of recovery.

"The crossing was commenced, taking about 15 yards interval between vehicles. Soon there came a check; the animals of one waggon, tired of pulling against the rush of ice-cold water, came to a standstill. The next waggon tried to pass to the right, but got into the deeper water, and stayed there until extra animals could be hitched in—this, in itself, proving a most difficult task.

"The column was again got moving, but several similar stops occurred, until a limbered waggon and a water-cart of the 14th Lancers got off the ledge to the left and were lost, two of the four drivers and all the mules being drowned.

"All the transport of the Regiment, except the mess-carts, crossed without trouble; these, being very light, had great difficulty in stemming the swift current, and one was carried off the ledge and swept down-stream.

"The driver of this cart was Pte. Harding, a good old soldier who was mentioned in despatches during the South African War for great gallantry ('Regimental History,' volume ii. pp. 129 and 207). He was accompanied in the cart by an Indian cook.

"Every one who saw them carried away thought they were certainly lost. Harding, however, kept his head, and, whilst being floated down the river at a great pace, sat and handled his reins as though he were driving along a hard road. The horse swam with the stream, the well-laden cart floated evenly on the water, and the whole lot went down-stream for about 800 yards before coming to a standstill. While in deep water, Harding gradually steered towards the bank, and at length guided the horse to where he could just scrape his feet along the shingly bottom, and come to a halt, standing.

"The horse was dead-beat, and it was feared that they would again be carried away by the rush of water, when an Indian Artillery driver rode into the river and reached the cart-horse's head. By seizing the rein and assisting and encouraging the horse, he eventually got it to shallower water, where aid was forthcoming to get the cart ashore.

"It was a most daring and gallant feat on the part of the Indian driver, and his conduct was brought to the notice of the Army Commander. Private Harding received the cheers and congratulations of all assembled there, and was recommended for a reward.

"This may not appear very exciting in print, but when it is remembered that the Tigris is about 300 yards wide, with such a volume of water rushing down that no communication by shouting is possible, except at very short distances, and that men and animals were struggling and drowning at the time, the writer thinks that all who read this account will agree that fording the Tigris with the transport afforded not the least of the many thrilling incidents of the closing days of the war."

CHAPTER XXII.

SCOPE AND MEANING OF THE MESOPOTAMIAN CAMPAIGN. THE SHARE IN IT OF THE THIRTEENTH HUSSARS.

THE bearing of the campaign in Mesopotamia upon the issues of the Great War has already been examined in the course of the narrative; but at the risk of some repetition it seems desirable to sum up here the achievements of the British in this part of the world, and to see what effect they had upon the general situation.

No one was in a better position to give a correct view of the subject in a few words than the Commander who had the thoroughly deserved good fortune to complete the overthrow of the Turks and the conquest of the country. General Marshall, in his despatch of the 1st February 1919, writes as follows:—

"The campaign in Mesopotamia has lasted just four years. From small beginnings, when Fort Fao was captured on November 6th, 1914, the ration strength of the force when Mosul was occupied had grown to some 420,000, including labour battalions.

"The area of territory of the Turkish Empire which has been conquered and occupied amounts to 114,000 square miles. The sphere of operations has included Trans-Caucasia and Trans-Caspia, and detachments furnished by the force are being employed over territory extending from Deir-as-Zor on the Euphrates, 330 miles north-west of Baghdad, to Merv in Trans-Caspia, some 1450 miles north-east of Baghdad.

"Actual captures since the beginning of the campaign amount to 45,500 prisoners and 250 guns, together with vast quantities of war material of all descriptions.

"These results have been achieved in a country destitute of shade in summer, and impassable owing to floods in wet weather, and are a lasting record of the gallantry and endurance of the officers and men, both British and Indian, who have fought uncomplainingly in spite of heat, thirst, rain, and discomfort, for four years in Mesopotamia."

This no doubt is a true and in some respects a striking presentment of the case; but it is a modest one, and more might have been said without exaggeration.

Mention has been made in earlier chapters of the scheme of world dominion conceived by the Germans, and of the part which Turkey was to play in it. Germany aimed not only at becoming the Paramount Power in Europe, but at the supreme control of the East. The two things were parts, and equally important parts, of the great conspiracy. They were, so to speak, the twin pillars upon which the whole vast superstructure of German dominion was to rest. For the Eastern part of the scheme Turkey was essential. Upon the geographical situation and the military power of Turkey, supported by German gold and German military science, supported also by the Sultan's influence as the spiritual head of Islam, everything depended. Through their vassal Turkey, the Germans were to strike down their most formidable enemy—England, whose fleet stood between them and the dominion of the world. This is not mere surmise. Germany made little secret of her intentions, or of the means by which they were to be fulfilled.

When Turkey proceeded to declare war against the Allies and to carry out her part in the German scheme, she was very soon able, after beating off an attack on the Dardanelles, to throw her whole weight into Asia. From the nature of things her military effort then resolved itself into an advance upon three fronts. On her left she struck at the Russians in the Caucasus, in the centre she struck at both Russians and English in Persia, and hoped to work through Persia upon Afghanistan and India. On her right she struck at England in Egypt. All these countries—the Caucasus, Persia, Afghanistan, the Indian Frontier, Egypt—were mainly or largely Mahomedan, and the Koran was to give weight to the blows of the Turkish sword.

Before very long Russia broke up, and England stood practically alone, so that in Western Asia it became a duel between her and

Turkey. The struggle was fought out on all three of the Turkish fronts, but mainly in the centre and on the Turkish right. On both these fronts England, though dangerously weak at first, grew stronger and stronger as time went on, until eventually she overwhelmed her enemy. Instead of being attacked and beaten in Egypt, she advanced from Egypt and crushed the Turkish armies in Palestine and Syria. Instead of being attacked and beaten in Persia and India, she advanced into Mesopotamia, and planted the British flag upon the citadels of Baghdad and Mosul. Under these tremendous blows the Turkish military power collapsed, and with it collapsed the Eastern part of the German world-scheme. Perhaps more than this might be said; for the fall of Turkey was no doubt a heavy shock and discouragement to Turkey's masters, struggling hard themselves to carry out their conquest of Europe. But it is enough to know that the British armies in Mesopotamia and Palestine brought Turkey to her knees.

It is not necessary to inquire which of the two had the leading share in this great feat of arms. They were portions of one Imperial Army, and though fighting hundreds of miles apart, with a vast desert between them, they were enabled, through England's command of the sea, to afford each other some support. The Mesopotamian Army more than once sent large forces round by sea to join the army in Palestine, and Allenby's victories in Palestine reacted with immense effect upon the situation in Mesopotamia. It was a signal instance of the influence of sea power in war, for it did much to neutralise the great advantage given to Turkey by her central position against her widely-parted enemies. The two of them, striking separately, yet in a sense together, from their common base, the sea, overthrew and ruined her.

It was the same sea power that enabled England to make each of them efficient in itself. Not only did the sea bring them their hundreds of thousands of fighting men, but the sea brought them everything needed to let their men fight with success. As to the Mesopotamian campaign, it was, as its very name implies, a river campaign, and the utilisation of the Turkish rivers, especially the Tigris, was the first essential of success. Nothing was so remarkable in all the history of the campaign as the way in which the Tigris was broken in and transformed from an almost insuperable obstacle into a great military highway. By a hard and sustained effort, impossible for any nation but the mistress of the seas, the turbulent undisciplined river was tamed and enlisted in the service of the invaders, and its carrying power was increased nearly a thousandfold. When Maude began his advance, its surface was swarming with steamboats that had found their way, thanks to the daring and skill of British seamen, from the Hughli and the Irawadi and the Thames, until they were numerous enough to carry a large army into the heart of the Turkish Empire, and to keep it supplied with all the multitudinous necessities of modern war.

The river steamers were supported by rapidly constructed railways and roads, and all manner of engineering works; new irrigation channels were made or ancient channels reopened; many of the wild Arab tribes were gradually brought to see the advantages of order and service; crops were raised and gathered; altogether the activities of our people were innumerable, and the conquest of a country almost as large as the United Kingdom went hand-in-hand with the measures necessary for making it prosperous and contented.

What its future will be no one can say. To take over such a country, with its turbulent population and unsettled frontiers and outlying responsibilities of many kinds, is a formidable task; but in any case the conquest of it, from such an enemy as the Turk, was no small contribution to the work done by the British Empire in the Great War. It was performed at a distance from England, among scenes unfamiliar to Englishmen, and, partly owing to Press restrictions, it did not greatly appeal to the country's attention, which was naturally enough fixed upon the progress of the conflict in Europe; but it was none the less a great achievement, and one of which all concerned in it may well be proud.

Among those concerned in it was the Regiment whose history forms the subject of this book, and there remains to be considered now the part which the Regiment played in the campaign.

The Thirteenth were perhaps fortunate in that they did not come to Mesopotamia until 1916, when the first half, and the most trying half, of the Mesopotamian Campaign was over. They had no part in the earlier fighting, when the British force was small and ill-found, and its difficulties great. They were spared the troubles and sufferings endured by the troops who conquered the lower part of the country; they did not march up with Townshend to the bloody field of Ctesiphon, or share in the desperate efforts to break through to his relief when he was shut up in Kut. When they landed at Basra all that was over. Horrified at the losses and sufferings entailed by its attempt—one of its customary attempts—to make a small force do the work of a large one, the nation had suddenly woken up to the needs of Mesopotamia; and, blaming every one but those chiefly to blame, itself and its statesmen, was now pouring in without stint Regiments and guns and river steamers and every kind of war material. It could not bring back to life the thousands of men sacrificed, or undo the injustice done to some of its best soldiers, but it was determined that the Mesopotamian force should for the future fight with reasonable chances of success; and though its attention was soon diverted again, it certainly made a fine effort, the result of which was to endure. The Thirteenth came in on the turn of the tide, and though they had before them two years of hard work and hard fighting, they were never to know the bitterness of defeat.

Nevertheless the work was severe, and the issue of the fighting was often doubtful for a time. That it uniformly ended in success, and eventually in complete triumph, was proof of very high qualities in the men who led and the men who followed. For the Turk is a stubborn enemy. He was once acknowledged, by general consent, as the best soldier in Europe; and if time has deprived him of some of his reputation, it still stands deservedly high.

As in the Palestine campaign, so in Mesopotamia, the Cavalry found its chance. There, as in all modern war, it was no doubt the Infantry and guns which mainly decided the fate of battles; but the Cavalry had much to do for the armies both before and after battle, in reconnaissance, in sweeping the country for supplies, in pursuit, and in the turning movements which at times brought about the enemy's ruin. Even in the actual shock of battle they were not wholly excluded from a share of fighting and honour. Some tributes to the value of their services in Mesopotamia have already been recorded in earlier chapters. There is one service which perhaps has not been sufficiently brought out. The whole plan of General Maude's attack upon the Turkish army which faced him on the north bank of the Tigris, when he began his advance in December 1916, was to push up the south bank, and keep threatening their rear from that side until they had so extended the line held by their troops that they were no longer strong enough to hold it all securely—to prevent his forcing a passage at some point and getting in behind their main body. As he put it in a letter to General Symons: "The Turks were very sticky and would not go for a long time, but we gradually stretched them and stretched them till owing to their casualties they were much weakened, and then we struck boldly at their tail across the Tigris. I had been hammering at this for two months, and the fact that they would not give up Sannaiyat, and consequently placed themselves in the dangerous position of having a line of communication in prolongation of their battle front, was their undoing."

And the Cavalry had much to do with this stretching process. In his despatch giving an account of the whole advance to Baghdad, General Maude writes:—

"The work of the Cavalry had been difficult. The flat terrain intersected with nalas obstructed movement without providing cover, and the state of the country after heavy rains made progress even for short distances laborious. The absence of water, too, away from the river, limited its radius of action. Nevertheless its reconnaissance work and the blows delivered against the enemy's communications helped in no small way to bring about that dissipation of his forces which was so essential to our success, and the pressure applied after the passage of the Tigris to the retreating enemy was instrumental in completing his final rout."

The Thirteenth, working from their camp on the Hai river, did their full share of the Cavalry work so described.

During the subsequent march on Baghdad, to quote again General Maude's letter to General Symons: "Your old Regiment did splendidly," and in the course of it, at Lajj, they had their first chance, while pressing the pursuit, of delivering a real Cavalry charge, mounted and sword in hand

This charge, "a brilliant charge," as General Maude called it, was the first of several which have been described in earlier chapters. At Tekrit, at Kulawand, and at Tuz Kermatli the Regiment got in with the sword; and if the final onslaught at Hadraniya was made on foot with the bayonet, it was none the less a hand-to-hand attack, after a gallop across the open to the foot of the bluff on which were the Turkish trenches. Nor is an assault with the bayonet less creditable than a mounted charge. Naturally a Cavalry Regiment will always long for that, and rejoice in it if it comes; but, as many leading soldiers have now recognised, Cavalry under present conditions, to be thoroughly efficient, must be able to fight on foot as well as on horseback. Ever since the American Civil War this has become clearer and clearer. The Thirteenth in Mesopotamia, as other Cavalry Regiments in this war, have in a measure reverted to their old rôle as Dragoons, to whom the firearm of the Infantry is as familiar as the sword.

Far from lessening the honour due to them, the fact that both at Lajj and at Hadraniya the Thirteenth showed they could fight either way, is very much to their credit. There was no lack of the Cavalry spirit. Their whole inclination was to charge on horseback, sword in hand. But, if necessary, they could do something more.

To conclude, perhaps the most valuable singly of all the services rendered by the Cavalry during the campaign, was the one of which the Hadraniya charge was only a part—namely, the long march round into the enemy's rear and the closing of his line of retreat, which led to the surrender of his whole force. The credit of this daring feat is due in the first place to Cassels and his own 11th Brigade, but the 7th Brigade too deserved much credit, for it may be doubted whether without their timely help Cassels and his men could have held their own against the repeated and desperate attacks of an enemy so superior in numbers.

Altogether, the Thirteenth Hussars did their duty well from the beginning to the end of their stay in Mesopotamia, which lasted for two years and a half. During that time they gained much honour at a heavy cost. Of their officers, eight were killed in action or died on service, two were disabled and taken prisoners, and fourteen were wounded. In other ranks the numbers were 90, 2, and 176. It is an honourable record.

CHAPTER XXIII.

RETURN TO ENGLAND—CONCLUSION.

THE Thirteenth remained in and about Mosul throughout the months of November and December 1918; but early in January 1919 they were ordered to return to Baghdad, and marching on the 11th of the month, arrived in Baghdad on the 29th.

There they remained in camp for some weeks, when the welcome order was received to embark for England. On the 10th March, the Regiment went on board a river steamer, and started on their journey down the Tigris. It was a large vessel, and with them were their old comrades of the Fourteenth Hussars. Transshipping at Basra to the *Etna*, and at Bombay to the *City of Sparta*, they arrived on the 29th April in the Mersey, and on the following day went ashore at Liverpool. On the 5th May they marched into the Cavalry Barracks at Aldershot, where they were to be quartered.

Here this History of 'The Thirteenth Hussars in the Great War' naturally comes to an end, for though the war was not formally concluded, the Regiment was now in England again, and its share in the military operations was over.

But so much has been written and spoken during the last few years regarding the value of Cavalry in modern war that, as remarked in the introductory chapter, it seems desirable, before leaving the subject of the services performed by the Regiment, to say a few words upon this question. It is one about which professional soldiers only can have an opinion of any value, and I therefore confine myself to quoting the opinion of professional soldiers.

ON THE WAY HOME

THE "CITY OF SPARTA" PASSING A JAPANESE DESTROYER IN THE SUEZ CANAL

The authorities whose views have been most definitely stated are Lord Haig, Lord Allenby, and the two Generals who commanded in Mesopotamia, Maude and Marshall.[1] The last two have been already quoted as to the Cavalry work in the advance to Baghdad, and beyond. General Allenby has written as follows regarding the operations in Palestine:—

"The Desert Mounted Corps took some 46,000 prisoners during the operations. The complete destruction of the VIIth and VIIIth Turkish armies depended mainly on the rapidity with which their communications were reached, and on quick decision in dealing with the enemy's columns as they attempted to escape. The vigorous handling of the Cavalry by its leaders, and the rapidity of its movement, overcame all attempts to delay its progress. The enemy's columns, after they had out-distanced the pursuing Infantry, were given no time to reorganise and fight their way through."

Of course, the Cavalry did not effect their work alone. General Allenby says that the breaking of the enemy's entrenched lines by the Infantry enabled the Cavalry to accomplish its mission. But this is no detraction from the merit of either. Both arms did their duty, and the result, in Palestine as in Mesopotamia, was what Henderson calls the most important operation of grand tactics—the surrounding and destruction of an enemy's army.

General Haig is perhaps the most important witness of all, not only because of the greatness of his command, but because the Western Front, with its immense system of trenches, stretching from end to end of the theatre of war, was the one where Cavalry was at special disadvantage. In his final despatch, summing up the features of the war, he deliberately takes up the question, and gives his answer. I quote the following words:—

"*The Value of Cavalry in Modern War.*

"17. From time to time, as the war of position dragged on, and the enemy's trench systems remained unbroken, while questions of man-power and the shortage of shipping became acute, the wisdom

[1] It may possibly be argued that both Lord Haig and Lord Allenby were Cavalry officers, and therefore perhaps inclined to think highly of the achievements of their own arm. They certainly were Cavalry officers, as was Lord French, a fact well worthy of note, but, like him, they were something more—great leaders of armies.

or necessity of maintaining any large force of mounted men was freely discussed. In the light of the full experience of the War, the decision to preserve the Cavalry Corps has been fully justified. It has been proved that Cavalry, whether used for shock effect under suitable conditions, or as mobile Infantry, have still an indispensable part to play in modern war. Moreover, it cannot safely be assumed that in all future wars, the flanks of the opposing forces will rest on neutral states or impassable obstacles. Whenever such a condition does not obtain, opportunities for the use of Cavalry must arise frequently."

General Haig proceeds to justify his opinion by examples, and concludes with a reference to the position of the British Cavalry on the morning of the Armistice which closed the war. "There is no doubt that, had the advance of the Cavalry been allowed to continue, the enemy's disorganised retreat would have been turned into a rout." In fact, the Armistice deprived the British Cavalry of the certainty of striking, in the last and greatest of all wars hitherto fought, and in that theatre of the war where they were at special disadvantage, the most stupendous blow ever struck by a body of armed horsemen since the history of the world began.

The British Cavalry consisted of Regiments similar to the Thirteenth Hussars. The story of such a Regiment is not therefore, as some think, the story of one unit of an arm which is now useless in war—as useless, it has been said, as bows and arrows. On the contrary, Cavalry has proved its usefulness up till now, and we may be sure that in the near future at all events, whether used on horseback for battle-shock, or as mobile Infantry, or in any other way of duty, it will still do good service to its country, as it has done in the past. The Cavalry spirit, please God, will never die in our people so long as war endures in the world, nor is there any reason to fear that we shall soon see the last of the famous squadrons in which that spirit is so gloriously embodied.

A GROUP OF OFFICERS OF THE REGIMENT TAKEN IN OCTOBER 1920 BY ELLIOTT & FRY

BACK ROW (Left to Right)—Lieut. A. H. WOOD, 2nd Lieut. H.R.H. PRINCE HENRY, Lieut. W. P. MADGIN, Lieut. M. C. KENNEDY, Lieut. W. W. N. DAVIES, Lieut. A. C. BARRINGTON, Lieut. C. F. COLLINS, Lieut. C. E. C. BOVEY

SITTING (First Row)—Major and Quartermaster A. COOKE, Bt. Major H. LL. JONES, D.S.O., Major E. H. STOCKER, Bt. Lt.-Col. E. F. TWIST, Lt.-Col. J. J. RICHARDSON, D.S.O., Bt. Lt.-Col. T. H. S. MARCHANT, D.S.O., Capt. J. H. HIND, Capt. J. G. OAKES, Capt. S. V. KENNEDY, M.C.

SITTING (Second Row)—Lieut. E. V. DUNBAR, Australian Forces; Lieut. W. H. HILLESS, Australian Forces.

APPENDIX I.

ROLL OF OFFICERS WHO BELONGED TO OR SERVED WITH THE REGIMENT BETWEEN AUGUST 1914 AND NOVEMBER 1918.

Lieutenant A. E. Annett.
„ R. R. Anson.
„ W. G. Baker.
„ E. P. Barrett.
„ J. L. M. Barrett.
2nd Lieutenant A. C. Barrington.
Lieutenant Hon. J. F. Best.[1]
„ J. W. Biggar.
Captain W. E. Bigge (Northumberland Hussars).
Captain W. Blyth.
Lieutenant E. Bristol.
„ F. H. Butler.
Major (Bt. Lieut.-Col.) E. J. Carter.
Lieutenant D. A. C. Chamberlain (R.A.M.C.)
2nd Lieutenant G. W. Cheeseman.[1]
Lieutenant J. I. Chrystall, M.C.
„ J. O. P. Clarkson.
„ B. W. D. Cochrane.
Captain G. G. Collett (R.A.M.C.)
Lieutenant C. F. Collins.
2nd Lieutenant R. C. Constable.[1]
Major and Q.M. A. Cooke.
Rev. H. Cooke, M.C.
Lieutenant E. G. Corner.
Captain F. C. Covell.
Major R. F. Cox.
Lieutenant W. P. Crawford-Greene.
Captain J. Cullenan (R.A.M.C.)
2nd Lieutenant W. W. N. Davies.
Major J. V. Dawson.
2nd Lieutenant H. G. Deerham.[1]
Captain M. H. C. Doll.
Lieutenant G. F. Earle.
„ A. J. C. Elkan.
Captain W. A. Elliott (R.A.M.C.)
„ J. W. D. Evans (21st Lancers).
„ W. H. Eve.
Lieutenant M. R. Farrer.
„ H. C. D. FitzGibbon, M.C.
2nd Lieutenant A. H. F. Fletcher.[2]
Lieutenant G. R. Foster.[1]
„ C. W. Gardner.
2nd Lieutenant W. R. Gibbons.[1]
Captain D. W. Godfree, M.C. (21st Lancers).
Lieutenant E. Goodman.
„ R. Gore.
„ C. H. Gowan, M.C.
Major R. S. Hamilton-Grace.
Lieutenant J. Hampson.
2nd Lieutenant W. B. Harrison.[1]
Lieutenant M. G. Hartigan, M.C.
2nd Lieutenant H. R. Hawkins.[1]
Captain R. C. Hill.

[1] Did not join Regiment for duty.
[2] Did not join Regiment for duty. Killed with R.F.C., May 22, 1917.

Captain J. H. Hind.
Lieutenant J. H. Hirsch.
„ E. T. Hollingdale.
Captain A. L. Horner (R.A.V.C.)
„ J. E. P. Howey (Bedford Yeomanry).
Captain Lord Huntingfield.[1]
„ T. K. Jackson.
„ J. A. Jeffrey, M.C.
Lieutenant C. W. Jemmett.
„ H. H. Johnson, M.M.
Captain L. C. Johnston (R.A.M.C.)
Major H. Ll. Jones, D.S.O.
Lieutenant H. R. Jones, M.C.
„ B. E. H. Judkins.
Major W. A. Kennard, D.S.O.[1]
Lieutenant M. C. Kennedy.
Captain S. V. Kennedy, M.C.
Lieutenant H. G. Keswick.
„ F. G. Lawrence.
„ T. E. Lawson-Smith.[2]
„ A. M. Le Patourel.
„ J. A. Lord.
„ J. H. Lucas.
Captain J. N. Lumley, M.C.
2nd Lieutenant G. Lynch-Staunton.
Captain H. MacDonald, M.C.
Lieutenant C. A. G. M'Lagan.
2nd Lieutenant W. Madgin.[3]
Major (Bt. Lieut.-Col.) T. H. S. Marchant, D.S.O.
Captain D. M. Methven (21st Lancers).
Lieutenant R. J. V. Michell.
„ N. L. Moon.
2nd Lieutenant H. A. Moore.[3]
Lieutenant J. F. Munster.
Captain N. Neill.[4]
Lieutenant W. G. Newman.
Captain H. G. T. Newton.
Lieutenant D. J. E. Norton, M.C.
„ W. J. L. Norwood.
Captain J. O. Oakes.
Lieutenant L. A. Ormrod.
Lieutenant L. Osmond.
„ L. G. Owen.
„ R. B. Pardon.
„ G. H. Parkes, M.C.
Captain F. N. Payne.
Lieutenant W. R. Pearson.
„ G. R. Pedder.
Lieut.-Col. W. Pepys.
Lieutenant E. F. Pinnington.
„ W. J. C. Pook.
Lieut.-Col. J. J. Richardson, D.S.O.
Lieutenant W. Riley-Smith.
Captain S. O. Robinson.
Lieutenant F. G. Roche (R.A.M.C.)
2nd Lieutenant E. V. Rolfe.
Captain W. G. Rose (R.A.M.C.)
Lieutenant G. R. Russell.
„ C. F. Ryder.[3]
Captain A. M. Sassoon, O.B.E., M.C.
„ V. H. Secker (14th Hussars).
Lieutenant T. Shanley.[3]
„ P. H. J. Sheil.
2nd Lieutenant A. J. R. M. Sibson.
„ R. W. M. G. Smith-Sligo.[3]
Major C. Steele.
Lieutenant D. A. Stirling.
Captain E. H. Stocker.
Lieutenant V. F. Symondson.[5]
Brig.-Gen. A. Symons, C.M.G.
Lieut.-Col. E. F. Twist.
Captain A. Vlasto (R.A.M.C.), M.C.
Lieutenant G. R. Watson-Smyth.
„ G. L. M. Welstead.
2nd Lieutenant C. A. Whitchurch.[3]
Lieutenant A. Williams.
Captain B. H. Williams.
Lieutenant T. Williams-Taylor.
„ C. A. F. Wingfield.
„ A. H. Wood.
Captain E. Wordley (R.A.M.C.)
2nd Lieutenant G. F. Wrigley.
Lieutenant E. Wright.

[1] Did not do duty with Regiment during war.
[2] Did not serve with Regiment during war. Killed on active service with 11th Hussars, 1914.
[3] Did not join Regiment for duty.
[4] Did not serve with Regiment during war. Killed on active service, 1914.
[5] Did not serve with Regiment during war. Killed with R.A.F., Nov. 13, 1918.

ROLL OF WARRANT OFFICERS, NON-COMMISSIONED OFFICERS, AND MEN, WHO SERVED WITH THE REGIMENT DURING THE WAR.

Those marked with an asterisk (*) did not serve in Mesopotamia with the Regiment.

7324 Private Abbott, W.
46265 ,, (Sad.) Adams, J.
24407 ,, Adams, J. N.
16435 ,, Adams, R. (*wounded in action*, Nov. 3, 1917).
321371 ,, Adderley, J.
15709 ,, Adrian, F.
4721 ,, Adrian, H.
21791 ,, Aiken, G. N.
10889 ,, Akers, G.
11030 ,, Alexander, A.
4763 S.S.M. Allan, W. J. (*twice mentioned in despatches*, Dec. 29, 1916, and Jan. 4, 1917; gazetted 2nd Lieutenant, Lincolnshire R., Nov. 3, 1917).
13940 Private Allaway, W.
24483 ,, Allen, A.
2336 Lance-Corporal Allidine, F.*
255670 Private Allot, W.
33146 ,, Allsop, E.
46238 ,, Allston, J. (*mentioned in despatches*, Aug. 27, 1918).
10659 ,, Allum, C. E.
21526 ,, Ames, F. T. (*wounded in action*, Mar. 5, 1917).
3499 ,, Ammon, W.*
27712 ,, Amphlett, W. S.
6859 Lance-Corporal Anderson, F.*
4770 Sergeant Anderson, G. (*killed in action*, Mar. 5, 1917).
1298 Private (S.S.) Anderson, W.*
15773 ,, Andrews, F. (*wounded in action*, Nov. 5, 1917).
23136 ,, Andrews, J. L. N.
6507 ,, Angier, W. B (*wounded in action*, Mar. 5, 1917).
3884 ,, Annand, G.
22369 ,, Annibal, G. L.
3939 ,, Ansell, J. G.*
1952 ,, Ansell, T. S.
16393 ,, Archer, F.
10942 Corporal Arlotte, E.*
46187 Sergeant Arnott, W. H.
5720 S.S.Far. Arthur, R.
46220 Private (S.S.) Ashcroft, A.
5574 ,, Ashdale, F.*
7815 Lance-Corporal Ashwood, T.
27837 Private Atkins, E.
255951 ,, Atkins, F.
3254 Lance-Sergeant Aukett, H. H. L. (*awarded* M.M., Dec. 1, 1918).
6613 Private Austin, J.*
4421 Lance-Corporal Austin, V.*
22509 Private Ayres, J. (*wounded in action*, July 18, 1915).*
16561 ,, Bache, H.
325891 ,, Baggott, A. E.

16289 Private Baker, A.*
255687 ,, Baker, J. E.
2715 ,, Baker, W.
3774 F.S.M. Bald, A. (*mentioned in despatches*, Aug. 27, 1918; *awarded* L.S. and G.C. Medal, April 1, 1917).
3130 Private Ball, D.
13660 ,, Ball, R. H.
4388 Corporal (S.S.) Ballard, A. C. W. (*died* Sept. 27, 1918).
4606 Sergeant Barham, W. (*wounded in action*, Feb. 4, 1917).
18943 Private Barker, J. H. (*wounded in action*, Nov. 3, 1917).
46254 Lance-Corporal Barling, F.
8809 Private Barlow, B.
4712 Sergeant Barnard, G. A. (*wounded in action*, Oct. 29, 1918).
23684 Lance-Corporal Barnfield, E.
32808 Private Barrett, A. S.
46250 Lance-Sergeant Barron, J.
8267 Lance-Corporal Bartlett, F. (*drowned in Tigris*, Oct. 28, 1918).
3117 Lance-Sergeant Bartlett, W. G. (*wounded in action*, Feb. 25, 1917).
19975 Private Bartley, J.
5300 Lance-Sergeant Batchelor, D.
5301 Corporal Batchelor, E.
281957 Private Bath, T.
27888 ,, Batten, A.
11694 ,, Beadle, A.
3977 Corporal Beadle, W. (*died*, Mar. 6, 1917, *of wounds received in action*, Mar. 5, 1917).
27907 Private Beales, G.
276662 ,, Beales, H. J. (*wounded in action*, Nov. 3, 1917).
4837 Private Bean, C.*
8138 ,, Beaty, W.
2712 ,, Beauchamp A. (*wounded in action*, Feb. 25, 1917; *mentioned in despatches*).
32800 ,, Beaver, E. F. (*wounded in action*, May 10, 1918).
6991 ,, Beaver, J.
41218 ,, Bedder, E.
24741 ,, Beesley, J. (*died*, May 20, 1917).
6871 Sergeant Bell, E. C. (*wounded in action*, Feb. 25, 1917).
3973 Private Bell, H. (*wounded in action*, Feb. 25, 1917; *died* Oct. 30, 1918).
4000 Lance-Corporal Bennett, D.
22761 Private Bennett, S. B.
325228 ,, Bennett, W. (*wounded in action*, Nov. 4, 1917).
7566 ,, Benoit, F. T.
281959 ,, Bent, A.
8564 ,, (S.S.) Berry, F.
15652 ,, Berry, F. (*wounded in action*, Dec. 3, 1917).
11959 ,, Berry, M. W.
22301 ,, Bertwistle, G.
10331 Sergeant Bettis, J. A.
11595 Private Bexon, L.
21481 ,, Biddiscombe, A.
10957 ,, Biggs, J. H. (*awarded* M.S.M., Dec. 24, 1918).
8894 ,, Birchenough, E.
325383 ,, Bird, H. H.
4430 ,, Bird, I.*
206277 ,, Bishop, A.
301559 ,, Blackburn, J. H.
3470 ,, Blake, J. G. A.
46209 ,, (S.S.) Blanchard, F. H.

1551 Private Blaney, J.
1130 Corporal (S.S.) Block, R.*
325366 Private Blundell, G. A.
31718 ,, Bocking, J.
1357 Lance-Corporal Bolton, W. (*wounded in action*, Mar. 5, 1917).
3928 Sergeant Bond, W.
37866 Private Bone, J.
1327 ,, Boorman, G.
12910 ,, Botfield, E. J.
15605 ,, Boulton, W. G.
4600 Lance-Corporal Bowie, G. W. (*awarded* D.C.M., Aug. 29, 1917).
32811 Private Bowley, C.
5718 Sergeant Bradley, W.
255016 ,, Bradshaw, A.
300436 Private Bramall, H.
9183 Corporal Bray, W.
1150 R.Q.M.S. Brearley, J. (*mentioned in despatches*, Aug. 14, 1917; *awarded* D.C.M., Aug. 29, 1917; *awarded* L.S. and G.C. Medal, April 1, 1918).
255770 Private Bree, D.
2711 Sergeant Brencher, L. P.
29343 Private Briston, J.
6893 ,, Bristow, W. F.*
41590 ,, Britnell, A. L.
3688 S.S.Far. Brockway, G.*
321377 Private Brookshaw, W.
27714 ,, Broomfield, F. (*drowned at sea*, April 15, 1917).
4112 ,, Brown, G. F.
23362 ,, Brown, G. W.
1176 ,, Brown, J.*
11458 ,, Brown, J. G.*
255875 ,, Brown, J. L.
24069 ,, Brown, N.
8475 Corporal (Sad.) Brown, W. R. (*wounded in action*, Feb. 25, 1917).
31902 Private Browning, W.
1404 Private Brunton, F.*
24751 ,, Buckland, E.
1710 ,, Buckley, J.
15801 ,, Bullen, E. R.
24739 ,, Bullock, T.
7514 Lance-Corporal Burder, H. A. (*wounded in action*, Nov. 5, 1917).
8575 Sergeant Burgess, G.
46196 Lance-Corporal Burgess, J. (*wounded in action*, Mar. 5, 1917).
6618 Private Burgess, T.*
6894 ,, Burgess, W.
46184 Corporal Burgess, W.
6918 Sergeant Burgess, W. H. (*wounded in action*, Mar. 5, 1917; *wounded in action*, May 10, 1918; *awarded* M.M., June 2, 1918).
9385 Private Burns, J.*
15316 Lance-Corporal Bush, F. K.
22850 ,, Bush, J.*
4794 Private Bushell, B. H.
29114 Lance-Corporal Butcher, R. A.
6768 Private Butler, F.*
41251 ,, Butler, H.
3259 ,, Butters, C.
6262 ,, Buttle, W.*
46253 ,, Byrne, J. (*wounded in action*, Nov. 5, 1917).
453 Sergeant Byrne, P.
13605 Lance-Corporal Calder, J.
32820 Private Callery, T.
6554 ,, Calver, F.*
24682 ,, Calvert, M.
3433 ,, Cameron, D.
595 ,, Campbell, P.* *alias* Devlin, P.
7135 Lance-Corporal Candy, R.*
382 Corporal Carlile, J. H.
3960 Sergeant Carnegie, T. (*awarded* M.S.M., April 29, 1917; *awarded* L.S. and G.C. Medal, April 1, 1917).

6177 Corporal Carpenter, J. (*mentioned in despatches*).
11281 Private Carradine, J.*
6913 ,, Carter, G.
46246 Lance-Corporal (Tptr.) Carter, W. P.
9206 Private Cartwright, A.
300025 ,, Cartwright, W. A.
285137 ,, Cave, N. P.
13516 ,, Challiner, H.*
860 ,, Chamberlain, A. V.
4509 ,, Chambers, H.*
31923 ,, Chamley, D.
6519 ,, Chapman, J.*
15828 ,, Chapman, J. T.*
46205 ,, Chapman, P.
1854 ,, (S.S.) Chapman, T.*
285444 Lance-Corporal Chaundy, E. G.
1970 Private Chinery, W. H.
3938 Sergeant Chipperfield, P. (*died* Mar. 6, 1917, *of wounds received in action*, Mar. 5, 1917).
3974 Private Chisholm, T.
6427 Sergeant Christie, H.
16094 Private Christie, J. G.*
256498 Lance-Corporal Church, W. G.
5886 R.S.M. Churchhouse, M.* (gazetted 2nd Lieutenant Lincolnshire R., June 16, 1915).
1018 Lance-Corporal Clancy, J. (*wounded in action*, April 27, 1918).
4275 Private Clare, J.
8270 Far.S. Clark, A. E. (*wounded in action*, Mar. 5, 1917).
2337 Private (S.S.) Clark, D. W.
6637 Sergeant Clark, S.* .
24488 Private Clarke, C.
235368 ,, (Tptr.) Clarke, F. J. W.
276664 ,, Clarke, G. W.
1666 ,, Clarke, W.*
15594 ,, Clarke, W. A.
6759 Private Cleary, E.
3132 Sergeant Clement, J. H.
9009 Private Cleveland, J. (*wounded in action*, Mar. 5, 1917; *awarded* M.S.M., April 29, 1917).
5791 Corporal (Tptr.) Coakes, J.*
10657 Private Cochrane, A. L.*
8769 ,, Coker, R. C.
24495 ,, Cole, J. (*died* Mar. 14, 1917, *of wounds received in action*, Mar. 5, 1917).
15846 ,, Cole, J. B. G.*
29633 ,, Cole, R. G.
46178 Far.Staff-S. Colenutt, A. (*awarded* L.S. and G.C. Medal, Oct. 1, 1918).
4851 Private Coles, S.*
6244 ,, Collard, E.
3628 ,, Collier, C.
1915 Sergeant Collins, P. (*wounded in action*, Mar. 5, 1917).
8395 Private Collins, S.
41414 ,, Comery, J.
6084 ,, Conolly, T.
9432 ,, Conroy, M. C.
24489 Corporal Cook, F.
11165 Private Cook, F. J.
6897 Sergeant Cook, G. (*died* July 27, 1916).
24753 Private Cook, H.
18130 ,, Cook, T. P.
29113 ,, Cooke, G.*
4226 ,, Cooke, J.
13739 ,, Cooper, C. (*wounded in action*, Oct. 19, 1917).
41111 ,, Cooper, J.
5577 Lance-Corporal Corbin, W. G. (*awarded* M.M., Nov. 16, 1917).
13599 Private Cordey, H.*
497 ,, Cornell, F.
4724 ,, Cornish, R.*
5805 ,, Corton, C.*

7409 Private Costello, D.
32832 ,, Cottrell, F. B.
4504 ,, (S.S.) Couch, J. F. (*awarded* D.C.M., Aug. 29, 1917; *died* July 4, 1918).
2351 ,, Course, R.*
5815 ,, Cox, E. J.
32822 ,, Cox, F.
14775 ,, Cozens, P.
10939 ,, Crabtree, E.
4523 ,, Craven, T. S. (*awarded* M.S.M., Dec. 24, 1918).
21486 ,, Creene, F. J.
32695 ,, Creeper, F. W.
16106 ,, Cressey, S. W.*
5801 ,, Crocker, S.
5079 ,, Croft, G.*
255719 ,, Crook, H.
21135 ,, Cross, G.
24421 ,, Crossinggum, B.
2898 ,, Crouch, T. (*killed in action*, Mar. 5, 1917).
10992 ,, Crowder, H.*
6731 ,, Cubitt, C. (*awarded* M.M., June 2, 1918).
24500 ,, Culleton, M.
255788 ,, Culpin, R.
21532 ,, Cumins, S. H.*
18285 ,, Cunningham, E.
6473 Sergeant Cunningham, J. (*wounded in action*, Feb. 15, 1917; *awarded* M.M., May 18, 1918).
255640 Private Cuthbert, A. G. V.
27842 ,, Cutler, H. J.
21206 ,, Danter, C. H.
5513 ,, Darby, J.
13176 ,, Dardis, R. (*wounded in action*, Feb. 25, 1917).
15830 ,, Davies, J. W.
21837 Private Davies, W. J. (*wounded in action*, Oct. 19, 1917).
9213 ,, Davis, H.
16942 ,, Davis, H. G. (*killed in action*, Feb. 25, 1917).
6505 ,, Davis, R. J.*
10971 ,, Davis, W. F. (*wounded in action*, Oct. 19, 1917).
2129 Lance-Corporal Day, G. A. (*wounded in action*, Feb. 4, 1917; *wounded in action*, Nov. 3, 1917).
281974 Private Deamon, E.
4117 Lance-Corporal Debman, S.*
46219 Lance-Sergeant Deeley, H. (*wounded in action*, Mar. 5, 1917).
22488 Private Dehner, F.*
3544 ,, Delaney, C. (*wounded in action*, Nov. 5, 1917).
205497 ,, Dent, W. D. (*died* Nov. 15, 1918).
8560 ,, Derry, F.
32841 ,, Dickens, J.
32838 ,, Dimblebee, G.
9221 ,, Disley, W.*
32842 ,, Dismore, A. H.
23179 ,, Dixey, W.
493 Lance-Sergeant Dobbie, C.
9369 Private Dobson, J.
6265 ,, Dolton, H.*
520 Corporal (S.S.) Donald, J.
3258 Private Donnachie, W.*
9116 ,, Donoghue, J. (*drowned in Tigris*, Oct. 28, 1918).
46182 S.S.M. Douthwaite, C. M. (*mentioned in despatches*, April 15, 1918).
5528 Private Dowdall, W.
3549 ,, Downie, A.*

3491 Private Downie, D. (*died* Oct. 3, 1916).
9038 ,, Doyle, J.
18361 ,, Driscoll, J.*
1984 ,, Drury, H. (*wounded in action*, Feb. 25, 1917).
26444 ,, (S.S.) Duckett, F.
24088 ,, Duke, S.
2169 ,, Dumbrill, G. J.
24073 ,, Dutton, B.
11124 Lance-Corporal Dutton, J. C.
21995 Private Dyer, E.
5440 ,, Dymock, J. E.
46216 ,, Dyne, H. A.
325214 ,, Dyson, H.
11899 ,, (S.S.) Earl, A. (*wounded in action*, May 10, 1918).
41427 ,, Earnshaw, A.
798 ,, Eaves, J.*
16463 ,, Edmunds, R.*
1173 O.R.Q.M.S. Edwardes, E. V.*
9634 Private Edwards, A. C.
46223 R.Q.M.S. Edwards, H. J. (*wounded in action*, Feb. 25, 1917; *mentioned in despatches*).
5444 Private Edwards, J.
9199 ,, Edwards, W. (*wounded in action*, Mar. 5, 1917).
3124 ,, Ellis, P.
21611 ,, Elphick, H. R.*
235756 ,, Elsworthy, A.
16845 ,, Enstone, C. (*wounded in action*, Mar. 5, 1917).
46239 Sergeant Ethell, J. R. (*wounded in action*, Mar. 5, 1917).
5447 Private Etherington, A.*
18185 ,, Evans, G. G. (*wounded in action*, Mar. 5, 1917).
424 ,, Ewen, H. M.*
206279 ,, (Tptr.) Ewing, R. C.
6957 Private Fahey, F.*
13840 ,, Fairbairn, F. (*wounded in action*, July 18, 1915).
4603 ,, Farmer, A.*
11749 ,, Farmer, R. S.
3855 S.S.Far. Farnden, H.*
8223 Private Faulkner, R.
321443 ,, Fear, C. J.
31943 ,, Fegan, J. G.
205407 ,, Fendom, A. S.
3252 Lance-Corporal Fenn, S.
32848 Private Fern, W. (*accidentally drowned*, July 17, 1918).
1530 Lance-Corporal Ferns, E.
9020 Private Fewell, J. E.
7012 Lance-Corporal Field, C.*
6881 Private Field, E.
206290 ,, Filewood, F.
18245 ,, Finch, F.
255654 ,, Fisher, W.
41 Sergeant Fitzpatrick, I.*
46235 Private Flanagan, F.
18343 ,, Flindell, D. (*died* Mar. 6, 1917, *of wounds received in action*, Mar. 5, 1917).
2052 ,, Flynn, M.*
3925 Corporal Forbes, J.
41598 Private Ford, F. G.
13531 ,, Ford, R.
4723 ,, Foreman, A.
321448 ,, Forsdick, B. H.
2263 Lance-Sergeant Foster, J. H. (*mentioned in despatches*, April 15, 1918).
27586 Private Fowler, E.
23463 ,, Fox, J.*
3964 Cook Sgt. Francis, A.*
21068 Private Francis, S. H
5882 ,, Francomb, A. W. (*killed in action*, Nov. 5, 1917).
46192 Sergeant Fraser, C. J.

8803 Private Free, S. W.
13062 ,, Freeman, G. H.
1137 ,, Freeman, O.
8119 ,, Freeman, W. (*wounded in action*, Mar. 5, 1917).
22872 ,, French, C.*
3940 ,, (Tptr.) French, H.*
32851 ,, French, R.
24368 ,, Frost, J.
5808 ,, Froud, A. G.
5703 Sergeant Frudd, E.*
1828 Cook Sgt. Fryer, E. H.
41678 Private Gabell, A. J.
5072 Sergeant Gage, F. R.
21316 Private Gage, H. J.
6229 ,, Gair, W. H.
4004 ,, Galbraith, D.
22803 ,, Galley, W. G.
5957 Lance-Corporal (Tptr.) Galloway, A. O.
5612 Private Garrod, E.*
1339 Lance-Corporal Garrod, G. A. (*wounded in action*, Mar. 5, 1917).
8268 Sergeant Garrod, S.*
6361 Lance-Corporal Garton, G. (*wounded in action*, Mar. 5, 1917).
4852 Lance-Corporal Gates, J.*
32855 Private Geaney, C.
15670 Lance-Corporal Gee, A
13604 Private Gent, J. H.
3887 ,, George, A.*
8392 Lance-Corporal (S.S.) George, J. (*wounded in action*, April 29, 1918).
11879 Private George, J.*
5802 Lance-Corporal George, R.*
6590 ,, George, W. S. (*died*, Mar. 16, 1917, *of wounds received in action*, Mar. 5, 1917).
24728 Private Gerrard, H. D. (*wounded in action*, Oct. 29, 1918).
46218 Sergeant Gilbert, W. (*prisoner of war*, Mar. 5, 1917).
6215 Sergeant Gilbert, W. J.*
325514 Private Giles, H. J.
30555 Sergeant (Tptr.) Gill, S. (*mentioned in despatches*, April 15, 1918).
46203 Private Gillies, R.
3545 Lance-Corporal Gilliver, F. (*wounded in action*, Mar. 5, 1917).
41255 Private Gilman, C. E.
23199 ,, Ginks, G.
2221 ,, Girling, F.*
29238 ,, Glass, A. W.
3113 ,, (Sad.) Glenister, A. E.
276670 ,, Glenn, G.
281980 ,, Glew, E.
8552 ,, Glover, W.
5417 S.Q.M.S. Goddard, D. F. B.
46199 Sergeant Goddard, E. (*wounded in action*, Feb. 25, 1917).
4507 Lance-Corporal Gold, A. F.
6471 Sergeant Gold, F.*
276669 Private Goodall, F.
6615 ,, Goodland, E. W.
3398 ,, Goold, H. H.
24568 ,, Goulding, E.
24569 ,, Goulding, T.
7612 Lance-Corporal Gowing, C. H.
4972 Lance-Corporal Grahame, I. C. J. (*wounded in action*, Feb. 25, 1917).
4850 Private Grant, J.*
6579 Sergeant Grant, R.*
46240 ,, Gray, J. (*mentioned in despatches*, Aug. 14, 1917; *killed in action*, Nov. 5, 1917).
13622 Private Gray, W.
32858 ,, Green, A.
4426 ,, Greene, P. C.
325325 ,, Greening, D.

255693 Private Grieff, R.
15736 ,, Griffen, H.
32857 ,, Griffen, H. H.
1534 ,, Griffin, —
15013 ,, Griffiths, J. E.
16505 ,, Grivell, C. (*wounded in action*, Feb. 25, 1917).
46175 ,, Groves, J.
13612 Lance-Corporal Gull, F.
24303 Private Gunton, T. (*wounded in action*, Feb. 4, 1917).
24494 ,, Guthrie, J. (*wounded in action*, Mar. 5, 1917).
7539 ,, Gutsell, S.
2004 Far.S. (S.S.) Gutsell, T.
15788 Private Habbitts, W. J.
10168 ,, Hack, B. G.
24456 ,, Haigh, J.
1128 S.Q.M.S. Haines, J.*
2348 ,, Haines, S. B. (*mentioned in despatches; awarded* D.C.M.)
21195 Private Haley, E.
46204 ,, Hall, E. A.
6263 ,, Hall, H.
24449 ,, Hall, J.
11416 ,, Hall, V.
18198 ,, Halstead, C. (*wounded in action*, Nov. 5, 1917).
4790 Lance-Corporal Ham, G.
2899 ,, (Tptr.) Hamilton, H. A. (*wounded in action*, Mar. 5, 1917).
301163 Private Hammond, W. (*accidentally drowned*, June 19, 1918).
6472 Sergeant Hammond, W. J. R.
27700 Private Hancock, T. F.
5808 ,, Hanglin, H.*
6708 ,, Hannah, I.*
5809 ,, Hanslip, W. (*wounded in action*, May 13, 1915).
19409 Private Harding, F.
22968 ,, Harding, H. J. (*wounded in action*, Feb. 25, 1917).
13902 ,, Harding, J. T. (*awarded* M.S.M.)
301071 ,, Hardyman, A. C.
24321 ,, Harker, T. E.
9277 ,, Harman, A.
21238 ,, Harris, C. (*wounded in action*, July 14, 1915; *wounded in action*, Dec. 5, 1917).
46243 Corporal Harris, E. (*wounded in action*, Mar. 5, 1917).
24577 Private Harris, L.
6667 Sergeant Harrison, A. (*killed in action*, Mar. 5, 1917).
2270 Private Hart, A.*
1138 ,, Hart, G. (*wounded in action*, Feb. 25, 1917).
46230 Corporal Hart, W.
8799 Private Hartopp, W.
6542 ,, Harvey, F. (*wounded in action*, Oct. 29, 1918).
16791 ,, Harvey, J. L. (*wounded in action*, April 28, 1918).
285650 ,, Hatwell, H. J.
1160 ,, Haughey, M. J.
16555 Lance-Corporal Hawkes, J. S.
32881 Private Hay, J.
14632 ,, Hayden, W. H.
7417 ,, Hayes, F.*
2352 Sergeant Hayes, J.
1637 Private Haynes, A.
15680 ,, Hayter, C. F.*
32863 ,, Hayworth, F. (*died* Nov. 14, 1918).
510 ,, Hazeltine, J. A.
13711 ,, (S.S.) Heard, A.
6550 ,, Heathman, A. E. (*wounded in action*, Feb. 25, 1917).

46227 Private Helme, N.
31946 ,, Hembrow, H.
7006 ,, Hemmings, H.
6523 Corporal (S.S.) Hemsworth, P. (*wounded in action*, Mar. 5, 1917).
425 Corporal Henderson, W.
4234 Private (S.S.) Henley, W.*
4605 ,, Henn, W. H.
13001 ,, Herbert, P. H. (*killed in action*, Mar. 5, 1917).
7846 ,, Hetherington, J.
5830 ,, Heyburn, G. E. (*wounded in action*, Mar. 5, 1917).
16431 ,, Hibbert, A. (*wounded in action*, Nov. 5, 1917).
24271 ,, Hickingbotham, J.
32872 ,, Hickles, G.
13043 ,, Higgs, A. F.
24382 ,, Higgs, J. W.
8269 S.Q.M.S. Higgs, W.
2897 Sergeant Hill, A. E.
4114 Private Hill, G. A. (*wounded in action*, Jan. 29, 1917).
9607 ,, Hill, H.
10802 R.S.M. Hill, R. C. (gazetted 2nd Lieutenant, Mar. 19, 1916).
1195 Sergeant Hill, W. J. (*mentioned in despatches; wounded in action*, Feb. 25, 1917; *wounded in action*, Oct. 29, 1918).
46479 Private Hilton, G. A.
18718 ,, Hirons, G. W.
1971 ,, Hiscox, S.*
5921 Corporal Hobbs, J. (*died* Aug. 1, 1916).
46256 S.S.M. Hockey, R. G.
301069 Private Hodder, C. H.
3859 Corporal Hodge, A.
31949 Private Hodge, F.
31950 ,, Hogan, J.
16668 Private Hogarth, F. W.
10385 ,, Hogben, A. S.*
2653 ,, Hogg, A. W. T.
22954 ,, Hogg, C.
32865 ,, Holbrook, W. J.
32869 ,, Holdsworth, J. W.
46226 ,, Holland, A. H.
13959 ,, Holland, F.
4608 ,, Holland, W.*
1162 ,, Hollaron, J.*
46191 Sergeant Holloway, H. S. (*wounded in action*, Feb. 15, 1917; *mentioned in despatches*, April 15, 1918).
3978 Sergeant Holloway, R. (*killed in action*, Oct. 29, 1918).
5306 Private Homer, H. A
32880 ,, Hone, R.
46237 ,, Hooper, S.
32879 ,, Hopcroft, T. E.
4422 ,, Hopkins, A. (*wounded in action*, Dec. 14, 1916).
24613 ,, Horrex, G. A.
8554 ,, Howard, H.
4913 Lance-Corporal Howard, J. (*died* July 22, 1915, *of wounds received in action*, July 22, 1915).
11912 Private Howard, J. S
301122 ,, Howard, O.
256561 Lance-Corporal Howden, H.
5812 Private Howland, S. M.
5305 ,, Howlett, S. (*wounded in action*, Nov. 4, 1917).
24051 ,, Howlett, T. H.
4148 ,, Hows, W. E. (*died* July 28, 1916).
15779 ,, Hubbard, H.
6768 ,, Hudson, A. E. (*wounded in action*, April 29, 1918).
206282 ,, Hudson, B.
46247 Lance-Sergeant Hudson, H. (*wounded in action*, Nov. 5, 1917).

8796 Private Humphrey, E. (*awarded* M.M., Dec. 1, 1918).
305706 ,, Humphrey, H. L.
15572 ,, Hunt, C. W.
13616 ,, Hunt, W. F. (*died* July 19, 1915, *of wounds received in action*, July 18, 1915).
4113 ,, Hunter, E.
32877 ,, Hurley, D.
4695 ,, Hurn, W.
6543 ,, Ives, W.
22420 ,, Jackson, G. W.*
255696 ,, Jackson, H. W.
8895 ,, Jackson, J.*
41457 ,, Jackson, J.
24493 ,, Jackson, W.
15628 ,, Jacobs, I. H.
22967 Lance-Corporal James, E. (*killed in action*, Mar. 5, 1917).
13828 Private James, H.
9021 ,, James, S., *alias* Roberts, H.
13417 ,, James-Moore, J. H. (*awarded* M.S.M., April 29, 1917).
24882 ,, Jarrett, S. W.
32884 ,, Jasper, F. G. (*killed in action*, Nov. 5, 1917).
18983 ,, Jeffrey, J. (*wounded in action*, Nov. 5, 1917).
3123 ,, Jelley, G.
285607 ,, Jiggins, W. F.
255905 ,, Johnson, A.
26790 ,, Johnson, G.
2717 Sergeant Johnson, J. (*wounded in action*, March 5, 1917).
9182 Private (S.S.) Johnson, W.
797 ,, Johnston, G.*
46255 Sergeant Johnstone, T.
9285 Private Jones, A. (*awarded* D.C.M., Mar. 5, 1917; *killed in action*, Mar. 5, 1917).
6154 Lance-Corporal Jones, C.*
11626 Private Jones, D. E.
9363 ,, Jones, J.
1163 ,, Jones, L.
18064 ,, Jones, R.*
285674 ,, Jones, S.
16995 ,, Jones, T. R. (*killed in action*, Mar. 5, 1917).
325391 ,, Jones, W. G.
3850 ,, Jordan, C. W. (*killed in action*, Mar. 5, 1917).
9283 ,, Jordan, M. (*wounded in action*, Dec. 17, 1916).
9247 ,, Jowett, R.*
255700 ,, Jukes, J. E.
13216 ,, Kane, M.
13242 ,, Keaney, P. (*prisoner of war*, April 23, 1917).
13460 ,, Keely, J.
206283 ,, Keeton, G.
41461 ,, Kelly, F. E.
7303 ,, Kelly, H.*
16533 ,, Kemp, J. W.
26448 ,, Kendall, R.
6164 Sergeant Kennard, F.
300482 Private Kennings, G.
11936 Lance-Corporal Kennington, C. H. (*killed in action*, Mar. 5, 1917).
3920 Private Kenny, J. (*wounded in action*, Feb. 25, 1917).
46188 Corporal Kent, G. F. S. (*wounded in action*, Mar. 5, 1917).
24478 Private Kerr, C. (*wounded in action*, April 28, 1918).
276631 Lance-Corporal Keyworth, J. H.

6850 Private Killacky, T. (*died* July 28, 1916).
6866 „ Killick, A. (*killed in action*, Feb. 4, 1917).
1200 „ Kimberley, W.
25067 „ King, A.*
2038 „ King, F.
15664 „ King, G. W.
6436 Sergeant King, J.
16228 Private King, W.
23059 „ Kingstone, F. C.
4792 „ Kirby, J. H.
27989 „ Kirby, T. L.
9156 „ Kirk, W. H.
7548 „ Kirkham, J.
11417 „ Kirkham, W.
4976 Lance-Corporal Kitcher, F. C. (*awarded* M.M., May 18, 1918).
3126 Sergeant Knapman, H. (*killed in action*, Mar. 5, 1917).
24308 Private Knight, A.
9945 „ Knight, B.
281994 „ Knight, L.
11641 „ Knight, W. G. (*wounded in action*, May 6, 1918).
4798 „ Kyte, A.*
4799 „ Kyte, F.
46213 Lance-Corporal Lacey, E. (*wounded in action*, Feb. 25, 1917).
46252 Private Ladley, A. E.
26789 „ Laggett, H.*
46172 S.S.M. (Rough Rider) Laing, H. (*awarded* L.S. and G.C. Medal, April 1, 1918).
11973 Private Lamb, W.
2511 Sergeant Lamb, W. F.
16936 Private Lambert, L. J.
4520 „ Landells, D.*
9739 „ Langford, J.
6253 „ Langley, A.
6882 S.S.M. Langley, W. J.
2513 Private Lapham, B.*
24502 Private Lapham, W. E.
2900 „ Larner, J. (*wounded in action*, Mar. 5, 1917).
4085 Sergeant Larter, E. L.*
46190 Corporal Lawless, T. G.
22591 Private Lawrence, G. (*wounded in action*, Feb. 25, 1917).
6899 „ Leddy, F.*
3493 „ Lee, A.*
26309 Lance-Corporal Lee, A.*
6630 Private Lee, F.*
276632 Lance-Sergeant Leeman, W. (*killed in action*, Oct. 29, 1918).
18749 Private Legge, S. T.
2713 „ Leplea, W.
21528 „ Levy, L. V.
2650 „ (S.S.) Lewin, H.*
2159 „ Lewis, J. W. (*wounded in action*, Jan. 14, 1917).
11629 „ Liddle, A. (*wounded in action*, Nov. 3, 1917).
4348 Lance-Corporal Lindsey, H. (*wounded in action*, Feb. 25, 1917; *wounded in action*, Nov. 5, 1917).
9627 Lance-Corporal Lingley, H.*
3929 Private Linton, J.
32892 „ Little, P. W. (*wounded in action*, Oct. 29, 1918).
24075 „ Llewellyn, J.
4796 „ Lock, F.*
3858 „ Locker, J. E. H.
6765 „ Lockerty, J.*
22903 Lance-Corporal Lomas, F. (*awarded* M.M., June 2, 1918).
6572 Sergeant Long, E. A.*
32893 Private Longstaff, J. G.
22450 Lance-Corporal Lonsdale, R.*

27616 Private Lovejoy, S. A. (*drowned at sea*, April 15, 1917).
1565 ,, Lovell, C.*
4609 Lance-Sergeant Lowder, E. (*wounded in action*, Feb. 25, 1917).
10809 Lance-Corporal (Tptr.) Lowe, J.
26310 Corporal Luetchford, W. J. (*wounded in action*, Nov. 5, 1917).
9567 Sergeant Luff, H.*
24902 Private Lyman, F. (*killed in action*, Mar. 5, 1917).
12158 ,, M'Allister, W.
10649 ,, M'Auliffe, J. G.
6240 S.Q.M.S. M'Bride, F A. (*awarded* D.C.M.)
1183 Private M'Cardy, P.
13537 ,, M'Cormick, A.
3807 ,, M'Creadie, W. (*died* Mar. 6, 1917, *of wounds received in action*, Mar. 5. 1917).
46207 ,, M'Donald, J.
13224 ,, M'Donnell, M. (*killed in action*, Nov. 5, 1917).
2563 ,, M'Dougall, M.
11253 ,, M'Govern, J. (*wounded in action*, Mar. 5, 1917).
1809 ,, M'Grath, E.
24900 ,, M'Gregor, E.
11636 ,, M'Ilwrath, J. C.
206284 ,, Mack, R. J.
32897 ,, M'Kay, J.
11958 ,, (S.S.) M'Kee, W. F.
5075 ,, M'Kenna, I.*
24845 ,, M'Kenzie, A. E.
11967 ,, Mackenzie, J.
13632 ,, M'Kirdy, J.
13099 ,, M'Nulty, P. (*killed in action*, Mar. 5, 1917).
3439 Private M'Queen, H. (*wounded in action*, Feb. 24, 1917).
31956 ,, M'Queen, R. C. B. (*wounded in action*, April 29, 1918).
256554 ,, Madelin, A.
9339 Acting-Sergeant Madge, C. J.
32909 Private Maggs, H. V.
7066 Lance-Corporal (Tptr.) Maguire, W. F. (*mentioned in despatches*).
8372 Private Mahoney, T.*
24886 ,, Main, D. (*wounded in action*, Mar. 5, 1917).
1908 ,, Mallindine, J. T.
24082 ,, Manby, E. (*wounded in action*, Nov. 5, 1917).
13704 ,, Mant, A.
27487 ,, Markham, H. J.
46179 ,, Marney, J. R.
5316 ,, Marr, W. J.
6624 ,, Marsh, E.*
4797 ,, (S.S.) Marsh, F. A. (*wounded in action*, Mar. 5, 1917).
31776 ,, (S.S.) Marshall, A. W.
6517 ,, Marshall, E.
41003 ,, Marshall, P.
9189 Lance-Corporal Martin, F. (*wounded in action*, July 18, 1915).
32898 Private Martin, J.
5581 ,, Martin, W. F. (*wounded in action*, Mar. 5, 1917).
15841 ,, Mason, B. F.
13541 ,, Mason, R.
6791 ,, Mason, Reuben (*killed in action*, Mar. 5, 1917).
13106 ,, Massey, C.
13112 ,, Massey, C. T.
3782 ,, Masters, W.*
32907 ,, Maton, L.

46224 Sergeant Matthews, W. (*awarded* M.M., Nov. 16, 1917; *killed in action*, April 28, 1918).
41723 Private Matthews, W. J.
46183 S.S.M. Medhurst, A.
22892 Private Mellanby, G. B.
13200 Lance-Corporal Mellor, A. (*wounded in action*, Feb. 25, 1917; *wounded in action*, May 10, 1918).
8605 Private Mercer, D.
24735 ,, Mercer, L.
15353 ,, Messenger, T. J.
7611 Lance-Corporal Metcalfe, I.*
301128 Private Miles, J.
1165 ,, Millar, T. A.
206305 ,, Millard, W. G.
8588 ,, Miller, C. A. (*wounded in action*, Feb. 25, 1917).
11960 ,, Miller, H. H. (*wounded in action*, Nov. 3, 1917).
3801 Lance-Corporal Miller, T. R. (*killed in action*, Mar. 5, 1917).
24892 Private Millican, S. (*accidentally shot*, Oct. 27, 1916).
41167 ,, Millis, S. H.
1554 ,, Mills, H.
23055 ,, Mills, J.
8587 ,, Minton, E. B.
6643 Corporal Mitchell, G.*
235634 Private Moaby, B.
8232 ,, Moggeridge, T.
5811 ,, Moir, F. (*wounded in action*, Mar. 5, 1917).
3128 Corporal Molyneux, F.
3852 Private Montague, F.
18767 ,, Moody, G. A.
27853 ,, Moon, J.
6880 ,, Moore, A.*
281999 ,, Moore, A. E.
15630 ,, Moore, E.
29866 Private Moore, J.
235344 ,, Moore, J.
32908 ,, Moore, J. T. T.
6705 Corporal Moore, L. J.
6397 Private Moore, P. (*died* Nov. 22, 1918).
46228 ,, Morgan, A. M.
1664 ,, Morgan, P. W. (*wounded in action*, Feb. 25, 1917).
1663 ,, Morgan, S. G.
3001 ,, Morgan, W.*
13135 ,, (S.S.) Morrey, J.
5432 ,, Morris, E.*
7561 ,, Morris, E.
6830 Corporal Morris, J. (*killed in action*, Mar. 5, 1917).
276679 Private Morris, J. E.
11916 ,, Morrison, J. (*prisoner of war*, Mar. 5, 1917).
3541 ,, Morrison, W.*
16961 ,, Mortimer, H. H. *awarded* M.M., Nov. 16, 1917).
5317 Sergeant Moss, E. E. (*awarded* M.M.)
24725 Private Moss, J. A.
15746 ,, Mothers, C.
46229 Corporal Mott, E.
46211 Private Moule, —
13219 ,, Mulcahy, P.
31825 ,, Munday, W.
9566 S.S.M. Mushett, E.
32914 Private Nash, T. E.
4857 Lance-Corporal Neal, G. C. (*wounded in action*, Feb. 25, 1917).
4435 Corporal Neale, A.*
31790 ,, Neale, A. C. (*wounded in action*, Feb. 25, 1917).
16407 Private New, A. J.*
4171 Corporal Newman, A. S. (*killed in action*, Nov. 5, 1917).
235478 Private Newman, F. A.

6989 Sergeant Newman, M.
31751 Private Newman, W. H. (*drowned in Tigris*, Oct. 28, 1918).
11259 ,, Newton, B.*
46234 ,, (Tptr.) Newton, W. J.
3809 ,, Nicholls, J.
27732 ,, Noone, W. F.
41566 ,, Norledge, F. G. (*drowned at sea*, April 15, 1917).
3490 ,, Norman, J.*
27798 ,, North, L. (*drowned at sea*, April 15, 1917).
27919 ,, Nunney, G. N.
9382 Lance-Corporal O'Connor, J. (*wounded in action*, Nov. 3, 1917).
11438 Private O'Connor, T. (*wounded in action*, Feb. 4, 1917).
31835 ,, O'Shea, M.
23686 ,, Owen, A. (*died* May 10, 1918, *of wounds received in action*, April 29, 1918; *awarded* M.M., May 18, 1918).
13090 ,, Owen, F.
11633 ,, Packer, W. H.
799 Sergeant Page, C. W.
13510 Private Palk, F. R.
4009 ,, (S.S.) Pankhurst, J. (*killed in action*, Mar. 5, 1917).
5522 Lance-Corporal Parker, A.*
41492 Private Parker, C.
29857 ,, Parker, J. E.
5302 ,, Parkes, C.
695 ,, Parr, J.*
3924 ,, Parsons, A.*
6161 ,, Parsons, F. G.
3125 Lance-Corporal Parsons, R.*
32918 Private Parton, J. A.
26121 ,, Partridge, W. J.*
18681 ,, Pate, A.*
23432 Private Patrick, V. (*wounded in action*, Oct. 29, 1918).
49091 ,, Pay, W. W. (*mentioned in despatches*).
7009 Corporal Payne, A. E.
32926 Private Payne, C.
1969 ,, Payne, F.
15676 ,, Payne, R.*
18332 ,, Payton, W. J.
7929 ,, Pearce, J.
2518 ,, Pearson, H.*
206306 ,, Pease, S. W.
22518 ,, Peet, G. (*wounded in action*, Mar. 5, 1917).
46249 ,, Penny, B.
7927 Far.S. Perkins, F.
24492 Private Perkins, G.
13082 Lance-Corporal Perkins, W. (*wounded in action*, Feb. 25, 1917).
24138 Private Peronne, B. R.*
26893 ,, Perrott, J. (*wounded in action*, Mar. 5, 1917).
4527 Corporal (Sad.) Perry, J.
10888 Private Peters, W.
46181 ,, Petman, R. D.
24553 ,, Petre, W.
3967 ,, Pettifor, H.
33312 ,, Phelps, H.
31959 ,, Phillips, J.
46245 ,, Pierce, J.
4420 ,, Piggott, J. C. (*wounded in action*, Feb. 1, 1917).
3744 Lance-Corporal Pinner, G.
27471 Private Piper, J. W.
1201 ,, Pitman, D. (*killed in action*, July 12, 1915).
4289 Corporal (Sad.) Plumbly, S. (*wounded in action*, Dec. 15, 1916).
24988 Private Plumbridge, E. W.

5811 Private Plumridge, C. F.* (*wounded in action*, Aug. 18, 1915).
5514 Lance-Corporal Poile, W.*
7746 Private Pomeroy, W. H. (*wounded in action*, Mar. 5, 1917).
26127 ,, Pook, A. F.*
46222 S.S.M. Popham, E. L. (*mentioned in despatches*, Aug. 14, 1917).
235772 Private Pople, C.
6838 Corporal Pople, G. (*wounded in action*, Mar. 5, 1917).
6221 Private Porter, R.
6235 ,, Posford, C.
46200 S.S.M. Potter, A. (*mentioned in despatches*, Aug. 14, 1917).
27823 Corporal Potter, S. (*wounded in action*, April 28, 1918).
6410 Private Powell, W. (*wounded in action*, Feb. 4, 1917; *wounded in action*, Oct. 29, 1918).
4505 ,, Pratt, F. W. (*wound-in action*, Dec. 17, 1916).
46180 ,, Pretsell, J.
240125 ,, Price, R.
80 ,, Pringle, W.
18089 ,, Prior, E. W.
2856 ,, Proctor, W.
2341 ,, Prosser, A. P. (*wounded in action*, Feb. 25, 1917).
4111 ,, Pullen, E. L.
13703 ,, Pullen, V.*
675 ,, (Tptr.) Purchase, J.
29262 ,, Purser, A.
24061 ,, Purvis, W.
16783 ,, Quinney, J.
11991 ,, Radford, W. C. N.
1801 ,, Raggett, A.*
4405 ,, Rainbird, L.
4429 Lance-Corporal Rainbird, W.*
6762 Private Raines, J.
16300 ,, Ralls, W. J.
6876 ,, Ralph, A. (*wounded in action*, Feb. 25, 1917).
4360 ,, Randall, F. E.
5319 ,, Randell, C.
1167 ,, Rapple, T.*
13476 Lance-Corporal Ratcliffe, J.
13436 Private Ratcliffe, R.*
4524 Lance-Corporal Ratty, G. (*wounded in action*, Mar. 5, 1917).
8885 Private Ratty, W.
9202 ,, Rawbone, S.*
18235 ,, Rawlinson, R.
46248 Corporal Rayment, J. W.
423 ,, Rayner, G. (*mentioned in despatches; awarded* D.C.M., Feb. 1, 1918).
4422 Lance-Corporal Reaves, A. E. (*wounded in action*, Nov. 5, 1917).
24931 Private Reaves, T.
41502 ,, Redman, R. G. A. (*drowned at sea*, April 15, 1917).
13865 ,, Rees, C. V.*
300480 ,, Reeves, G.
235876 ,, Reeves, W.
11127 ,, Reid, C. W.*
2253 Lance-Sergeant Reid, J.
235715 Private Rendle, A. R. C.
205542 ,, Revolta, G. R. E.
15784 ,, Reynolds, F. G. (*wounded in action*, July 18, 1915).
3808 Lance-Sergeant Richards, A. N. (*awarded* M.S.M., Oct. 16, 1918).
11436 Private (S.S.) Richards, J.
46187 Sergeant Richardson, A.
22931 Lance-Corporal Richardson, R. W.
28341 Private Ricketts, W.

46242 Private Ridley, G. E.
206307 ,, Risdon, A.
32933 ,, Roberts, A.
16278 ,, Roberts, F.
9014 ,, Roberts, G.*
8076 ,, Roberts, J. H.
6166 ,, Roberts, J. L. (*awarded* D.C.M., 1917; *died* April 10, 1917, *of wounds received in action*, April 8, 1917).
7821 Corporal Roberts, W. H. (*wounded in action*, Feb. 25, 1917).
11911 Private Robertson, J.
8551 ,, Robertson, N.
1198 ,, Robins, F.
31771 ,, Robinson, W. H.
11504 Lance-Corporal Rochester, L. M.
27991 Private Rockcliffe, F. W.
32939 ,, Rogers, F. R.
1947 ,, Rogers, H.
4118 Sergeant Rogers, J. H. (*mentioned twice in despatches; awarded* M.S.M., Dec. 24, 1918).
32937 Private Rogers, W.
276683 ,, Rook, E.
1361 ,, Rostron, J.
31960 ,, Rowley, L. (*wounded in action*, April 28, 1918).
206114 ,, Rubins, H.
2936 ,, Russell, G.
9045 ,, Ryan, M. (*wounded in action*, Mar. 5, 1917).
9191 Lance-Corporal Ryder, J. W.
27739 Private Salter, W. A.
49118 ,, Sambrook, T. W.
32954 ,, Sampson, C. P.
6170 ,, Saunders, E.
1188 Lance-Corporal Saunders, J.
25010 Private Savill, A. G.
10656 ,, Sawyer, F. J.
41804 ,, Sawyer, J. G. (*drowned at sea*, April 15, 1917).
4343 Lance-Corporal (Tptr.) Scarr, C. H. T.
11937 Private Schofield, L. B.
3704 ,, Schubert, H.
1426 ,, Schubert, J. (*died* July 26, 1916).
4570 ,, Scobell, —
5970 ,, Scott, A. C.
46197 ,, Scott, J.
4342 ,, Scott, S. A.
13567 ,, Scott, W. J.
1576 ,, Seagull, G. A.
24316 ,, Searing, H.
10972 ,, Searle, E.
26261 ,, Sebiour, F.
9026 ,, Seekins, H.
46186 R.S.M. Seekins, S. (*awarded* M.C., Feb. 7, 1918).
1189 Corporal Sexton, L.
6753 Private Shadgeth, H.*
41785 ,, Shannon, J.
5785 ,, Shaw, A. J. (*wounded in action*, May 6, 1918).
24071 ,, Sheen, W. (*wounded in action*, Mar. 5, 1917).
321441 ,, Shepherd, S. W.
4344 ,, Sheppard, J. O.
32964 ,, Sheppard, T. G.
32955 ,, Sherlock, S.
5585 ,, Shoebridge, J.
6202 ,, Shorter, A. E. (*wounded in action*, Feb. 1, 1917).
46241 Lance-Corporal Shrimpton, W. J.
5734 Sergeant Sidwell, W. V.
46185 Private Sims, W. H.
751 Lance-Corporal Singleton, J. E.
32963 Private Sinnott, H.

282017 Private Sisson, J. M.
285535 ,, Slatter, J. H.
4858 ,, Slaughter, W.*
30441 ,, Slavin, R. (*drowned at sea*, April 15, 1917).
255710 ,, Sleath, F. A.
46231 Corporal Smith, A.
46206 Private Smith, A. C.
655 Sergeant (Sad.) Smith, B.*
9024 Private Smith, C. J.
46221 Sergeant Smith, C. W. (*wounded in action*, Feb. 4, 1917).
5582 Private Smith, E. H.
3855 ,, Smith, G.
6249 ,, Smith, H.*, *alias* Parsons.
5606 Corporal (S.S.) Smith, J.*
2058 Private Smith, J.*
5803 ,, Smith, J. (*wounded in action*, Feb. 25, 1917).
8294 ,, (S.S.) Smith, J. (*wounded in action*, Feb. 4, 1917).
41524 ,, Smith, J. W.
7037 ,, (S.S.) Smith, P.
276688 ,, Smith, S. (*died* Oct. 8, 1918).
5572 ,, Smith, T.
10252 ,, Smith, T. S.
531 ,, Smith, W.*
41669 ,, Smith, W. J. (*drowned at sea*, April 15, 1917).
24355 ,, Smith-Mutlow, P.
6392 ,, Smithers, J. (*mentioned in despatches*).
256178 ,, Snell, A. S.
301229 ,, (S.S.) Snell, G.
4105 ,, Soley, A.
32944 ,, Solomon, E.
18825 ,, Soper, R. (*died* July 27, 1916).
256557 Private Souter, G.
46212 Lance-Sergeant Spanton, F. (*prisoner of war*, Mar. 5, 1917).
2174 Private Sparrow, A. (*wounded in action*, Feb. 1, 1917).
32960 ,, Spear, F.
24363 ,, Spenceley, T. J.
10861 Sergeant Spencer, F.
8050 Private Spicer, C.*
6010 Lance-Corporal (Tptr.) Spicer, J. A. (*killed in action*, Mar 5, 1917).
5315 Private Spicer, W. E.
3851 Sergeant Spokes, H. S.
5512 Private Stacey, W.*
6608 ,, Stamp, W. G.
5573 ,, Stannett, G.
46173 S.S.Far. Stanton, W. (*mentioned in despatches*).
2207 Lance-Corporal Starbrook, W. D.
282027 Private Starbuck, F.
13470 ,, Steedman, P.
6217 ,, Steer, F. (*accidentally killed*, Feb. 9, 1916).
5721 Corporal Stent, C. J.
4115 Private Stephens, E.*
23500 ,, Stephens, W. P.
6495 ,, Stephenson, F. (*died* Dec. 19, 1917).
2901 Lance-Corporal Stevens, J. (*mentioned in despatches*, Aug. 14, 1917; *awarded* D.C.M., Aug. 29, 1917).
23537 Private Stevens, W. G.
7009 Sergeant Stevenson, E.*
5586 Private Steward, H. W. (*wounded in action*, Nov. 5, 1917).
4939 Lance-Corporal Stewart, J.*
29236 Private Stewart, W. A. E.
10970 Lance-Corporal Stillwell, F.
24680 Private Stobbs, H.

1190 Acting-Sergeant Stoneham, D. (*wounded in action*, July 18, 1915).
13583 Private Stoneman, S. H.
18206 ,, Stones, J. L. K. (*killed in action*, Mar. 5, 1917).
16213 ,, Strange, C. A.
8765 ,, Strange, G. E.
4859 Sergeant Strawbridge, S. G. (*awarded* D.C.M., Aug. 26, 1918).
46225 Private Street, H. J.
3620 ,, Stroud, G.
9039 ,, Stubbs, R. (*wounded in action*, Nov. 5, 1917).
24490 ,, Sturgess, J.
2135 Lance-Corporal Styles, F.
16808 Private Styles, H.
10722 Sergeant (Tptr.) Styles, J. (*killed in action*, Mar. 5, 1917).
6624 Lance-Corporal Sullivan, —*
46166 Private Swaddling, J.
26138 ,, Swindell, W.
3702 Lance-Corporal Swindle, F.
49090 Private Sykes, T.
256531 ,, Sykes, T.
26039 Corporal Taber, A.
27902 Private (S.S.) Tait, A.
3715 S.Q.M.S. Tarr, S. (*awarded* M.S.M.)
16508 Private Tarr, W. G. (*died* July 27, 1916).
255891 ,, Tarry, G.
46194 S.Q.M.S. Tassie, W. D. (*awarded* D.C.M., Feb. 11, 1917).
19433 Private Tatner, J. G. (*wounded in action*, April 29, 1918).
1998 Lance-Sergeant Taylor, A. (*wounded in action*, Mar. 5, 1917).
1871 Private Taylor, A. F.
13900 Private Taylor, A. J.*
15661 ,, Taylor, B. G.
18475 ,, Taylor, C. E. (*wounded in action*, Dec. 17, 1916; *wounded in action*, Mar. 5, 1917; *wounded in in action*, Nov. 5, 1917).
3127 Lance-Sergeant Taylor, J. A.
796 Lance-Corporal Taylor, M.
9218 Private Teague, P. J. (*died* Sept. 25, 1918).
6514 S.Q.M.S. Tegg, F. J. (*killed in action*, Nov. 5, 1917).
9479 Private Tell, N. E.
3515 R.Q.M.S. Tennant, G. F.
18737 Private Terry, A. E.
4428 Sergeant Tetheridge, W. H. (*awarded* D.C.M., Nov. 16, 1917).
11642 Private Thackray, G. (*killed in action*, Mar. 5, 1917).
27566 ,, Thatcher, A.
325624 ,, Thatcher, C.
46208 Corporal Thomas, F. (*wounded in action*, Feb. 25, 1917).
24897 Private Thomas, H. G.
1562 ,, Thomas, W.*
3804 Corporal Thomas, W. H. W. (*wounded in action*, Mar. 5, 1917).
206170 Private Thomason, F.
8220 ,, Thompson, G.
13419 ,, Thompson, J.*
24726 ,, Thompson, J. W. (*wounded in action*, May 10, 1918).
3883 ,, Thomson, W. (*died* Nov. 6, 1917, *of wounds received in action*, Nov. 5, 1917).
4703 S.Q.M.S. Thomson, W. K. (*died* July 25, 1916).

10989 Private Thorne, A. E.*
32971 ,, Thornell, S. V. (*died* May 8, 1918, *of wounds received in action*, May 6, 1918).
31992 ,, Thornton, J.
27644 ,, Thornton, T.
3002 Lance-Corporal Thorogood, G.*
9044 Private Thorpe, J.*
276691 ,, Thorpe, J. W.
7918 ,, Thrackay, B.
41539 ,, Thresher, W.
3494 ,, Thursbly, C. B.
20705 ,, Thurston, T.
11553 ,, Tilling, S.*
32968 Lance-Corporal Tindale, J.
13692 Private Tindle, W. H. (*killed in action*, May 3, 1917).
5309 ,, Tipper, W.*
3853 ,, Titmus, A.*
23138 ,, Tolman, S. F.
235403 ,, Tombs, H. H.
3131 ,, Tooke, A. E.
9286 Lance-Corporal Tookey, E. F.
27808 Private Towers, B.
10887 Corporal Townsend, J. (*wounded in action*, May 10, 1918).
11254 Private Tracey, J. (*killed in action*, July 12, 1915).
5439 ,, Trafani, A. (*wounded in action*, Mar. 5, 1917).
7549 ,, Tranter, R. (*wounded in action*, Oct. 29, 1918).
13978 ,, Treanor, J.
33313 ,, Tribe, E.
6596 ,, Trowbridge, F.*
13832 ,, Tucker, E.
11139 ,, Tucker, J. P.
205691 ,, Tull, V.
22535 Private Tunnicliffe, J. (*killed in action*, Nov. 3, 1917).
3129 ,, (S.S.) Tuppin, T.*
13685 ,, Turner, C.
5074 ,, Turner, E.
8218 ,, Turner, E. Y. (*wounded in action*, Feb. 25, 1917).
13474 ,, Turner, G.
9145 ,, Turner, G. T.
15619 Lance-Corporal Turner, H. E. (*mentioned in despatches*)
27910 Private Turner, J.
2992 ,, Turner, T.
5861 ,, Turpin, B. G.
206158 ,, Tutte, E.
26806 ,, Twigg, D. (*killed in action*, Feb. 25, 1917).
21527 Lance-Corporal Tyler, L. I.
13529 Private Ullock, W.*
235489 ,, (S.S.) Vaughan, A. S. (*wounded in action*, May 10, 1918).
235451 ,, Vaughan, R.
4855 ,, Venning, J.
4424 Lance-Sergeant Venning, J.C. (*wounded in action*, Dec. 14, 1916).
6248 Sergeant Vernon, W.
8224 Private Vickers, J. (*died* Nov. 6, 1918).
1171 ,, Vigus, F. G.
4003 Acting-Sergeant Vile, W. H.
6915 Lance-Corporal Vinall, A. (*awarded* D.C.M.; *died* Oct. 31, 1918, *of wounds received in action*, Oct. 29, 1918).
7394 Private Vincent, J. H.
4491 Lance-Corporal Viner F. (*awarded* M.M., Dec. 1, 1918).
5516 Private Viney, A. V. (*wounded in action*, Nov. 5, 1917).

3232 Private Wade, G.
2357 ,, Walker, A.
18347 ,, Walker, S.
276623 ,, Wall, C. N.
4529 ,, Wallace, J.*
22635 ,, Waller, E.
3786 Lance-Corporal Waller, H.*
9568 Private Walley, W.*
5303 ,, Wallhead, A. (*killed in action*, Mar. 5, 1917).
4423 ,, Walmsley, F.
26067 ,, Walsh, M.
15861 ,, Walters, S.
2116 Arm.S.S. Walton, H.
5453 Private Ward, A.*
24863 ,, Ward, A. J. F. (*wounded in action*, Nov. 5, 1917).
5912 Sergeant Ward, B. J. (*awarded* M.S.M.)
281934 Corporal Wardle, S.
4107 Sergeant Ware, H.*
41665 Private Warner, S. B.
6582 ,, Warner, W. J.
27755 ,, Watchman, A. E.
11514 ,, Watkins, A. H.
9192 Lance-Corporal Watkins, A. W. (*wounded in action*, Mar. 5, 1917; *awarded the* D.C.M., Aug. 29, 1917; *mentioned in despatches*, Aug. 14, 1917).
46214 Private Watson, H.
1938 ,, Way, F. W.
20919 ,, Wearing, W.
16847 ,, Webb, A.
1548 Lance-Corporal Webb, A. F. (*wounded in action*, Feb. 1, 1917).
5614 Private Webb, C. A.
16517 ,, Webb, F.
18727 ,, Webb, H. J.
6378 ,, Webber, G.*
16707 ,, Webber, W.
276624 Private Webster, W. W.
23453 ,, Wedd, A.
24435 ,, Welch, A.
10982 ,, Wells, A. V.
9234 Lance-Corporal Went, A.*
32976 Private Wernham, J.
4922 ,, West, W. W.
2177 ,, Westall, F.*
5551 Sergeant Westbrook, A. (*wounded in action*, Mar. 5, 1917).
325502 Private Westbury, A.
15683 Sergeant Westerman, R. A. (*awarded* M.S.M., Dec. 24, 1918).
6746 Private Weston, A.*
23501 ,, Wheatcroft, P.
10646 ,, Wheatley, A. W.*
27481 ,, Whebby, G. H.
4900 Sergeant Wheeler, A.
1907 Lance-Corporal White, E.
6247 Private White, E. J.
16441 ,, White, H. N.
2282 ,, White, H. W.
21954 ,, White, J.
1148 ,, White, S.*
1416 ,, Whitelock, S. D.
5433 Lance-Corporal Whitington, F. (*awarded* M.M., Dec. 1, 1918).
9282 Private Whitmore, C.*
9227 ,, Whitney, W.
2519 ,, Whittington, W.
9279 Lance-Corporal Whittle, J. (*wounded in action*, April 8, 1917).
3710 Sergeant Whyte, R.*
15087 Private Wickham, W. F. (*wounded in action*, Feb. 25, 1917).
206298 ,, Wilkins, W. L.
4349 ,, Wilkinson, F.*
24614 ,, Williams, D. G. (*wounded in action*, Nov. 5, 1917).

276627 Private Williams, E.
2256 „ Williams, G.
13120 „ Williams, J.
31990 „ Williams, J.
24756 „ Williams, W. J.
3350 Sergeant Willis, J.
30615 Private Willis, J.
6956 „ Willoughby, P.*
41778 „ Wilson, D. J.
13061 „ Wilson, F. (*wounded in action*, Mar. 5, 1917).
7464 „ Wilson, J.
3501 Sergeant Wiltshire, T.
10965 Private Wimpey, W. A.
2902 „ Winfield, J. W.
1192 „ Winter, F. (*wounded in action*, Oct. 29, 1918).
6493 „ Wiseby, A. W.
46202 S.S.M. Wood, S.
205786 Private Woodhead, F.
27440 „ Woodland, W. J.
1172 „ Woods, J.
8228 Lance-Corporal Woollaston, A.*
282045 Private Woolley, L.
8240 „ Wordley, C.
7003 Private (Tptr.) Wren, F. M. (*died of wounds*, Mar. 5, 1917, *received in action*, Mar. 5, 1917).
13611 „ Wren, P. (*wounded in action*, Mar. 5, 1917).
6604 Lance-Corporal Wright, D. C. (*wounded in action*, Feb. 4, 1917; *awarded* M.S.M., April 29, 1917).
10668 Private Wright, F.*
24907 „ Wright, F.
18363 „ Wright, H. C.
1753 „ Wright, L.*
2710 „ Wright, W.*
573 S.S. (Sad.) Wykes, W.
206086 Lance-Corporal Wyncoll, C. H. A. (*mentioned in despatches*).
8352 Private Yates, T W.
13614 „ Yelverton, D.
2200 Corporal (Sad.) Youells, J. W. (*wounded in action*, Feb. 25, 1917).
2042 Private Young, H. (*died* May 11, 1918).
26955 „ Young, J. A. G.*

APPENDIX II.

CASUALTIES.

All the casualties occurred with the Regiment unless otherwise statea

OFFICERS.

KILLED.

Lieutenant T. E. Lawson-Smith (With 11th Hussars.)	Near Messines	Nov. 1, 1914.
Captain Norman Neill (Brigade-Major, 7th British Cavalry Brigade.)	At Zwarteleen	Nov. 6, 1914.
2nd Lieutenant J. F. Munster	At Shumran Bend	Feb. 4, 1917.
Captain William H. Eve	At Lajj	Mar. 5, 1917.
2nd Lieutenant G. Lynch-Staunton	,,	,,
,, E. V. Rolfe	,,	,,
,, J. O. P. Clarkson (Attached 16th M.G. Squadron, 7th Cavalry Brigade.)	Near Jaffer's Tomb	Mar. 10, 1917.
Captain S. O. Robinson	At Tekrit	Nov. 5, 1917
,, H. Macdonald, M.C. (Assistant Political Officer.)	At Amadia, Kurdistan	July 14, 1919.

DIED OF WOUNDS.

Captain H. A. de P. O'Kelly (With 18th Hussars.)	At Meteren	May 19, 1915.

Other Causes.

Brevet Major R. S. Hamilton-Grace (Result of a motor accident, G.S.O. 2nd Hqs. Cav. Corps.)	At Burgues	Aug. 4, 1915.
Captain H. G. T. Newton (Accidentally drowned.)	In River Tigris	April 25, 1917.
Captain F. Norman Payne (Cholera. Staff Captain to Military Governor, Baghdad.)	At Baghdad	Nov. 14, 1917.
Major W. A. Kennard, D.S.O. (Of Pneumonia.)	At Etaples	Dec. 1918.

Wounded.

Brevet Major H. Ll. Jones, D.S.O. (With 4th Dragoon Guards.)	In France	Oct. 28, 1914.
Lieutenant B. H. Williams (With 11th Hussars.)	,,	Oct. 31, 1914.
Lieutenant G. R. Watson-Smyth	Near Lillers	July 14, 1915.
,, M. H. C. Doll	Near Authuille	Sept. 11, 1915.
Captain S. O. Robinson	At Shumran Bend	Feb. 4, 1917.
Lieutenant T. Williams-Taylor	,,	,,
,, J. A. Lord	At Imam Mahdi	Feb. 25, 1917.
Lieut.-Colonel J. J. Richardson, D.S.O.	At Lajj	Mar. 5, 1917.
Brevet Lieut.-Colonel E. F. Twist	,,	,,
Brevet-Major J. V. Dawson	,,	,,
Lieutenant G. L. Welstead	,,	,,
,, G. R. Pedder	,,	,,
,, H. C. D. FitzGibbon, M.C.	Near Deli Abbas	April 8, 1917.
2nd Lieutenant C. A. G. M'Lagan	At Tuz Kermatli	April 21, 1918.
,, G. R. Russell	At Hadraniyeh	Oct. 29, 1918.
Captain J. W. D. Evans (21st Lancers) Attached.	,,	,,

Wounded and Prisoners.

Lieutenant E. F. Pinnington	At Lajj	Mar. 5, 1917.
,, T. Williams-Taylor	At Tekrit	Nov. 5, 1917.

OTHER RANKS.

Killed.

France.

11254	Private J. Tracey	At Noyelles les Vermelles	July 12, 1915.
1201	,, D. Pitman	,, ,,	,,

Mesopotamia.

6866	Private A. Killick	At Shumran Bend	Feb. 4, 1917.
26806	,, D. Twigg	At Imam Mahdi	Feb. 25, 1917.
16942	,, H. G. Davis	,,	,,
10722	Sergeant J. S. Styles (Tptr.)	At Lajj	Mar. 5, 1917.
4770	,, G. Anderson	,,	,,
3126	,, H. Knapman	,,	,,
6667	,, A. Harrison	,,	,,
6830	Corporal J. Morris	,,	,,
22967	Lance-Corporal E. James	,,	,,
11936	,, C. H. Kennington	,,	,,
3801	,, T. R. Miller	,,	,,
6791	Private Reuben Mason	,,	,,
11642	,, G. Thackray	,,	,,
2898	,, T. Crouch	,,	,,
9285	,, Alfred Jones, D.C.M.	,,	,,
13099	,, P. M'Nulty	,,	,,
16995	,, T. R. Jones	,,	,,
13692	,, W. H. Tindle	,,	,,
13001	,, P. H. Herbert	,,	,,
5303	,, A. Wallhead	,,	,,
3850	,, C. W. Jordan	,,	,,
24902	,, F. Lyman	,,	,,
4009	,, J. Pankhurst (S.S.)	,,	,,
18206	,, J. L. K. Stones	,,	,,
22535	,, J. Tunnicliffe	Near Daur	Nov. 3, 1917.
46240	Sergeant J. Gray	At Tekrit	Nov. 5, 1917.
4171	,, A. S. Newman	,,	,,
5882	Private A. W. Francomb	,,	,,
46224	Sergeant W. Matthews, M.M.	At Tuz Kermatli	April 28, 1918.
3978	,, R. Holloway	At Hadraniyeh	Oct. 29, 1918.
276632	Lance-Sergeant W. Leeman	,,	,,

DIED OF WOUNDS.

13616	Private W. F. Hunt	At Nœux les Mines	July 19, 1915.
4913	Lance-Corporal J. Howard	In France	July 22, 1915.
7003	Private F. M. Wren (Tptr.)	In Mesopotamia	Mar. 5, 1917.
3977	Corporal W. Beadle	,,	Mar. 6, 1917.
3938	Sergeant P. Chipperfield	,,	,,
3807	Private W. M'Creadie	,,	,,
18343	,, D. Flindell	,,	,,
24495	,, J. Cole		Mar. 14, 1917.
6590	Acting Corporal W. S. George	,,	Mar. 16, 1917.

6166	Private J. L. Roberts, D.C.M.	In Mesopotamia	April 10, 1917.
3883	„ W. Thomson	„	Nov. 6, 1917.
32971	„ S. V. Thornell	„	May 8, 1918.
23686	„ A. Owen, M.M.	„	May 10, 1918.
6915	Corporal A. Vinall, D.C.M.	„	Oct. 31, 1918.

Deaths accepted for Official Purposes.

6010	Tptr. J. A. Spicer	At Lajj	Mar. 5, 1917.
6514	S.S.M. F. J. Tegg	At Tekrit	Nov. 5, 1917.
13224	Private M. M'Donnell	„	„
32884	„ F. G. Jasper	„	„

Drowned in the River Tigris during Operations.

31751	Private W. H. Newman	At Hadraniyeh	Oct. 28, 1918.
8267	Lance-Corporal F. Bartlett	„	„
9116	Private J. Donoghue	„	„

Other Causes.

6217	Private F. Steer	At Le Plouy, France	Feb. 9, 1916.
4703	S.Q.M.S. W. K. Thomson	At Basrah	July 25, 1916.
1426	Private J. S. Schubert	„	July 26, 1916.
6897	Sergeant G. Cook	„	July 27, 1916.
16508	Private W. G. Tarr	„	„
18825	„ R. Soper	At Makina	„
6850	„ T. Killacky	At Basrah	July 28, 1916.
4148	„ W. E. Hows	„	„
5921	Corporal J. Hobbs	„	Aug. 1, 1916.
3491	Private D. Downie	„	Oct. 3, 1916.
24892	„ S. Millican	Accidentally shot at sea, "H.T." *Huntsgreen*	Oct. 27, 1916.
27798	„ L. North	At sea, "H.T." *Cameronia*	April 15, 1917.
41566	„ F. G. Norledge	„ „	„
41699	„ W. J. Smith	„ „	„
41502	„ R. G. A. Redman	„ „	„
27616	„ S. A. Lovejoy	„ „	„
27714	„ F. Broomfield	„ „	„
30441	„ R. Slavin	„ „	„
41804	„ J. G. Sawyer	„ „	„
24741	„ J. Beesley	At Baghdad	May 20, 1917.
6495	„ F. Stephenson	In Mesopotamia	Dec. 19, 1917.
2042	„ H. Young	„	May 11, 1918.
301163	„ W. Hammond	In River Tigris	June 19, 1918.

4504	Private J. F. Couch, D.C.M. (S.S.)	At Chaldari	July 4, 1918.
32848	„ W. Fern	„	July 17, 1918.
9218	„ P. J. Teague	In Mesopotamia	Sept. 25, 1918.
4388	Corporal A. C. W. Ballard (S.S.)	At Baghdad	Sept. 27, 1918.
276688	Private S. Smith	In Mesopotamia	Oct. 8, 1918.
3973	„ H. Bell	At Basrah	Oct. 30, 1918.
8224	„ J. Vickers	In Mesopotamia	Nov. 6, 1918.
32863	„ F. Hayworth	At Baghdad	Nov. 14, 1918.
205497	„ W. D. Dent	„	Nov. 15, 1918.
5397	„ P. Moore	„	Nov. 22, 1918.

WOUNDED.

21238	Private C. Harris	Nœux Les Mines	July 14, 1915.
22509	„ J. Ayres	„	July 18, 1915.
13840	„ F. Fairbairn	„	„
9189	Lance-Corporal F. Martin	„	„
15784	Private F. G. Reynolds	„	„
1190	Acting Sergeant D. Stoneham	„	„
5811	Private C. F. Plumridge	„	Aug. 18, 1915.
4422	„ A. Hopkins	Kalahaji Fahan	Dec. 14, 1916.
4424	Lance-Corporal J. G. Venning	„	„
4289	Sad. Corporal S. Plumbly	Atab	Dec. 15, 1916.
9283	Private M. Jordan	Bassouia	Dec. 17, 1916.
4505	„ F. Pratt	„	„
18475	„ C. E. Taylor	„	„
2159	„ J. W. Lewis	Sheik Suliman	Jan. 14, 1917.
4114	„ G. A. Hill	Bassouia	Jan. 29, 1917.
4420	„ J. C. Piggott	Reconnaissance, Bassouia	Feb. 1, 1917.
6202	„ A. E. Shorter	„ „	„
2174	„ A. Sparrow	„ „	„
1548	„ A. F. Webb	„ „	„
4606	Sergeant W. Barham	Shumran Bend	Feb. 4, 1917.
2129	Lance-Corporal G. A. Day	„	„
24303	Private T. Gunton	„	„
11438	„ T. O'Connor	„	„
6410	„ W. Powell	„	„
46221	Sergeant C. W. Smith	„	„
8294	Lance-Corporal (S.S.) J. Smith	„	„
6604	Lance-Corporal D. C. Wright	„	„

6473	Sergeant J. Cunningham .	Nahr Massag .	Feb. 15, 1917.
46191	„ H. S. Holloway .	„ . .	„
3439	Private H. M'Queen . .	Imam Mahdi . .	Feb. 24, 1917.
3117	Lance - Sergeant W. G. Bartlett	„ . .	Feb. 25, 1917.
2172	Private A. Beauchamp .	„ . .	„
6871	Sergeant E. C. Bell . .	„ . .	„
3973	Private H. Bell . . .	„ . .	„
8475	Corporal Sad. W. R. Brown	„ . .	„
13176	Private R. Dardis . .	„ . .	„
1984	„ H. Drury . .	„ . .	„
46223	S.S.M. H. J. Edwards .	„ . .	„
46199	S.Q.M.S. E. Goddard . .	„ . .	„
4972	Lance-Corporal I. C. J. Grahame	„ . .	„
16505	Private C. Grivell . .	„ . .	„
22968	„ H. J. Harding .	„ . .	„
1138	„ G. Hart . . .	„ . .	„
6550	„ A. E. Heathman .	„ . .	„
1195	Sergeant W. J. Hill . .	„ . .	„
3920	Private J. Kenny . .	„ . .	„
46213	Lance-Corporal E. Lacy .	„ . .	„
22591	Private G. Lawrence . .	„ . .	„
4348	Lance-Corporal H. Lindsey.	„ . .	„
4609	Sergeant E. Lowder . .	„ . .	„
13200	Lance-Corporal A. Mellor .	„ . .	„
8588	Private C. A. Miller . .	„ . .	„
4857	Lance-Corporal G. C. Neal	„ . .	„
31790	Corporal A. C. Neale . .	„ . .	„
13082	Lance-Corporal W. Perkins	„ . .	„
2341	Private A. P. Prosser . .	„ . .	„
6876	„ A. Ralph . .	„ . .	„
7821	Lance - Corporal W. H. Roberts	„ . .	„
5803	Private J. Smith . .	„ . .	„
46208	Corporal F. Thomas . .	„ . .	„
8218	Private E. Y. Turner . .	„ . .	„
15087	„ W. F. Wickham .	„ . .	„
2200	Corporal Sad. J. W. Youells	„ . .	„
21526	Private F. T. Ames . .	Lajj . . .	Mar. 5, 1917.
6507	„ W. B. Angier . .	„ . . .	„
1357	Lance-Corporal W. Bolton .	„ . . .	„
46196	„ J. Burgess .	„ . . .	„
6918	Sergeant W. H. Burgess .	„ . . .	„
8270	Farrier Sergeant A. E. Clark	„ . . .	„
9009	Private J. Cleveland . .	„ . . .	„
1915	Sergeant P. Collins . .	„ . . .	„
46219	„ H. Deeley . .	„ . . .	„

9199	Private W. Edwards	Lajj	Mar. 5, 1917.
16845	„ C. Enstone	„	„
46239	Sergeant J. R. Ethell	„	„
18185	Private G. G. Evans	„	„
8119	„ W. Freeman	„	„
1339	Lance-Corporal G. A. Garrod	„	„
6361	Lance-Corporal G. Garton	„	„
3545	„ F. Gilliver	„	„
24494	Private J. Guthrie	„	„
2899	Lance-Corporal (Tptr.) H. A. Hamilton	„	„
46243	Corporal E. Harris	„	„
6523	„ (S.S.) P. Hemsworth	„	„
5830	Private G. E. Heyburn	„	„
2717	Sergeant J. Johnson	„	„
46188	Corporal G. F. S. Kent	„	„
2900	Private J. Larner	„	„
11253	„ J. M'Govern	„	„
24886	„ D. Main	„	„
4979	„ (S.S.) F. A. Marsh	„	„
5581	„ W. F. Martin	„	„
5811	„ F. Moir	„	„
1664	„ W. P. Morgan	„	„
22518	„ G. Peet	„	„
26893	„ J. Perrott	„	„
7746	„ W. H. Pomeroy	„	„
6838	Corporal G. Pople	„	„
4524	Lance-Corporal G. Ratty	„	„
9045	Private M. Ryan	„	„
24071	„ W. Sheen	„	„
1998	Lance-Sergeant A. Taylor	„	„
18475	Private C. E. Taylor	„	„
3804	Corporal W. H. W. Thomas	„	„
5439	Private A. Trafani	„	„
9192	Lance-Corporal A. W. Watkins	„	„
5551	Sergeant A. Westbrook	„	„
13061	Private F. Wilson	„	„
13611	„ P. Wren	„	„
9279	„ J. Whittle	Deli Abbas	April 8, 1917.
13739	„ C. Cooper	Kurdarrah River	Oct. 19, 1917.
21837	„ W. J. Davies	„	„
10971	„ W. F. Davis	„	„
16435	„ R. Adams	Tekrit	Nov. 3, 1917.
18943	„ J. H. Barker	„	„
276662	„ H. J. Beales	„	„

2129	Lance-Corporal G. A. Day .	Tekrit . . .	Nov. 3, 1917.
11629	Private A. Liddle . .	,, ...	,,
11960	,, H. H. Miller . .	,, ...	,,
9382	Lance-Corporal J. O'Connor	,, ...	,,
3973	Private H. Bell . . .	,, . . .	Nov. 4, 1917.
325228	,, W. Bennett . .	,, ...	,,
5305	,, S. Howlett . .	,, ...	,,
15773	,, F. Andrews . .	,, . . .	Nov. 5, 1917.
7514	Trumpter H. A. Burder .	,, ...	,,
46253	Private J. Byrne . .	,, ...	,,
3544	,, C. Delaney . .	,, ...	,,
18198	,, C. Halstead . .	,, ...	,,
16431	,, A. Hibbert . .	,, ...	,,
46247	Lance-Sergeant H. Hudson	,, ...	,,
18983	Private J. Jeffrey . .	,, ...	,,
4348	Lance-Corporal H. Lindsey	,, ...	,,
26310	Corporal W. J. Luetchford .	,, ...	,,
24082	Private E. Manby . .	,, ...	,,
4422	Lance - Corporal A. E. Reaves	,, ...	,,
5586	Private H. W. Steward .	,, ...	,,
9039	,, R. Stubbs . .	,, ...	,,
18475	,, C. E. Taylor . .	,, ...	,,
5516	,, A. V. Viney . .	,, ...	,,
24863	,, A. J. F. Ward .	,, ...	,,
24614	,, D. G. Williams .	,, ...	,,
15652	,, F. Berry . .	Chai Khana . .	Dec. 3, 1917.
21238	,, C. Harris . .	,, . .	Dec. 5, 1917.
1018	Lance-Corporal J. Clancy .	Khastradalah or Tuz .	April 28, 1918.
16791	Private J. L. Harvey . .	,, ,, .	,,
24478	,, C. Kerr . . .	,, ,, .	,,
27823	Corporal S. Potter . .	,, ,, .	,,
31960	Private L. Rowley . .	,, ,, .	,,
8392	Lance - Corporal (S.S.) J. George	,, ,, .	April 29, 1918.
6768	Private A. E. Hudson .	,, ,, .	,,
19433	,, J. G. Tatner . .	,, ,, .	,,
31956	,, R. C. B. M'Queen .	,, ,, .	May 1, 1918.
1164	,, W. G. Knight .	Kirkuk . . .	May 6, 1918.
5785	,, A. J. Shaw . .	,, ...	,,
32800	,, E. F. Beaver . .	Daramah . . .	May 10, 1918.
6918	Sergeant W. H. Burgess .	,, ...	,,
11899	Private (S S.) A. Earl . .	,, ...	,,
13200	Lance-Corporal A. Mellor .	,, ...	,,
24726	Private J. W. Thompson .	,, ...	,,
10887	Corporal J. Townsend .	,, ...	,,
235489	Private (S.S.) A. S. Vaughan	,, ...	,,
4712	Sergeant G. Barnard . .	Hadraniyeh . .	Oct. 29, 1918.

24728	Private H. D. Gerrard .	Hadraniyeh . .	Oct. 29, 1918.
1195	Sergeant W. J. Hill . .	,, . .	,,
32892	Private P. Little . .	,, . .	,,
23432	,, V. Patrick . .	,, . .	,,
6410	,, W. Powell . .	,, . .	,,
7549	,, R. Tranter . .	,, . .	,,
1192	,, F. Winter . .	,, . .	,,
6542	,, F. Harvey . .	Crossing Ford at Hadraniyeh	,,

PRISONERS OF WAR.

46218	Sergeant W. Gilbert . .	Lajj	March 5, 1917.
46212	Lance-Sergeant F. Spanton .	,, ...	,,
11916	Private J. Morrison . .	,, . . .	,,
13242	,, P. Keaney . .	Dishdari . . .	April 23, 1917.

APPENDIX III.

HONOURS AND REWARDS.

Companion of the Bath.

Brig.-General J. T. Wigan.

Companion of St Michael and St George.

Brig.-General J. T. Wigan.
Colonel Sir A. Leetham.
„ A. Symons.

Commander of British Empire.

Brig.-General W. C. Smithson, D.S.O.
Lieut.-Colonel H. Phillips.
Colonel H. J. Blagrove.

Distinguished Service Order.

Brig.-General J. T. Wigan.
„ E. A. Wiggin.
Lieut.-Colonel J. J. Richardson.
Bt. Lieut.-Colonel T. H. S. Marchant.
„ J. G. Rees.
Bt. Colonel W. Pepys.
Bt. Major H. Ll. Jones.
Captain W. A. Kennard.

Bar to Distinguished Service Order.

Lieut.-Colonel J. J. Richardson.
Bt. Lieut.-Colonel J. G. Rees.

2nd Bar to Distinguished Service Order.

Bt. Lieut.-Colonel J. G. Rees.

Member of British Empire.

Lieutenant R. C. Hill.

Military Cross.

Captain C. C. Dangar.
,, J. N. Lumley.
,, D. W. Godfree, 21st Lancers (*attd.*)
Lieutenant A. M. Sassoon.
,, S. V. Kennedy.
,, J. A. Jeffrey.
,, J. I. Chrystall.
,, D. J. E. Norton.
,, H. C. D. FitzGibbon.
,, H. R. Jones.
,, M. G. Hartigan.
,, B. H. Williams.
,, C. H. Gowan.
2nd Lieutenant H. MacDonald.
,, G. H. Parkes.
46186 R.S.M. S. F. Seekins.

Legion of Honour (French).

Brig.-General E. A. Wiggin, D.S.O.
Bt. Major R. S. Hamilton-Grace (*deceased*).
,, H. Ll. Jones, D.S.O.

Croix de Guerre (French).

Lieut.-Colonel J. J. Richardson, D.S.O.
Lieutenant D. A. Stirling.

Silver Medal for Military Valour (Italian).

Lieutenant H. C. D. FitzGibbon, M.C.

Distinguished Service Medal (American).

Captain F. C. Covell.

St Stanislaus, 2nd Class (Russian).

Brig.-General A. H. M. Taylor, D.S.O.
Lieut.-Colonel H. Phillips.

Order of St Anne, 3rd Class (Russian).

Captain H. E. Spencer.

Order of Star of Roumania (with Swords).

Major (Bt. Lieut.-Colonel) E. J. Carter.

Order of the White Eagle, 5th Class with Swords (Serbia).

Lieut.-Colonel J. H. Tremayne, D.S.O.

Distinguished Conduct Medal.

46194 S.Q.M.S. W. D. Tassie.

9285 Private Alfred Jones (*killed in action*).

4600 Acting Corporal G. W. Bowie.

1150 S.S.M. J. Brearley.

4504 S.S. J. F. Crouch (*deceased*).

6166 Private J. L. Roberts „

2901 Lance-Corporal J. Stevens.

9192 „ A. W. Watkins.

4428 Sergeant W. H. Tetheridge.

4859 „ S. G. Strawbridge.

423 Corporal G. Rayner.

6915 „ A. Vinall (*died of wounds*).

6240 S.Q.M.S. F. A. M'Bride.

2348 „ S. B. Haines.

Military Medal.

46224 Sergeant W. Matthews (*killed in action*).

5577 Lance-Corporal W. G. Corbin.

16961 Private H. H. Mortimer.

6473 Sergeant J. Cunningham.

4976 Lance-Corporal F. Kitcher.

23686 Private A. Owen (*died of wounds*).

6731 „ C. Cubitt.

6918 Sergeant W. H. Burgess.

22903 Corporal F. Lomas.

3254 Sergeant H. H. L. Aukett.

8796 Private E. Humphrey.

4491 Lance-Corporal F. Viner.

5433 Corporal F. Whitington.

5317 Sergeant E. E. Moss.

Meritorious Service Medal.

3960 Sergeant T. Carnegie.

9009 Private J. Cleveland.

13417 „ J. H. James-Moore.

6604 Lance-Corporal D. C. Wright.

3715 R.Q.M.S. S. Tarr.

3808 Lance-Sergeant A. N. Richards.

4523 Sergeant T. S. Craven.

10937 Private J. H. Biggs.

46198 S.Q.M.S. B. J. Ward.

4118 Sergeant J. H. Rogers.

15683 „ R. A. Westerman.

Page 363. 4504. *For* "Crouch" *read* "Couch."

13902 Private J. T. Harding.
9714 Sergeant W. Holden.

Bronze Medal for Military Valour (Italian).

46200 S.S.M. A. Potter.
15619 Lance-Corporal H. E. Turner.

Croix de Virtute Militara, 2nd Class, for Bravery (Roumanian).

46258 Private J. Allston.

Medaille Barbatie se Credente, 3rd Class (Roumanian).

4523 Sergeant T. S. Craven.

MENTIONED IN DESPATCHES.

[1] Colonel A. Symons, C.M.G.
[2] Lieut.-Colonel J. J. Richardson, D.S.O
[1] Bt. Lieut.-Colonel E. F. Twist.
[1] Major R. F. Cox.
[1] Captain A. M. Sassoon, M.C.
Major (Bt. Lieut.-Colonel) F. G. Bayley.
Brig.-General J. T. Wigan, D.S.O.
[1] Captain S. O. Robinson (*killed in action*).
Bt. Major C. Steele.
Captain J. N. Lumley, M.C.
" J. A. Jeffrey, M.C.
Acting Captain C. H. Gowan, M.C.
Lieutenant J. L. M. Barrett.
" J. I. Chrystall, M.C.
" A. Williams.
" D. A. Stirling.
" G. L. Welstead.
" B. H. Williams.
Major and Quartermaster A. Cooke.
[3] Lieut.-Colonel W. Pepys.
Major T. R. Farquhar.
Captain and Quartermaster R. W. F. Ashworth.
[1] Bt. Major R. S. Hamilton-Grace (*deceased*).
[3] Major W. A. Kennard, D.S.O.
[3] Bt. Major H. Ll. Jones, D.S.O.
[3] Bt. Lieut.-Colonel T. H. S. Marchant, D.S.O.
Captain S. V. Kennedy, M.C.
Lieutenant M. H. C. Doll.
Captain D. W. Godfree, M.C., 21st Lancers (*attd.*)

[1] Twice. [2] Four times. [3] Three times.

Lieutenant A. Le Patourel.
Captain J. H. Hind.
[1] Bt. Lieut.-Colonel J. G. Rees, D.S.O.

46223 R.Q.M.S. H. J. Edwards
1150 S.S.M. J. Brearley, D.C.M.
46222 ,, E. L. Popham.
46200 ,, A. Potter.
46240 Sergeant J. Gray (*killed in action*).
9192 Lance-Corporal A. W. Watkins, D.C.M.
2901 ,, J. Stevens, D.C.M.
[2] 4118 Sergeant J. H. Rogers.
46258 Private J. Allston.
3774 Far.S.M. A. Bald.
46182 S.S.M. C. M. Douthwaite.
2263 Sergeant J. Foster.
30555 Sergeant-Trumpeter S. Gill.
46191 Sergeant H. S. Holloway.
2348 S.Q.M.S. S. B. Haines
423 Corporal G. Rayner.
46173 Acting F.Q.M.S. W. Stanton.
206086 Lance-Corporal C. H. A. Wyncoll.
6177 Corporal J. Carpenter.
6392 Private J. Smithers.
1195 Sergeant W. J. Hill.
7066 ,, W. M'Guire.
15619 Corporal H. E. Turner.
[2] 4763 S.S.M. (A/R.S.M.) W. J. Allan (11th Hussars).
2712 Private A. Beauchamp.
49102 ,, (A./Farr.-Sergeant) W. W. Pay (attached R.A.V.C.)

[1] Three times. [2] Twice.

PROMOTION.

Lieut.-Colonel A. Symons, C.M.G. . . .	Brevet Colonel.
Major E. F. Twist	Brevet Lieut.-Colonel.
,, T. H. S. Marchant, D.S.O. . . .	,,
Captain R. S. Hamilton-Grace (*deceased*) .	Brevet Major.
Major E. J. Carter	Brevet Lieut.-Colonel.
,, F. G. Bayley	,,
Captain H. Ll. Jones, D.S.O.	Brevet Major.
,, C. Steele	,,
,, H. E. Spencer	,,
,, J. V. Dawson	,,
,, J. G. Rees, D.S.O.	Brevet Lieut.-Colonel.
Lieutenant and Quartermaster A. Cooke .	Hon. Captain. Hon. Major.
Captain and Quartermaster R. W. F. Ashworth	,,

APPENDIX IV.

OFFICERS WHO LEFT INDIA FOR FRANCE WITH THE REGIMENT ON 17TH NOVEMBER 1914.

Lieut.-Colonel A. Symons, *Commanding*.
Major R. F. Cox.
,, T. H. S. Marchant.
Captain W. H. Eve.
,, C. Steele.
,, J. N. Lumley, *Adjutant*.
Lieutenant H. G. T. Newton.
,, A. M. Sassoon.
,, S. V. Kennedy.
,, J. V. Dawson.
,, J. A. Jeffrey.
,, G. R. Watson-Smyth.
2nd Lieutenant R. Gore.
,, M. C. Kennedy.
,, J. L. M. Barrett.
,, A. E. Annett.
,, C. H. Gowan.
,, J. I. Chrystall.
Lieutenant and Quartermaster A. Cooke, Quartermaster.

Other Ranks, 499.

ATTACHED.

Captain A. L. Horner, A.V.C.
,, G. G. Collett, R.A.M.C.

THE FOLLOWING OFFICERS WERE IN ENGLAND AT THE OUTBREAK OF WAR AND REJOINED THE REGIMENT ON THE DATES AGAINST THEIR NAMES.

Lieut.-Colonel J. J. Richardson, D.S.O.	May 1, 1915.
Bt. Lieut.-Colonel E. F. Twist	December 18, 1914.
Temporary Lieut.-Colonel W. Pepys	,, ,,
,, ,, E. J. Carter	,, ,,
Bt. Major H. Ll. Jones, D.S.O.	June 22, 1915.
Captain J. H. Hind	December 20, 1914.
,, S. O. Robinson	July 26, 1915.
,, J. O. Oakes	June 21, 1915.

THE FOLLOWING SERVING OFFICERS WERE ATTACHED TO UNITS OTHER THAN THE REGIMENT.

Major W. A. Kennard, D.S.O.	Northumberland Hussars.
Bt. Major R. S. Hamilton-Grace	Staff of 2nd Cavalry Brigade.
,, H. Ll. Jones, D.S.O.	4th Dragoon Guards (13th Hussars, June 22, 1915).
Captain E. H. Stocker	Staff of Yeomanry Brigade.
,, N. Neill	Staff of 7th B. Cavalry Brigade.
,, J. H. Hind	2nd Life Guards (13th Hussars, December 20, 1914).
,, S. O. Robinson	5th Dragoon Guards (13th Hussars, July 26, 1915).
,, J. O. Oakes	11th Hussars (13th Hussars, June 22, 1915).
Lieutenant T. E. Lawson-Smith	11th Hussars.
,, C. A. F. Wingfield	2nd Life Guards (13th Hussars, January 2, 1915).
,, V. F. Symondson	11th Hussars, and R.A.F.
,, G. R. Pedder	11th Hussars (13th Hussars, January 2, 1915).

THE FOLLOWING OFFICERS WHO HAD LEFT THE REGIMENT REJOINED THE REGIMENT OR OTHER UNITS AS FOLLOWS.

Brig.-General E. A. Wiggin, D.S.O.	Commanding Mounted Brigade.
,, A. H. M. Taylor, D.S.O.	Commanding London Mounted Brigade.

Brig.-General J. T. Wigan, C.B., C.M G., D.S.O.	Commanding Yeomanry Brigade.
„ W. C. Smithson, C.B.E., D.S.O.	Commanding 2nd Line Mounted Brigade.
„ E. N. Pedder . . .	Commanding 2nd Line Mounted Brigade.
„ C. Williams . . .	Commanding Mounted Brigade.
Colonel Sir F. Maclean	Special Appointment.
„ Sir A. Leetham	War Office.
„ H. J. Blagrove, C.B . . .	Commanded Prisoners of War Camp.
Lieut.-Colonel H. Phillips . . .	Special Appointment.
„ A. W. B. Spencer . .	Tank Corps.
„ W. Pepys, D.S.O. . .	5th Dragoon Guards, Yorkshire Hussars, and Tank Corps.
Bt. Lieut.-Colonel F. G. Bayley . .	E.S.O.
„ „ J. G. Rees, D.S.O. .	Welsh Horse.
Major J. F. Church	1st Indian Field Squadron, Indian Cavalry Corps.
„ H. E. Spencer	11th Hussars.
„ E. W. Denny, D.S.O. . . .	Brigade Major, Yeomanry Brigade.
„ J. H. Tremayne, D.S.O. . .	Derbyshire Yeomanry.
„ F. H. Wise (*deceased*) . . .	Remount Department.
„ H. J. Jeddere-Fisher . . .	Queen's Regiment.
„ J. C. Metcalfe (*killed in action*) .	Cheshire Regiment.
Captain C. C. Dangar, M.C. (*deceased*) .	12th Reserve Cavalry Regiment, E.E.F.
„ Hon. M. F. S. Howard . .	8th Hussars.
„ G. Halswelle	1st Life Guards.
„ J. D. Lyons	Royal Horse Guards.
„ E. W. M'Arthur.	12th Reserve Cavalry Regiment.
„ H. A. de P. O'Kelly (*killed in action*)	18th Hussars.
„ S. Wybrants	Brigade Major, Yeomanry Brigade.
„ R. W. F. Ashworth . . .	Adjutant, No. 1 Base Remount Depot.
„ T. K. Jackson	11th Hussars, 13th Hussars, December 18, 1914.
„ G. H. Hodgkinson . . .	Staff Captain, Yeomanry Brigade.
„ L. S. Battye	Bristol Cavalry Depot.
„ G. Rupert	Special Appointment.
„ Lord Huntingfield . . .	3rd Reserve Cavalry Regiment.
„ R. H. J. M. Bolton . . .	3rd Reserve Cavalry Regiment.
Lieutenant W. V. Holt	3rd Signal Squadron R.E.
„ B. H. Williams . . .	11th Hussars, 13th Hussars, December 18, 1914.
„ E. W. Anderson (*killed in action*)	Scottish Rifles.
„ H. Norfolk (*killed in action*) .	Worcester Yeomanry.

THE FOLLOWING OFFICERS OF THE SPECIAL RESERVE THIRTEENTH HUSSARS JOINED THE REGIMENT AS FOLLOWS.

Lieutenant M. H. C. Doll	from Dec. 18, 1914,	to May 1, 1917.
„ H. C. D. FitzGibbon, M.C.	„ April 11, 1916,	„ Nov. 11, 1918.
„ F. C. Covell	„ Feb. 3, 1916,	„ June 27, 1916.
„ B. E. H. Judkins	„ June 2, 1916,	„ Sept. 15, 1916.
„ J. A. Lord	„ Dec. 28, 1916,	„ June 24, 1917.
„ N. L. Moon	„ Feb. 6, 1916,	„ July 18, 1916.
„ J. F. Munster (*killed in action*)	„ Feb. 3, 1916,	„ Feb. 4, 1917.
2nd Lieutenant H. G. Keswick	„ May 5, 1917,	„ Nov. 11, 1918.
„ C. A. G. M'Lagan	„ Dec. 22, 1917,	„ April 28, 1918.
„ B. W. D. Cochrane	„ Nov. 18, 1917,	„ Jan. 11, 1918.

THE FOLLOWING OFFICERS OF OTHER UNITS JOINED THE REGIMENT FOR DUTY.

Captain G. G. Collett	R.A.M.C.
„ A. L. Horner	R.A.V.C.
„ J. Vlasto, M.C.	R.A.M.C.
„ D. M. Methven	21st Lancers.
„ D. W. Godfree, M.C.	„
„ J. W. D. Evans	„
„ W. E. Bigge	Northumberland Hussars.
„ J. E. P. Howey	Bedford Yeomanry.
„ V. H. Secker	14th Hussars.
„ E. Wordley	R.A.M.C.
„ W. A. Elliott	„
„ W. G. Rose	„
„ J. Cullenan	„
Lieutenant L. C. Johnston	„
„ F. G. Roche	„
„ D. A. C. Chamberlain	„
The Rev. H. R. Cooke, M.C.	A.C.D.

THE FOLLOWING TEMPORARY OFFICERS WERE ATTACHED TO THE THIRTEENTH HUSSARS AND JOINED THE REGIMENT.

Lieutenant E. Bristol	December 18, 1914.
„ J. W. Biggar	May 24, 1917.
„ W. P. Crawford-Greene	July 27, 1915.
„ G. F. Earle	March 12, 1917.

Lieutenant A. C. J. Elkan . . . . . August 17, 1917.
" M. G. Hartigan, M.C. . . . February 19, 1917.
" H. R. Jones, M.C. . . . . May 5, 1917.
" C. W. Jemmett . . . . . October 17, 1917.
" A. Le Patourel . . . . . March 12, 1917.
" R. J. V. Michell . . . . March 26, 1916.
" H. MacDonald, M.C. . . . . November 12, 1917.
" W. J. L. Norwood . . . . February 16, 1918.
" W. G. Newman . . . . . November 23, 1918.
" L. J. Owen . . . . . May 5, 1917.
" L. Osmond . . . . . November 12, 1917.
" G. H. Parkes, M.C. . . . . January 9, 1918.
" W. C. J. Pook . . . . . February 16, 1918.
" W. R. Pearson . . . . . October 25, 1915.
" R. B. Pardon . . . . . December 23, 1916.
" E. F. Pinnington . . . . February 19, 1917.
" P. H. J. Shiel . . . . . March 12, 1917.
" E. Goodman . . . . . June 2, 1918.
" H. H. Johnson, M.M. . . . December 19, 1917.
" F. G. Lawrence . . . . . March 29, 1918.
" M. R. Farrer . . . . . February 27, 1918.
" F. H. Butler . . . . . December 19, 1917.
2nd Lieutenant F. Norman Payne (*deceased*) . December 23, 1916.
" E. V. Rolfe (*killed in action*) . February 19, 1917.
" R. R. Anson . . . . December 23, 1916.
" J. H. Lucas . . . . January 9, 1918.
" E. P. Barrett . . . . February 10, 1918.
" G. R. Russell . . . . February 27, 1918.
" E. Wright . . . . . March 7, 1918.
" W. G. Baker . . . . June 2, 1918.
" C. W. Gardner . . . . May 1, 1918.
" E. T. Hollingdale . . . June 2, 1918.
" E. G. Corner . . . . June 18, 1918.
" G. Lynch-Staunton (*killed in action*) . . . . February 21, 1917.

THE FOLLOWING COMMISSIONED OFFICERS JOINED THE REGIMENT DURING THE WAR.

Captain D. J. E. Norton, M.C. . . . . July 28, 1915.
Lieutenant C. A. F. Wingfield . . . . August 12, 1914.
" J. H. Hirsch . . . . . October 14, 1915.
" D. A. Stirling . . . . . October 14, 1915.
" L. A. Ormrod . . . . . December 18, 1914.
" T. Williams-Taylor . . . . October 16, 1915.
" W. Blyth, D.C.M. . . . . August 27, 1915.
" A. Williams . . . . . February 4, 1916.

Lieutenant R. C. Hill	March 19, 1916.
„ G. L. M. Welstead . . .	February 21, 1917.
„ J. Hampson	February 21, 1917.
2nd Lieutenant J. O. P. Clarkson (*killed in action*)	October 25, 1915.
„ A. H. Wood	April 6, 1918.
„ C. F. Collins	April 6, 1918.
„ A. C. Barrington . . .	February 19, 1919.
„ A. J. R. M. Sibson . . .	February 19, 1919.
„ W. W. N. Davies . . .	February 19, 1919.

During the period of the War—*i.e.*, from August 4, 1914, to November 11, 1918—108 officers and 1348 other ranks did duty with the Regiment.

APPENDIX V.

All Saints' Garrison Church, Aldershot.

2nd AUGUST 1920.

Service held at the

DEDICATION AND UNVEILING OF THE MEMORIAL TABLET

In Memory of the Officers, N.C.O.'s, and Men of the 13th Hussars who gave their lives in the Great War, 1914 - 1918.

"Dead March" (*Chopin*).

OPENING SENTENCES.

I am the Resurrection and the Life, saith the Lord; he that believeth in Me, though he were dead, yet shall he live; and whosoever liveth and believeth in Me, shall never die. (John xi. 25, 26.)

Greater love hath no man than this, that a man lay down his life for his friends. (John xv. 13.)

I heard a voice from heaven saying unto me, Write, From henceforth blessed are the dead that die in the Lord: even so, saith the Spirit, for they rest from their labours.

PSALM XXIII.

Dominus regit me.

1. The Lord is my Shepherd: therefore can I lack nothing.

2. He shall feed me in a green pasture: and lead me forth beside the waters of comfort.

(*Photograph by Gale and Polden Ltd., Aldershot*)

THE MEMORIAL TABLET IN ALL SAINTS' GARRISON CHURCH, ALDERSHOT

DEDICATED AND UNVEILED, 2ND AUGUST 1920

3. He shall convert my soul: and bring me forth in the paths of righteousness, for His Name's sake.

4. Yea, though I walk through the valley of the shadow of death, I will fear no evil: for Thou art with me, Thy rod and Thy staff comfort me.

5. Thou shalt prepare a table before me against them that trouble me: Thou hast anointed my head with oil, and my cup shall be full.

6. But Thy loving-kindness and mercy shall follow me all the days of my life: and I will dwell in the house of the Lord for ever.

(*Gloria.*)

THE LESSON.

WISDOM iii. 1-6.

But the souls of the righteous are in the hand of God, and there shall no torment touch them. In the sight of the unwise they seemed to die; and their departure is taken for misery, and their going from us to be utter destruction: but they are in peace. For though they be punished in the sight of men, yet is their hope full of immortality. And having been a little chastised, they shall be greatly rewarded; for God proved them and found them worthy for Himself. As gold in the furnace hath He tried them, and received them as a burnt-offering.

HYMN.

"Fight the good fight." (A.M. 540.)

PRAYERS.

Almighty God, with whom do live the spirits of them that depart hence in the Lord, and with whom the souls of the faithful, after they are delivered from the burden of the flesh, are in joy and felicity; We give Thee hearty thanks, for that it hath pleased Thee to take our brethren into Thy merciful keeping: beseeching Thee that we, with all those that are departed in the true faith of Thy Holy Name, may have our perfect consummation and bliss, both in body and soul, in Thy eternal and everlasting glory; through Jesus Christ our Lord. AMEN.

Almighty God, we commend to Thy loving kindness the souls of Thy servants, the men of this Regiment, whom we now commemorate before Thee. Accept, O Lord, the offering of their self-sacrifice, and grant to them with all Thy faithful servants a place of refreshment and peace where the light of Thy countenance shines for ever, and where all tears are wiped away; through Jesus Christ our Lord. AMEN.

Comfort, O Lord, we pray Thee, all who are mourning the loss of those near and dear to them. Be with them in their sorrow. Support them with the knowledge of Thy love. Teach them to rest and lean on Thee. Give them faith to look beyond the troubles of this present time, and to know that neither life nor death can separate us from the love of God which is in Christ Jesus our Lord, to whom with the Father and the Holy Ghost be all honour and glory, now and for ever. AMEN.

THE GRACE.

Rehearsal of the Names of those to be Commemorated.

Unveiling Ceremony.

By Lieut.-Col. J. J. RICHARDSON, D.S.O., Commanding 13th Hussars.

Dedication of the Memorial Tablet.

By the Rev. F. I. ANDERSON, C.M.G., S.C.F., C. of E., Aldershot, Chaplain to the King.

In the name of the Father, and of the Son, and of the Holy Ghost, we dedicate this Memorial to the glory of God, and in memory of the Men of this Regiment who have fallen in the great War. May their example inspire us to courage in the greater war against all evil. May their memory ever burn brightly in those who here or elsewhere remember their deeds, and, strengthened by their fellowship, look forward to reunion with them in the inheritance of the saints in light.

We give Thee thanks, O Lord, Father Almighty, for all those Thy servants who waxed valiant in fight and wrought righteousness, and counting not their life dear unto themselves, laid it down for their friends; and we pray Thee that, having fought a good fight and finished their course with joy, they may rejoice evermore with them that have come out of great tribulation, and, having washed their robes and made them white in the blood of the Lamb, stand before Thy Throne and serve Thee day and night for ever; through Him that overcame and is set down on Thy Throne, Jesus Christ our Lord, who liveth and reigneth with Thee and the Holy Ghost now and ever and world without end. AMEN.

The Address.

Most merciful Father, Who hast been pleased to take unto Thyself the souls of those whom we now commemorate before Thee, grant unto us who are still in our pilgrimage and who walk as yet by faith that, having served Thee with constancy on earth, we may be joined hereafter with Thy blessed saints in glory everlasting, through Jesus Christ our Lord. AMEN.

THE LORD'S PRAYER.

THE BENEDICTION.

The Lord grant to us, together with all His faithful servants, rest and peace in Christ Jesus our Lord. AMEN.

The National Anthem.

The Last Post by Massed Trumpeters.

HYMN.

O God, our help in ages past,
  Our hope for years to come,
Our shelter from the stormy blast,
  And our eternal home;

Beneath the shadow of Thy Throne
  Thy Saints have dwelt secure;
Sufficient is Thine Arm alone
  And our defence is sure.

O God, our help in ages past,
  Our hope for years to come,
Be Thou our guard while troubles last,
  And our eternal home. Amen.

Reveille by Massed Trumpeters.

APPENDIX VI.

REGIMENTAL ASSOCIATIONS.

Thirteenth Hussars Association.

92 VICTORIA STREET, WESTMINSTER, S.W. 1.

TELEPHONE: 3708 VICTORIA.

Cheques and Postal Orders to be made payable to the Secretary, Thirteenth Hussars Association, and crossed Cox & Co.

Secretary—

MISS FLORENCE CROSS.

President—

LIEUT.-GENERAL SIR ROBERT BADEN-POWELL, BART., K.C.B., C.V.O.

The objects of the Association are :—

To relieve distress amongst ex-men of the Regiment or their dependants.
To assist and advise them as to establishing themselves in civil life on discharge from the Army.
To advise them with regard to the care and education of their children.
To help them or their families to obtain hospital or convalescent treatment when required.
To provide a centre in London to which they may apply when in need of help or advice.

The Association is controlled by a Committee composed of past and present officers and non-commissioned officers of the Regiment, and all applications made to it are treated as confidential.

Application may be made either personally or by letter to the Secretary at the above address.

Donations and Subscriptions to the Funds of the Association will be gratefully received, and should be sent to the Secretary.

Thirteenth Hussars Old Comrades Association.

This Association—originally called the "Old Comrades Dinner Club"—was formed in accordance with a suggestion made at a meeting held at the Office of the 13th Hussars Association, on Saturday, 1st March 1913, under the Presidency of Lieut.-Colonel E. A. Wiggin, when a large number of past and present officers of the Regiment were present, also a good number of late N.C.O.'s and men. A committee was formed, and it was decided to start a Club of Old Comrades of the Regiment under the title of "The Old Comrades Dinner Club," for the purpose of holding an Annual Dinner.

The Annual Dinner is held on the fourth Saturday in October.

All retired officers, non-commissioned officers, and men whose addresses are known have been invited to enrol themselves as members.

Men of the Regiment of good character on taking their discharge are eligible for election, and are requested to send their names and addresses to the Hon. Secretary.

The Annual Subscription is 2s. 6d.

Invitations to the dinner are sent to members and (amongst others) to all Chelsea Pensioners of the Regiment who come as guests of the Association.

The Association has been the means of assisting many past members of the Regiment to obtain employment, and is very largely resorted to by men in need of help and advice.

Donations and subscriptions to the funds of the Association should be sent to the Hon. Secretary, from whom all particulars as to membership and otherwise can be obtained.

Hon. Secretary—

MR FRED. E. M. THOMPSON,

14 BLANDFORD STREET, MANCHESTER SQUARE, LONDON, W. 1.

INDEX.

www.ingramcontent.com/pod-product-compliance
Lightning Source LLC
LaVergne TN
LVHW060012210926
848481LV00013B/3695
9781843425366